Springer Series in
Nuclear
and **Particle Physics**

Springer Series in **Nuclear** and **Particle Physics**

Editors: Mary K. Gaillard · J. Maxwell Irvine · Vera Lüth · Achim Richter

A.N.Antonov   P.E.Hodgson   I.Zh.Petkov

Nucleon Correlations in Nuclei

With 96 Figures

Springer-Verlag

Berlin Heidelberg New York
London Paris Tokyo
Hong Kong Barcelona
Budapest

Prof. Dr. Sc. Anton Nikolaev Antonov
Institute of Nuclear Research and Nuclear Energy
Bulgarian Academy of Sciences
Boul. Tzarigradsko Shosse 72, Sofia 1784, Bulgaria

Dr. Peter Edward Hodgson
Nuclear Physics Laboratory, University of Oxford
Keble Road, Oxford OX1-3RH, UK

Prof. Dr. Sc. Ivan Zhelyazkov Petkov
Institute of Nuclear Research and Nuclear Energy
Bulgarian Academy of Sciences
Boul. Tzarigradsko Shosse 72, Sofia 1784, Bulgaria

Library of Congress Cataloging-in-Publication Data, Antonov, A.N. Nucleon correlations in nuclei / A.N. Antonov, P.E. Hodgson, I.Zh. Petkov. p. cm. — (Springer series in nuclear and particle physics) Includes bibliographical references and index.
1. Nucleon–nucleon interactions. 2. Nuclear structure. I. Hodgson, P.E. (Peter E.) II. Petkov, I.Zh. III.

ISBN-13: 978-3-642-77768-4   e-ISBN-13: 978-3-642-77766-0
DOI: 10.1007/978-3-642-77766-0

Title. IV. Series. QC793.5.N828A58 1993 539.7'212—dc20   92-27020

© Springer-Verlag Berlin Heidelberg 1993

Softcover reprint of the hardcover 1st edition 1993

Typesetting: Macmillan India Ltd., Bangalore 25

56/3140/SPS–543210 – Printed on acid-free paper

Preface

In recent years there has been growing interest in the nucleon–nucleon correlations inside nuclei. In many respects the motions of the nucleons can be very well described by an overall mean field, so that the motion of each nucleon is governed by the mean field due to all the other nucleons. This concept underlies the Fermi-gas, Hartree–Fock and shell models and has enabled a range of nuclear properties to be calculated, often to surprising accuracy. It gradually became clear, however, that these mean-field models are limited by the effects due to the very strong interactions between the nucleons that occur at short distances; these are the short-range correlations. They are responsible for instance for the high-momentum components in the nucleon momentum distribution, and prevent the simultaneous description of the nuclear density and momentum distributions by the same mean field. It thus becomes necessary to develop methods for including the effects of nucleon correlations in nuclei, and these are the main subject of this book.

Some related problems of nuclear structure were discussed in an earlier book by the same authors: *Nucleon Momentum and Density Distributions in Nuclei* (Clarendon Press, Oxford, 1988). The main aim of that book was to study the effects of nucleon–nucleon correlations, both short-range and tensor, on the nucleon momentum distribution, which is particularly sensitive to these correlations, and on the nucleon density distribution. This was done by developing a theory of the relation between these two distributions for the nuclear ground state based on the Hohenberg–Kohn theorem. In addition many other theoretical methods going beyond the Hartree–Fock theory were described, in particular the mass operator expansion in nuclear matter, the exp(S)-method, the Jastrow method and Brueckner correlation theory, the phenomenological method of two-body correlation operators, the natural orbital method, the generator coordinate method and the coherent density fluctuation model. Comparisons between these theories and the experimental data available up to the end of 1986 were made especially for those related to the high-momentum behaviour of the nucleon momentum distribution.

The main motivation of the present book is to give a more detailed account of the theory of the nucleon correlations, and to apply it to analyse the wealth of new experimental data. Nucleon correlations are closely related to clustering in nuclei, and there have been many new studies related to this phenomenon. Mean-field theories have been extensively developed, in particular by making use of the dispersion relations that connect the real and imaginary parts of the mean field. There have been many detailed studies of deep-hole states in nuclei using nucleon transfer and knock-out reactions. The occupation numbers and wave functions of the single-particle states have been studied in detail. Many of these data show substantial deviations from mean field calculations, and require the development of more sophisticated theories that include the effects of short-range and tensor correlations.

The nuclear theories discussed in detail include the complex mean field analysis of bound and scattering states, unified by the dispersion relations, and a critical analysis of the Hartree–Fock method on the basis of the natural orbital representation. Particular attention is paid to the effects of correlations on the characteristics of nuclear states including the spectral functions for deep-hole states and the natural orbital occupation numbers. The sum rules for the spectral functions and the methods used to determine the occupation numbers are described, and also the applications of the natural orbital method, the nucleon momentum distributions at finite temperatures and new correlation methods for the nucleon and cluster momentum distributions. An account is given of the y-scaling method for extracting information on the nucleon momentum distribution from the data on the inelastic scattering of electrons.

The application of the generator coordinate method and the coherent density fluctuation model to calculate the energies and density distributions of the excited monopole states is described, as are the problems associated with understanding the longitudinal and transverse response functions in quasi-elastic electron scattering. A section is devoted to the theories used to describe deep-inelastic lepton–nucleus scattering and another to photonuclear processes at energies above the giant resonance region. Theoretical studies of the effects of nucleon–nucleon correlations on the cross-sections of elastic and deep-inelastic proton scattering at intermediate energies are described, and also the effects of nucleon–nucleon correlations on alpha-particle and heavy-ion elastic scattering.

Much of this theoretical work has been stimulated by new experimental data, in particular electron and proton inelastic scattering, nucleon transfer and photonuclear reactions, and alpha-particle and heavy-ion scattering. Throughout the book the theories are compared with the latest experimental data, enabling their usefulness and range of validity to be assessed. This has given important new information on the effects of the nucleon correlations in nuclei. Many previous reviews have been devoted to pairing effects and random phase approximation calculations, and this work is only briefly mentioned. The existence of the quark substructure is recognized, but only through the effects that cannot be explained in terms of nucleon–nucleon interactions. Few-body systems, which allow a more microscopic approach, are also excluded. Our treatment throughout is essentially nonrelativistic although some relativistic effects are discussed when we consider scattering at higher energies.

Mean-field and beyond mean-field theories are discussed in Part 1 (Chapters 1–4). Chapter 1 contains examples of independent-particle models and methods. The complex mean-field theories are described in Chapter 2. Different correlation methods applied to infinite nuclear matter and finite systems are considered in Chapter 3. In Chapter 4 the basic assumptions and formalism of the generator coordinate method, the coherent density fluctuation model and the natural orbital method are given.

Effects of nucleon–nucleon correlations on characteristics of nuclear states are considered in Part 2 (Chapters 5–8). Spectral functions, widths of deep-hole nuclear states and related sum rules are analysed in Chapter 5. Natural orbitals and occupation numbers of single-particle states are considered in Chapter 6. Particular attention is paid to the study of nucleon momentum and density distributions as well as cluster momentum distributions in nuclei (Chapter 7). Energies and density distributions in the ground and excited monopole states are discussed in Chapter 8 together with various correlation methods.

Nucleon–nucleon correlation effects are also studied in processes of particle–, photo– and nucleus–nucleus interactions and these form the subject of Part 3 (Chapters 9–12). Elastic and quasielastic electron– and deep-inelastic lepton–nucleus scattering are described in Chapter 9, photonuclear processes in Chapter 10, elastic and deep-inelastic intermediate energy proton–nucleus scattering in Chapter 11 and alpha-particle and heavy ion elastic scattering on nuclei in Chapter 12.

This work shows that nucleon–nucleon correlations are important for the correct description of nuclear structure and nuclear reactions, so that a realistic explanation of nuclear properties both at low and high energies has to be based on the methods going beyond the mean-field approximation.

We thank our colleagues, particularly Dr. V.A. Nikolaev, Dr. Chr.V. Christov, I.S. Bonev, E.N. Nikolov, Dr. M.V. Stoitsov, Dr. L.P. Kaptari, Dr. G.A. Lalazissis, Dr. F. Malaguti, Dr. A. Uguzzoni and Dr. E. Verondini, who have collaborated with us. They have contributed in important respects to the work described here and we thank them for allowing us to use material which we have published together.

We thank the Royal Society of London and the Bulgarian Academy of Sciences for supporting the exchange agreement that has made our co-operation possible.

We are also grateful to all colleagues, authors, and publishers who have permitted us to reproduce illustrations from their work.

Sofia and Oxford,   A.N. Antonov
January 1993   P.E. Hodgson
                                                                 I.Zh. Petkov

Contents

Part 1
Mean-field and Beyond Mean-field Nuclear Theoretical Methods

1 Independent-particle Models

In this chapter we present briefly the formalism of the Fermi-gas model (Sect. 1.1), the shell model (Sect. 1.2) and the Hartree–Fock method (Sect. 1.3). These examples of models and methods in the framework of the mean-field approximation, as well as the one-body density matrix, and the density and nucleon momentum distributions are necessary to understand how the nucleon–nucleon correlations are described in the remainder of the book.

1.1 The Fermi-gas Model

The Fermi-gas model is the simplest of the independent-particle models. It is assumed that the nucleons are moving freely in a volume Ω. The mean free path of a nucleon in the nuclear medium is comparable with the size of the nucleus [1.1–4]. The single-particle wave function in the independent-particle Fermi-gas model is the plane wave:

$$\varphi_\alpha(r) = \Omega^{-1/2} \exp(i k_\alpha \cdot r), \tag{1.1}$$

where k_α is the nucleon momentum. The total wave function of the system is the Slater determinant:

$$\Psi_F(\{r_i\}) = \hat{\mathscr{A}} \prod_{i=1}^{A} \varphi_{\alpha_i}(r_i), \tag{1.2}$$

where $\hat{\mathscr{A}}$ is the antisymmetrization operator. If the normalization volume is a cube with sides of length L, then

$$k = \frac{2\pi}{L}(n_x, n_y, n_z), \tag{1.3}$$

where n_x, n_y and n_z are positive or negative integers.

The total energy of the independent-particle Fermi-gas is:

$$E_0 = \left\langle \Psi_F \left| -\frac{\hbar^2}{2m} \sum_i \nabla_i^2 \right| \Psi_F \right\rangle, \tag{1.4}$$

or [1.4b]:

$$E_0 = -\frac{\hbar^2}{2m} \int [\nabla_r^2 \rho(r, r')]_{r'=r} \, dr, \tag{1.5}$$

where the one-body density matrix is

$$\rho(r, r') = \int \Psi_F^*(r, r_2, \ldots) \Psi_F(r', r_2, \ldots) dr_2 \ldots dr_A. \tag{1.6}$$

For plane waves this has the form:

$$\rho(r, r') = \frac{4}{\Omega} \sum_{k < k_F} \exp(-ik \cdot r') \exp(ik \cdot r). \tag{1.7}$$

The factor 4 in (1.7) accounts for the spin and isospin degeneracy of each state. In the ground state of the system the single-particle levels are occupied up to the Fermi-level which has the momentum k_F and energy ε_F. The set of all occupied levels is called the Fermi-sea. If this sea contains A nucleons then

$$A = 4 \sum_{|k| < k_F} 1 \rightarrow 4\Omega \int_{|k| \leq k_F} \frac{dk}{(2\pi)^3} = \frac{16\pi}{3} (k_F/2\pi)^3 \Omega. \tag{1.8}$$

This enables the density $\rho_0 = A/\Omega$ to be expressed in terms of the Fermi-momentum:

$$\rho_0 = \frac{16\pi}{3} (k_F/2\pi)^3. \tag{1.9}$$

It follows from (1.9) that

$$k_F = (3\pi^2 \rho_0/2)^{1/3}. \tag{1.10}$$

Equations (1.9) and (1.10) give the important relation between the Fermi-momentum and the density of the system.

In analogy with the substitution used in (1.8) the one-body density matrix (1.7) can be obtained also in the form:

$$\rho(r, r') = 3\rho_0 \frac{j_1(k_F|r' - r|)}{k_F|r' - r|}. \tag{1.11}$$

The Fermi-energy can be expressed in terms of the Fermi-momentum and the density:

$$\varepsilon_F = \frac{\hbar^2 k_F^2}{2m} = \frac{\hbar^2}{2m} (3\pi^2 \rho_0/2)^{2/3}. \tag{1.12}$$

The total energy of the system can be determined using (1.5), (1.7), (1.8) and (1.10):

$$E_0 = \frac{3}{5} A \frac{\hbar^2 k_F^2}{2m}. \tag{1.13}$$

The energy per nucleon is then

$$\frac{E_0}{A} = \frac{3}{5} \frac{\hbar^2 k_F^2}{2m} = \frac{3}{5} \varepsilon_F. \tag{1.14}$$

If one uses the values of the Fermi-momentum of nuclear matter of normal density (approximately $k_F \simeq 1.38 \text{ fm}^{-1}$ [1.4a]), then

$$\varepsilon_F \approx 39 \text{ MeV}, \tag{1.15}$$

which leads to $E_0/A \approx 24 \text{ MeV}$. The positive value of E_0/A shows that the noninteracting Fermi-gas model cannot describe the bound state of the system of nucleons. The model can be improved [1.4a] by considering a gas of non-interacting nucleons moving in a potential well with volume Ω and depth

$$V_0 = \varepsilon_F + S, \tag{1.16}$$

where S is the separation energy of a nucleon ($\sim 8 \text{ MeV}$). Thus from (1.15) and (1.16) $V_0 \simeq 47 \text{ MeV}$.

As can be seen from (1.11) the density distribution of the non-interacting Fermi-gas is constant:

$$\rho(r) = \rho(r, r' = r) = \rho_0 = \frac{A}{\Omega} = \text{const.} \tag{1.17}$$

The nucleon momentum distribution, which is expressed by the one-body density matrix:

$$n(k) = \int \rho(r, r') \exp[ik(r - r')] \, dr \, dr' \tag{1.18}$$

results in the form

$$n(k) = 4\Omega\Theta(k_F - |k|), \tag{1.19}$$

where the step-function

$$\Theta(k_F - |k|) = \begin{cases} 1, & |k| < k_F \\ 0, & |k| > k_F \end{cases} \tag{1.20}$$

determines the occupation numbers in the non-interacting Fermi-gas.

The inclusion of simple (often schematic) interaction between the nucleons can be used as a first step towards a more realistic Fermi-gas model. In [1.5] a dilute Fermi-gas with a repulsive interaction is considered. The inclusion of the interaction leads to to partial depletion of the levels below the Fermi-level

$$\frac{n(k < k_F)}{4\Omega} < 1 \tag{1.21}$$

and partial occupation of the levels above it

$$\frac{n(k > k_F)}{4\Omega} > 1. \tag{1.22}$$

It is found also that

$$n(k) \xrightarrow[k/k_F \to \infty]{} k^{-4} \tag{1.23}$$

when the conditions $k \gg k_F$ and $ka < 1$ (where a is the scattering length) are satisfied simultaneously. The results obtained for $n(k)$ show the effects of the nucleon–nucleon hard core repulsion when a has the value of the core size ($a \simeq 0.4$ fm). There is a discontinuity of the momentum distribution at k_F of an amount $[1 - 4(k_F a/\pi)^2 \ln 2]$. It has been shown by *Migdal* [1.6] that the existence of a discontinuity in the momentum distribution at $k = k_F$ is a property of infinite fermion systems with an arbitrary interaction between the nucleons.

The results from [1.5] have been confirmed by the studies of *Belyakov* [1.7] and *Sartor* and *Mahaux* [1.8] of dilute fermion systems with repulsive forces between the particles.

It is pointed out in [1.1] that the correlations induced by the nucleon–nucleon interaction have small effects on the nuclear binding energy but they smear out the form of the nucleon momentum distribution. The latter shows a resemblance with the finite temperature Fermi-gas momentum distribution. In this case the wave function contains high-momentum components. This is confirmed by the experimental data from various particle–nucleus interactions at higher energies (see Part 3).

1.2 The Shell Model

Due to the action of the Pauli principle the nucleus is not a dense system of nucleons and the strong nucleon–nucleon forces are reduced by the fact that the nucleons are quite far apart. The interactions due to the singular force at small nucleon–nucleon distances are infrequent and this makes the idea of the independent motion of the nucleons acceptable as a first approximation to the many-body problem [1.9].

Considering the nucleus as a non-relativistic quantum-mechanical system with two-body interactions between the nucleons, the nuclear problem is reduced to the solution of the Schrödinger equation:

$$\hat{H}\Psi(1, \ldots, A) = E\Psi(1, \ldots, A), \tag{1.24}$$

where

$$\hat{H} = \sum_{i=1}^{A} \left(-\frac{\hbar^2}{2m}\nabla_i^2 \right) + \sum_{i<j}^{A} v(i,j) \tag{1.25}$$

is the Hamiltonian of the system.

The mean field $V(i)$ in which the ith nucleon is moving can be introduced by the decomposition

$$H = T + \sum_{i}^{A} V(i) + \sum_{i<j}^{A} v(i,j) - \sum_{i}^{A} V(i) \tag{1.26}$$

$$\equiv \hat{H}_0 + \hat{V}_{\text{res}},$$

where

$$\hat{V}_{res} \equiv \sum_{i<j}^{A} v(i,j) - \sum_{i}^{A} V(i) \qquad (1.27)$$

is the residual interaction.

The main assumption of the shell model is to neglect the effects of the residual interaction:

$$\hat{H} \simeq \hat{H}_0 = \sum_{i}^{A} \left(-\frac{\hbar^2}{2m} \nabla_i^2 \right) + \sum_{i} V(i). \qquad (1.28)$$

This reduces (1.24) to the simpler equation

$$\sum_{i=1}^{A} \left\{ -\frac{\hbar^2}{2m} \nabla_i^2 + V(i) \right\} \Psi_{SM} = E \Psi_{SM}. \qquad (1.29)$$

The nuclear shell-model many-body wave function Ψ_{SM} is an antisymmetrized product of the single-particle functions $\varphi_k(i)$, which are solutions of the single-particle equations:

$$\left[-\frac{\hbar^2}{2m} \nabla_i^2 + V(i) \right] \varphi_k(i) = \varepsilon_k \varphi_k(i). \qquad (1.30)$$

The sum $E_0 = \sum_k \varepsilon_k$ is the ground state energy of the system.

In the case of spherical symmetry of the mean field (V) the angular momentum l is a good quantum number and the wave function $\varphi_{nlm}(r)$ can be written in the form:

$$\varphi_{nlm}(r) = \varphi_{nl}(r) Y_{lm}(\theta, \varphi), \qquad (1.31)$$

where the radial function $\varphi_{nl}(r)$ satisfies the equation [1.4]:

$$\frac{d^2 \varphi_{nl}}{dr^2} + \frac{2}{r} \frac{d\varphi_{nl}}{dr} + \frac{2m}{\hbar^2} \left[E_{nl} - V(r) - \frac{\hbar^2}{2m} \frac{l(l+1)}{r^2} \right] \varphi_{nl} = 0 \qquad (1.32)$$

and $Y_{lm}(\theta, \varphi)$ is the spherical harmonic. The function $\varphi_{nl}(r)$ has to be finite when $r \to 0$ and equal to zero at $r \to \infty$.

For light nuclei the potential $V(r)$ is usually approximated by the three-dimensional isotropic harmonic oscillator potential:

$$V(r) = -V_0 + \tfrac{1}{2} m \omega^2 r^2. \qquad (1.33)$$

The normalized radial wave function in this case is [1.4a]

$$\varphi_{nl}(r) = C_{nl} \exp(-\alpha^2 r^2/2) \cdot r^l \cdot F(1-n, l+3/2, \alpha^2 r^2), \qquad (1.34)$$

where F is the degenerate hypergeometric function and

$$C_{nl} = \frac{1}{\Gamma(l+3/2)} \left(\frac{2\Gamma(l+n+1/2)}{\Gamma(n)} \right)^{1/2} \cdot \alpha^{l+3/2}, \quad \alpha = (m\omega/\hbar)^{1/2}. \qquad (1.35)$$

Another simple form of the potential $V(r)$ is the spherical potential well with a finite depth

$$V(r) = \begin{cases} -V_0, & r < R \\ 0, & r > R \end{cases} \tag{1.36}$$

or with infinite walls:

$$V(r) = \begin{cases} -V_0, & r < R \\ \infty, & r > R. \end{cases} \tag{1.37}$$

For both (1.36) and (1.37), the solution of (1.32) in the internal region is

$$\varphi(r) = Cj_l(kr), \quad r < R \tag{1.38}$$

where $j_l(kr)$ is the spherical Bessel function. In the case (1.37) the boundary condition is

$$j_l(kR) = 0 \tag{1.39}$$

and the roots of the function $j_l(x)$ determine the energies of the states.

The third mean-field potential often used is the Saxon–Woods potential:

$$V(r) = -V_0\{1 + \exp[(r - R)/a]\}^{-1}. \tag{1.40}$$

The parameters of the mean-field potentials can be obtained by fitting the experimental data on single-particle energies, density distributions and radii.

The realistic potentials have to include the spin–orbit interaction $\sim \hat{l} \cdot \hat{s}$. For the harmonic oscillator and square-well forms an additional term depending on $\hat{l}^2$ is needed in order to ensure that the magic numbers are reproduced:

$$V = V(r) + f(r)\hat{l} \cdot \hat{s} + D\hat{l}^2. \tag{1.41}$$

The abilities of the shell model to describe the energy levels, the magnetic moments, the quadrupole moments, spins and parities, etc. are discussed widely in the literature (e.g. [1.1, 1.4, 1.9, 1.10]). Here we consider the one-body density matrix corresponding to Ψ_{SM}:

$$\rho(r, r') = \sum_{k=1}^{A} \varphi_k^*(r')\varphi_k(r). \tag{1.42}$$

The density and nucleon momentum distributions are respectively:

$$\rho(r) = \sum_{k=1}^{A} |\varphi_k(r)|^2, \tag{1.43}$$

and

$$n(k) = \rho(k, k' = k) = \sum_{i}^{A} |\tilde{\varphi}_i(k)|^2, \tag{1.44}$$

where $\tilde{\varphi}(k)$ is the Fourier transform of $\varphi(r)$.

It has to be emphasized that the local density distribution in the case of the harmonic oscillator potential (1.33) has the Gaussian asymptotic form at $r \to \infty$.

This is in contradiction with the behaviour of $\rho(r)$ exhibited in the experimental data $(e^{-\alpha r})$.

The nucleon momentum distribution obtained from the shell model which is discussed extensively in this book as a quantity sensitive to the short-range nucleon–nucleon correlations turns out to be quite unrealistic. The shell-model momentum distributions $n(k)$ have steep slopes when k increases in contradiction with the large high-momentum components obtained in the methods that include the nucleon–nucleon correlations and also those obtained from the available experimental data. This question is discussed thoroughly in Chap. 7.

1.3 The Hartree–Fock Approximation

The mean-field potential can be derived by the Hartree–Fock method [1.11, 1.12] from the two-body nucleon–nucleon interaction $\sum_{i<j} v_{ij}$ on the basis of the variational principle:

$$\delta E[\Psi] = 0, \tag{1.45}$$

where

$$E[\Psi] = \langle \Psi|\hat{H}|\Psi\rangle / \langle \Psi|\Psi\rangle, \tag{1.46}$$

and

$$\hat{H} = \sum_i t_i + \sum_{i<j} v(i,j) \tag{1.47}$$

is the Hamiltonian of the system. The variational space of functions consists of a set of Slater determinants built up from single-particle functions $\varphi_k(i)$. The lowest value of the energy E of the A-nucleon system can be found by varying $E[\Psi]$ with respect to the single-particle functions $\varphi_k(i)$. This procedure leads to the Hartree–Fock system of equations which has the following form in the coordinate space for a local two-body potential $v(r_i, r_j)$ not depending on spin and isospin variables [1.2, 1.9, 1.13]:

$$-\frac{\hbar^2}{2m}\nabla_k^2\,\varphi_k(r_k) + \sum_{j=1}^{A}\int \varphi_j^*(r_j)v(r_j,r_k)\,\varphi_j(r_j)\varphi_k(r_k)\,dr_j$$

$$-\sum_{j=1}^{A}\int \varphi_j^*(r_j)v(r_j,r_k)\,\varphi_j(r_k)\varphi_k(r_j)\,dr_j = \varepsilon_k\varphi_k(r_k), \quad k = 1,2,\ldots,A. \tag{1.48}$$

The system (1.48) can be written also in the form:

$$-\frac{\hbar^2}{2m}\nabla_k^2\varphi_k(r_k) + V_H(r_k)\varphi_k(r_k) - \int V_F(r_k,r_j)\varphi_k(r_j)\,dr_j = \varepsilon_k\varphi_k(k_k), \tag{1.49}$$

where

$$V_H(r_k) \equiv \sum_{j=1}^{A}\int \varphi_j^*(r_j)v(r_j,r_k)\varphi_j(r_j)\,dr_j \tag{1.50}$$

and

$$V_{\mathrm{F}}(r_k, r_j) \equiv \sum_{j=1}^{A} \varphi_j^*(r_j) v(r_j, r_k) \varphi_j(r_k) \tag{1.51}$$

are the local Hartree- and non-local Fock-contributions to the mean-field potential. The latter can be found in a self-consistent way by solving the equations (1.48) or (1.49) by iteration starting with a set of phenomenological shell-model wave functions. The Hartree–Fock basis of single-particle functions obtained from (1.48) and (1.49) minimizes the ground state expectation of the Hamiltonian:

$$E_{\mathrm{HF}} = \langle \Psi_{\mathrm{HF}} | \hat{H} | \Psi_{\mathrm{HF}} \rangle. \tag{1.52}$$

The wavefunction Ψ_{HF} in (1.52) is a Slater determinant built from the A deepest bound orbitals φ_j. In the Hartree–Fock ground state of the system all A single-particle states are occupied up to the highest state which is called the Fermi level. The ground state energy is

$$E_{\mathrm{HF}} = \sum_{k=1}^{A} \varepsilon_k - \frac{1}{2} \sum_{k, j=1}^{A} \langle kj | v | \widetilde{kj} \rangle, \tag{1.53}$$

where

$$\begin{aligned} \langle kj | v | \widetilde{kj} \rangle = &\iint \varphi_k^*(r) \varphi_j^*(r') v(r, r') \varphi_k(r) \varphi_j(r') \mathrm{d}r \, \mathrm{d}r' \\ &- \iint \varphi_k^*(r) \varphi_j^*(r') v(r, r') \varphi_j(r) \varphi_k(r') \mathrm{d}r \, \mathrm{d}r'. \end{aligned} \tag{1.54}$$

The one-body density matrix has the following form in the Hartree–Fock approximation:

$$\rho_{\mathrm{HF}}(r, r') = \sum_{\alpha}^{\alpha_{\mathrm{F}}} \varphi_\alpha^*(r') \varphi_\alpha(r), \tag{1.55}$$

where α_{F} denotes the Fermi level. The density and nucleon momentum distributions are respectively:

$$\rho_{\mathrm{HF}}(r) = \sum_{\alpha}^{\alpha_{\mathrm{F}}} |\varphi_\alpha(r)|^2 \tag{1.56}$$

and

$$n_{\mathrm{HF}}(k) = \sum_{\alpha}^{\alpha_{\mathrm{F}}} |\tilde{\varphi}_\alpha(k)|^2, \tag{1.57}$$

where $\tilde{\varphi}_\alpha(k)$ is the Fourier transform of $\varphi_\alpha(r)$.

The Hartree–Fock and the Hartree–Fock-type mean-field methods make it possible for many nuclear characteristics to be calculated from given nucleon–nucleon interactions. In the case of the hard repulsive core in the nucleon–nucleon interaction this scheme is not, however, applicable because the matrix elements diverge. Usually this difficulty is overcome by using an effective interaction with parameters adjusted to fit particular experimental data. In the Brueckner–Hartree–Fock method which goes beyond the limits of the Hartree–Fock approximation (see Sect. 3.2) the nucleon–nucleon interaction is renormalized by the g-matrix expansion. Alternatively, higher order many-body terms are taken into account by introducing a density dependence. The density-dependent Hartree–Fock method (see e.g. *Negele* [1.14], *Bartz* et al. [1.13], *Svenne* [1.15], *Ring* and *Schuck* [1.9]) and its application to calculate nuclear characteristics, such as binding energies, density distributions, radii and nucleon momentum distributions is discussed in Sect. 3.2. In general, the use of effective forces and especially the effective forces of *Skyrme* [1.16] and *Gogny* [1.17] for the binding energies, nuclear density distributions and mean-square radii in the ground state of spherical nuclei leads to a good agreement with the experimental data on electron- and proton-nucleus scattering. The results for the nucleon momentum distributions show, however, large deviations from the high-momentum components of $n(k)$ (at $k > 2\,\mathrm{fm}^{-1}$) found using more realistic correlation methods (Chaps. 3 and 4) and from the experimental data for various nuclei. The Hartree–Fock-momentum distributions are close to the shell-model predictions with a steep fall-off for large momenta. The main reason for this behaviour of $n(k)$ is that the determinant Hartree–Fock ground state nuclear wave function includes the Pauli correlations but does not include the important part of the short-range, tensor and other nucleon–nucleon correlations. These correlations which are related to the characteristic features of the nucleon–nucleon forces at small distances lead to the high-momentum part of the nucleon momentum distribution (Chap. 7) as well as to the depletion of the Fermi sea (Chap. 6), to the form of the hole-state spectral functions which is far from the shell-model predictions (Chap. 5) and to the important effects on different particle- and ion–nucleus scattering cross-sections discussed in Part 3 of this book. An analysis of the limits of the Hartree–Fock approximation in nuclear theory is given in Sect. 3.1.

2 Complex Mean-field Analyses

In this chapter the complex nuclear potential, or optical potential, is considered. This potential describes the motion of a nucleon, bound or unbound, in the mean field of all the other nucleons comprising the nucleus. The field due to the sum of all the individual nucleon–nucleon interactions is thus represented by a simple one-body potential. This greatly simplifies the calculation of a wide range of nuclear structure and nuclear reaction phenomena. In this sense this chapter is related to the review of the mean-field approaches given in Chap. 1. It contains more details about the application of this mean-field approach to the study of the bound single-particle states and the scattering states. On the other hand, this approach contains the possibility of going beyond the Hartree–Fock approximation. In this sense this chapter presents one of the ways for the introduction of the consistent correlation methods (beyond the mean-field approximation) and has an intermediate place in Part 1 of this book between the review of the independent-particle models (Chap. 1) and that of the correlation methods given in Chaps. 3 and 4.

In Sect. 2.1 we give briefly the idea of the introduction of the optical potential and its place in the many-body nuclear theory.

The knowledge of the complex one-body potential can be unified in two different but connected ways. Firstly, the potential is defined for both negative and positive energies, so that it may be used to calculate the properties of bound single-particle states and also the scattering of unbound nucleons by nuclei (Sect. 2.2). Secondly, the real and imaginary part of the potential can be no longer treated as separate but are connected by the dispersion relations (Sect. 2.3) which are also applied to unify the data at negative and positive energies.

2.1 The Complex Potential. Formal Considerations

The idea of the introduction of the optical potential arose as an attempt to reduce the many-body problem for the elastic scattering of a particle on a system of bound particles to a single-particle problem and thus to avoid explicit consideration of the non-elastic processes.

We shall consider briefly the formal theory of the generalized optical potential (see, e.g. [2.1–6]).

The state of a system consisting of $A + 1$ particles (a nucleus with A nucleons and an incident particle) is described by means of the many-particle Schrödinger equation

$$(E - \hat{H})\Psi = 0 \tag{2.1}$$

with

$$\hat{H} = \hat{H}_0 + \hat{V} + \hat{H}(\xi), \tag{2.2}$$

where $\hat{H}_0$ is the kinetic energy operator for the incident nucleon and $\hat{V}$ is the interaction potential $\hat{V} = \sum_{i=1}^{A} v(r, R_i)$, where v is the two-particle interaction between the initial particle with radius-vector r and the nucleons of the target with radius-vectors R_i. $\hat{H}(\xi)$ is the internal Hamiltonian of the target nucleus. If the eigenstates of $\hat{H}(\xi)$ are $\phi_\alpha(\xi)$ with energies ε_α and $\varepsilon_0 = 0$, then the total wave function Ψ can be written as an expansion

$$\Psi = \sum_\alpha \psi_\alpha \phi_\alpha. \tag{2.3}$$

If only the elastic channel is open, the open-channel part of Ψ is $\psi_0 \phi_0$. Using the projection operators $\hat{P}$ and $\hat{Q}$ which project on and off the open channels:

$$\hat{P}\Psi = \psi_0 \phi_0, \quad \hat{Q}\Psi = (1 - \hat{P})\Psi \tag{2.4}$$

it can be shown (e.g. [2.4]) that the equation for ψ_0 has the form:

$$\left[E - \hat{H}_0 - \langle \phi_0 | \hat{V} | \phi_0 \rangle - \left\langle \phi_0 \left| \hat{V}\hat{Q} \frac{1}{E - \hat{H}_{QQ}} \hat{Q}\hat{V} \right| \phi_0 \right\rangle \right] \psi_0 = 0, \tag{2.5}$$

where

$$\hat{H}_{QQ} = \hat{Q}\hat{H}\hat{Q}. \tag{2.6}$$

The expression

$$\hat{V}_{\text{opt.}} = \langle \phi_0 | \hat{V} | \phi_0 \rangle + \left\langle \phi_0 \left| \hat{V}\hat{Q} \frac{1}{E - \hat{H}_{QQ}} \hat{Q}\hat{v} \right| \phi_0 \right\rangle \tag{2.7}$$

defines the generalized optical potential operator. In the case of a local interaction V and if the exchange effects are neglected the term $\langle \phi_0 | \hat{V} | \phi_0 \rangle$ is a single-particle potential which is local in configuration space and gives the interaction between the incident nucleon and the target nucleus in its ground state. The second term accounts for the transitions to different intermediate excited states of the nucleus. This term is non-local in configuration space.

In this way the complicated operator $\hat{V}_{\text{opt}}$ leads formally to the reduction of the many-particle problem to a single-particle one (Eq. (2.5)). The expression (2.7) is complex and therefore can be written in the form

$$\hat{V}_{\text{opt}} = \hat{V} + i\hat{W}. \tag{2.8}$$

The imaginary term $\hat{W}$ determines the part of the incident particles which go in the non-elastic channels, that are responsible for the "absorption" of the particles by the target nucleus:

$$\mathrm{div}\left\{\frac{\hbar}{2im}(\psi_0\nabla\psi_0^* - \psi_0^*\nabla\psi_0)\right\} = \frac{2}{\hbar}W\psi_0^*\psi_0. \tag{2.9}$$

An important property of the optical potential is its relation to the correlation function of the nucleus. If the two-particle interaction v is expressed by the effective interaction t (related to the amplitude of the scattering of the particle on a single nucleon in nuclear medium) and the mass number A is replaced by $A - 1$, then the optical potential takes the form:

$$V_{\mathrm{opt}} = V_{00}' + \sum_{n\neq0} V_{0n}'(E - H_0 - E_n - V_{nn})^{-1}V_{n0}'. \tag{2.10}$$

In (2.10):

$$V_{nn} = (A - 1)\langle n|t|n\rangle, \tag{2.11}$$

$|n\rangle$ being a nuclear eigenfunction and E_n the nuclear excitation energies.

In the applications of the optical potential the effective interaction t is often replaced by the two-particle scattering amplitude ($t \to t_0$, impulse approximation). Then the first term in (2.10) is obtained in the form [2.6]

$$V_{00} \approx (A - 1)\rho(r)t_0(0), \tag{2.12}$$

where $\rho(r)$ is the one-body density distribution in the target and $t_0(0) = -2\pi f(0)/m$, $f(0)$ being the scattering amplitude at $\theta° = 0°$.

The nucleon–nucleon correlations are contained in the second term of Eq. (2.10). At high energies E it can be assumed that the dependence on the index n in the propagator of the expression

$$\Delta V_{\mathrm{opt}} = \sum_{n\neq0} V_{0n}(E - H_0 - E_n - V_{nn})^{-1}V_{n0} \tag{2.13}$$

can be neglected. E_n and V_{nn} can then be replaced by average quantities $\bar{E}$ and $\bar{V}$, respectively. In this case ΔV_{opt} (2.13) can be expressed in the momentum representation by the quantity:

$$K(\boldsymbol{q}_1, \boldsymbol{q}_2) \equiv \sum_{n\neq0} V_{0n}(\boldsymbol{q}_1) V_{n0}(\boldsymbol{q}_2)$$

$$= (A - 1)^2 t(\boldsymbol{q}_1)t(\boldsymbol{q}_2)K_0(\boldsymbol{q}_1, \boldsymbol{q}_2), \tag{2.14}$$

where

$$K_0(\boldsymbol{q}_1, \boldsymbol{q}_2) = (A - 1)/A[\rho(\boldsymbol{q}_1, \boldsymbol{q}_2) - \rho(\boldsymbol{q}_1)\rho(\boldsymbol{q}_2)] + O\left(\frac{1}{A}\right) \tag{2.15}$$

is the nuclear correlation function.

In (2.15):

$$\rho(\boldsymbol{q}_1, \boldsymbol{q}_2) = \iint d\boldsymbol{r}_1\,d\boldsymbol{r}_2\,\rho(r_1, r_2)\exp[-i(\boldsymbol{q}_1\cdot\boldsymbol{r}_1 + \boldsymbol{q}_2\cdot\boldsymbol{r}_2)], \tag{2.16}$$

where $\rho(r_1, r_2)$ is the one-body density matrix and

$$\rho(q) = \int dr\, \rho(r) \exp(-i q \cdot r). \tag{2.17}$$

The difference $\Delta\rho = \rho(r_1, r_2) - \rho(r_1)\rho(r_2)$ which is the Fourier transform of the main term in (2.15) shows that the nucleon–nucleon correlations are related to the Pauli principle and to the character of the nucleon–nucleon forces at small distances. For instance, $\Delta\rho$ is different from zero in the case of hard core nucleon–nucleon interactions independently of the Pauli-principle.

2.2 The Complex Potential for Bound and Scattering States

i) Bound Single-particle States

A nucleon bound in a one-body potential has a series of allowed eigenstates characterized by their energies and quantum numbers. Such states are found in nuclei by studies of one-nucleon transfer and knockout reactions and these enable their characteristics to be determined. It is found to good accuracy that the same potential gives all the energies and quantum numbers of the single-particle states in a particular nucleus [2.7].

The potential has central and spin-orbit terms

$$V(r) = U_u f_u(r) + U_s \left(\frac{\hbar}{m_\pi c}\right)^2 \cdot \frac{1}{r} \frac{d f_s(r)}{dr} L \cdot \sigma, \tag{2.18}$$

where U_u and U_s are the central and spin–orbit potential depths. The radial variation of the central part of the potential follows rather closely the nuclear density distribution, and this is conveniently parametrised by the Saxon–Woods form factor

$$f_u(r) = \{1 + \exp[(r - R)/a_n]\}^{-1}, \tag{2.19}$$

where $R = r_u A^{1/3}$ and a_n are the radius and surface-diffuseness parameters. The radial variation of the spin–orbit term is chosen to have the surface-peaked form $(d f_s(r)/dr)/r$, where $f_s(r)$ also has the Saxon–Woods form. $L \cdot \sigma$ term in (2.18) is the spin–orbit operator. Inclusion of the spin–orbit term splits the single-particle states in accordance with experiment and describes correctly the magic numbers of nucleons. Typical values of these parameters are $U_u \approx 50\,\text{MeV}$, $U_s = 7\,\text{MeV}$, $r_u = 1.25\,\text{fm}$, $a_n = 0.65\,\text{fm}$, $r_s = 1.1\,\text{fm}$ and $a_s = 0.65\,\text{fm}$.

Many detailed studies have shown that these parameters give a good account of the energies of single-particle states for a range of nuclei [2.8, 2.9]. In addition, the wave functions of these states ψ_α can be combined with the occupation numbers n_α that can also be found from analysis of nucleon transfer reactions to give the nuclear density distributions

$$\rho(r) = \sum_\alpha n_\alpha |\psi_\alpha|^2, \tag{2.20}$$

where the sum runs over all occupied states. The Eq. (2.20) can be used to calculate the charge distribution by including only the protons and the matter distribution by including both neutrons and protons. In these calculations the charge and matter distributions of the nucleons are folded in, and several other corrections are included [2.10–12].

It is usual in these analyses to fix the potential parameters to suitable average values and to determine the values of the potential depths for each state. It is then found that they vary with energy in a way that will be discussed in detail later.

Experimental studies of these bound single-particle states show that in general they are split into a number of fragments, and this effect is greatest away from the closed shells. It is still possible to define the energy of the state as the centroid of the fragment distribution, and in addition the width of the distribution may be connected with the imaginary part of the potential $W(r)$. For an unfragmented state the width is

$$\Gamma(E) = 2\langle \psi_\alpha | W(r, E) | \psi_\alpha \rangle. \tag{2.21}$$

In the case of a fragmented state this must be replaced by a sum of similar expressions for all the fragments.

Nucleon transfer reactions are most appropriate for the study of single-particle states near the Fermi surface, and in this region the individual fragments may be resolved and identified. Deeper states may be studied by knock-out reactions such as (e, e′p) and (p, 2p) and it is then not possible to resolve the fragments. The energy and width of the single-particle state are then identified with the mean energy and width of the fragment distribution, after allowance is made for the energy resolution of the experimental apparatus.

ii) Scattering States

The differential elastic scattering cross-sections and polarizations for the interaction of nucleons by nuclei have been measured for many nuclei and energies. These data may be analyzed using an one-body potential whose real part has the same form (2.18) as that used for bound states with the addition of an imaginary potential

$$W(r) = Wg(r), \tag{2.22}$$

where W is the depth of the imaginary part of the potential and $g(r)$ is the form factor that may have the volume form $g_v(r) = f(r)$ or the surface derivative form

$$g_d(r) = 4a_d \frac{\mathrm{d}f(r)}{\mathrm{d}r}. \tag{2.23}$$

The factor $4a_d$ in (2.23) is introduced to ensure that the maximum value of $g_d(r)$ is unity. The subscript v and d indicate the parameters of the volume and surface derivative form factor parameters.

The optimum values of the parameters of the potential may be found by varying them until the best fit to a particular data set is obtained. Many analyses have been made in this way, and the resulting parameters tabulated [2.13]. The method of analysis has been discussed by *Hodgson* [2.14].

A standard optical model analysis is one where the form factor parameters are fixed to average values, and the data fitted by variation of the potential depths only. It is then found that good fits can be obtained with potentials that vary smoothly with energy and show little variation from one nucleus to another. The optical potential thus provides a concise way of describing a vast amount of experimental data and also makes possible quite reliable calculations of cross-sections and polarizations for nuclei and energies for which no data are available. More fundamentally, it shows that one may usefully define a nuclear mean field valid for all nuclei.

This concept, however, has obvious limitations. The potential must be affected to some extent by the differences in structure among nuclei, but these effects are quite small and so the overall usefulness of the mean-field concept remains. As the experimental data improved in quality, some of these effects became apparent. The real potential was found to have a small but definite dependence on the nuclear asymmetry parameter $(N - Z)/A$, and there are some effects of shell structure, particularly on the imaginary potential. In the case of light nuclei, and for analyses at low energies, it was often not possible to obtain an adequate fit even if all the parameters of the potential were varied.

Even for medium and heavy nuclei, standard analyses sometimes failed to give an optimum fit to the data. These anomalous cases were of several types:
a) Analyses of the elastic scattering of 30, 40 and 61.4 MeV protons by ^{208}Pb and several other nuclei showed that an improved fit to the differential cross-section is obtained by adding a surface-peaked potential of Saxon–Woods derivative form to the standard Saxon–Woods potential [2.15].
b) Analyses of the elastic scattering of 7 and 22 MeV neutrons by ^{208}Pb gave significantly different optimum values of the form factor parameters at these two energies [2.16, 2.17]. It was found possible to fit these and other data from 0 to 24 MeV with potentials having energy-dependent form factor parameters

$$r_u = 1.302 - 0.0055 \cdot E, \qquad r_w = 1.383 - 0.0042 \cdot E,$$

$$a_u = 0.162 + 0.019 \cdot E, \qquad a_I = 0.7.$$

If these data are analysed with energy-independent form factors the overall fits are significantly poorer, and the optimum real potential departs from a linear energy dependence at low energies. The volume integral of the real potential shows an anomalous departure from linearity in the region of the Fermi surface.
c) The values of the total neutron cross-section calculated at around 1 MeV from a standard potential fitted to data above 5 MeV are significantly higher than the measured values.
d) In addition to these anomalies at lower energies there is evidence that at higher energies the Saxon–Woods form factor for the real part of the optical potential is inadequate [2.18].

Faced with such departures from the standard model, two responses are possible. The phenomenological response is to introduce more adjustable parameters by allowing the existing parameters to vary with energy, or by adding different form factors, or by allowing the potential to depend on some features of nuclear structure. Alternatively, one can seek a more fundamental understanding of the optical potential in the hope that this will enable the anomalies to be understood without the need to increase the number of adjustable parameters. As will be shown later, this hope has indeed been fulfilled by the use of dispersion relations.

iii) The Nuclear Mean Field

The results of the analyses of bound and scattering states described above may be combined to give the behaviour of the nuclear mean field from negative to positive energies.

The real potential depth U in MeV is found to vary with energy. The overall dependence on energy E (in MeV) for protons can be expressed as a quadratic function [2.19]:

$$U = 52.4 - (0.37 \pm 0.02) \cdot E + (0.0007 \pm 0.0001) \cdot E^2$$
$$+ 24 \, (N - Z)/A + 0.4 \, Z/A^{1/3}. \tag{2.24}$$

The last two terms describe the slight dependence of the potential on the nuclear asymmetry parameter, and the effect of the energy dependence in the presence of the Coulomb field. This smoothly varying part of the potential is identified as the Hartree–Fock field, and its energy dependence is attributed to the nonlocality of the potential.

In addition to this almost linear dependence there is evidence for non-linear behaviour around the Fermi surface at about -8 MeV. This effect is called the Fermi surface anomaly.

The spin–orbit potential U_s has a much weaker energy dependence that can be represented over the whole range of energies by the linear expression [2.20]:

$$U_s = (6.5 - 0.023 \cdot E) \, \text{MeV}. \tag{2.25}$$

The energy dependence of the imaginary potential is more difficult to determine because the form factor has predominantly the surface form at low energies and changes continuously to the volume form at higher energies. Furthermore, few analyses have been made with both forms included because it is then difficult to obtain the strength of either of them with reasonable accuracy.

Analyses with only the volume form fix the imaginary potential at positive energies with fair precision, and at negative energies there are some data available on the widths of single-particle states. Theoretical arguments indicate that the strength of the potential is symmetric about the Fermi energy, and this is consistent with the experimental data. The energy variations of both the volume and surface imaginary potentials for neutrons on ^{40}Ca are shown in Fig.

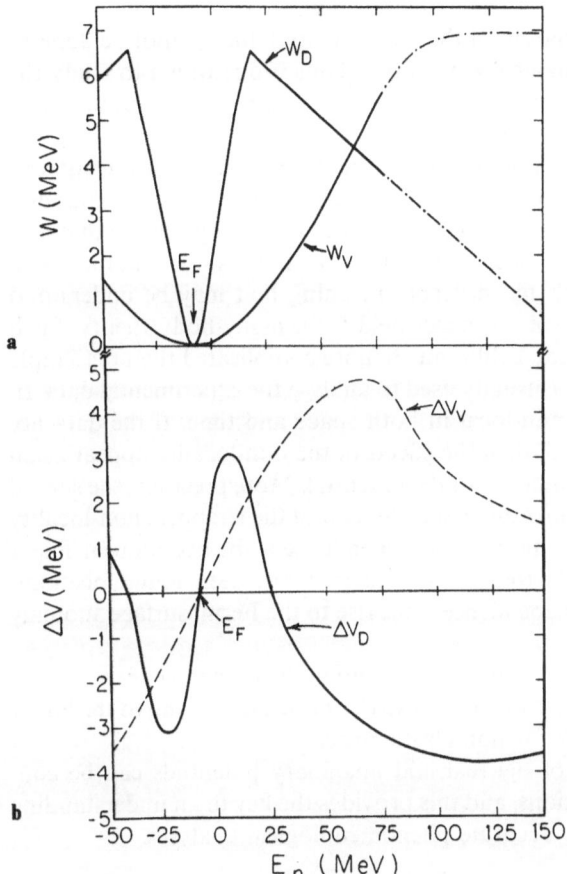

Fig. 2.1. (a) Energy variation of the volume (W_v) and surface (W_D) parts of the imaginary potential for neutrons on ^{40}Ca. (b) The corresponding corrections to the real potential calculated from the dispersion relations assuming that the energy dependence of the imaginary potential is symmetric about the Fermi energy E_F [2.21]

2.1, together with the corrections to the real potential that are obtained using the dispersion relations described in the following Sect. 2.3.

The difficulty of defining the imaginary potential is connected with its greater sensitivity to nuclear structure. As it takes into account the sum of all non-elastic processes it depends on the strength of those processes, and hence on the number of open reaction channels. Since the level density is notably smaller for nuclei near closed shells one can expect them to have smaller values of the imaginary potential than other nuclei, and this is indeed found. Thus the imaginary potential depends quite markedly on nuclear structure. The real potential, on the other hand, depends rather weakly on nuclear structure, the only established effect being the dependence on the asymmetry parameter shown in Eq. (2.24). Thus while it is possible to combine data from many nuclei to give an overall energy variation of the real potential, this cannot be done to the same accuracy for the imaginary potential. Thus if one wants to study the relation between the real and the imaginary potential, as is done later, this must be done for each nucleus separately.

It is notable that the concept of the nuclear mean field unifies the phenomena associated with bound nucleons in single-particle states and with unbound nucleons in scattering states. It is the same complex effective potential that acts on nucleons at all energies. The phenomenological analyses show that this mean field is energy dependent; this has a deeper meaning that may be understood with the help of calculations of the mean field from many-body theory. Such calculations show that the mean field is much more complicated than the simple phenomenological expressions usually used to analyse the experimental data. In particular, it is found to be non-local in both space and time. If the data are analysed with a local potential, then the effects of the non-locality appear as an energy dependence of the parameters of the potential. More precisely, the spatial non-locality give rise to a momentum dependence and the temporal non-locality to an energy dependence. The momentum dependence is the overall near-linear energy dependence of the Hartree–Fock field that has been found phenomenologically, and the energy dependence gives rise to the Fermi surface anomaly already mentioned [2.22].

Theoretical studies of the imaginary potential show that, as expected, it increases with energy as we go away from the Fermi surface, owing to the larger number of channels available for non-elastic processes.

These energy variations of the real and imaginary potentials can be connected by the dispersion relations, and this provides the key to an understanding of the anomalies revealed by accurate phenomenological analyses.

2.3 Dispersion Relations Methods

i) Basic Relationships

As shown in Sect. 2.1 the optical potential is a complex non-local function $v(k, E)$ of the momentum k and the energy E. It is an analytic function of the energy and therefore satisfies the dispersion relation (DR)

$$v(k, E) = \frac{1}{2\pi i} \int_C \frac{v(k, E')}{E' - E} \, dE' , \qquad (2.26)$$

where the contour C encloses the singularity at $E' = E$. The physical meaning of the dispersion relation is associated with the requirement of causality. If we impose the condition that the wave scattered by a potential cannot be emitted before the incident wave arrives, then the DR follows as a necessary consequence [2.23].

The complex optical potential is separated into real and imaginary parts (Eq. (2.8)):

$$v(k, E) = V(k, E) + i W(k, E). \tag{2.27}$$

Evaluating the contour integral and separating into real and imaginary parts then gives the DR

$$V(k, E) = \frac{P}{\pi} \int_{-\infty}^{\infty} \frac{W(k, E')}{E' - E} dE' \tag{2.28}$$

and

$$W(k, E) = -\frac{P}{\pi} \int_{-\infty}^{\infty} \frac{V(k, E')}{E' - E} dE', \tag{2.29}$$

where P indicates principal value.

These dispersion relations connect the real and imaginary parts of the potential, whereas in the phenomenological analyses they are treated as independent of each other. They therefore provide a way of unifying the potential over the whole energy range. The early attempts to apply them to the optical potential were made by *Passatore* [2.24]. An obvious difficulty is the range of integration, that far exceeds the range available experimentally. It is only relatively recently that a way has been found to use the DR to obtain significant results.

The first step is to separate the real potential into a part that depends only on the momentum and a part depending only on the energy. Then Eq. (2.28) becomes

$$V(k, E) = V_{HF}(k) + \frac{P}{\pi} \int_{-\infty}^{\infty} \frac{W(E')}{E' - E} dE'. \tag{2.30}$$

A more detailed discussion of this DR is given by *Mahaux* and *Ngô* [2.25].

The momentum-dependent part $V_{HF}(k)$ may be identified with the Hartree–Fock field and the energy-dependent part with the Fermi surface anomaly. The deviation of the real potential $V(k, E)$ from the Hartree–Fock field is related to the subject of this book and is discussed in detail in Chaps. 3 and 4.

The difficulty due to the range of the energy integration may be overcome in three ways. Firstly by separating the imaginary potential into two parts, secondly by using subtracted DR and thirdly by using the symmetry of the imaginary potential about the Fermi energy. These will now be discussed. The symbol P, indicating principal part, will be omitted from subsequent formulae.

The first method uses the different energy dependence of the surface-peaked and volume components of the imaginary potential

$$W(r) = W_d(r) + W_v(r). \tag{2.31}$$

The surface-peaked potential $W_d(r)$ dominates at low energies and falls to zero at large energies, so that the integral in (2.30) converges. The volume term $W_v(r)$ has the same radial dependence as the Hartree–Fock potential $V_{HF}(k)$ and so may be absorbed into it, possibly producing the small departure from linearity already mentioned. The effect of the DR (2.30) is thus to add to the real potential a small surface-peaked component derived from the surface-peaked component of the imaginary potential. We thus obtain an explanation of the anomaly a) mentioned in Sect. 2.2. Furthermore, since the addition of a surface-peaked potential leads to an increase of the radius of the equivalent volume potential, the anomaly b) is also explained.

The method of subtracted DR uses the expression

$$V(k, E) - V(k, E_F) = \frac{E - E_F}{\pi} \int\limits_{-\infty}^{\infty} \frac{W(E')}{(E' - E)(E' - E_F)} \, dE', \tag{2.32}$$

where E_F is the Fermi energy. This integral converges sufficiently rapidly for it to be evaluated, and may therefore be applied to potentials obtained in analyses using a volume potential only. It is therefore convenient to use this method for potentials obtained by phenomenological analyses in which the form factor parameters do not vary with energy. In such analyses all the effects of the Fermi surface anomaly appear in the potential depth. The alternative method, just described, shows the origin of the anomaly in an increased radial dependence due to the surface-peaked part of the imaginary potential.

If the imaginary potential is symmetric about the Fermi energy, Eq. (2.30) becomes

$$V(k, E) = V_{HF}(k) + \frac{2}{\pi}(E - E_F) \int\limits_{E_F}^{\infty} \frac{W(E') \, dE'}{(E' - E_F)^2 + (E - E_F)^2}. \tag{2.33}$$

This integral has convergence properties similar to (2.32) and so provides an alternative method of analysis.

The phenomenological optical model analyses determine certain moments of the potential with greater accuracy than the individual parameters. It is therefore often useful to work with radially integrated DR. Particularly useful is the volume integral per nucleon defined by

$$J_V = \frac{4\pi}{A} \int\limits_{0}^{\infty} V(r) \cdot r^2 \, dr \tag{2.34}$$

and a similar relation for the imaginary part of the potential.

The corresponding DR is

$$J_V(E) = J_V^{HF}(k) + \frac{1}{\pi} \int\limits_{-\infty}^{\infty} \frac{J_w(E')}{E' - E} \, dE'. \tag{2.35}$$

In general one can define the qth moment of the potential by

$$J_V^{(q)} = \frac{4\pi}{A} \int\limits_0^\infty V(r) r^q \, dr, \tag{2.36}$$

which satisfies the DR

$$J_V^{(q)}(E) = J_V^{(q)\text{HF}}(k) + \frac{1}{\pi} \int\limits_{-\infty}^\infty \frac{J_w^{(q)}(E')}{E' - E} \, dE'. \tag{2.37}$$

Subtracted DR for the radially integrated potentials can be defined as before.

The above discussion already shows, in a qualitative way, how the application of the DR gives additional insight into the optical potential and relates the real and imaginary parts in a way that enables us to understand the anomalies found in standard optical model analyses, particularly those centred on the Fermi surface.

The physical reason for the Fermi surface effect is the coupling between the elastic and non-elastic channels. If there were no excited states of the target nucleus and no reaction channels, the imaginary potential would fall to zero and the Fermi surface effect would vanish. If for simplicity only inelastic channels are considered, then the elastic scattering can take place via virtual nuclear excitations and this affects the elastic scattering and hence the phenomenological optical potential. This effect is greatest when the energy is comparable with that of the excited nuclear states and since the coupling is greatest to the low-lying states the effect is greatest around the Fermi surface. The effects of inelastic scattering can be studied by considering the elastic and inelastic scattering together using the coupled-channels formalism. If this is used to generate an elastic scattering with cross-section, which is then fitted phenomenologically with an optical potential, the results show the effects of the coupling to the inelastic channels. Several analyses made in this way have confirmed that the coupling to inelastic channels does produce changes in the real potential similar to those associated with the Fermi surface effect [2.26].

The DR have been used to analyse many sets of accurate experimental data, and some of these analyses are described below.

ii) Dispersive Optical Model Analyses

The DR (2.28) have been used by *Ahmad* and *Haider* [2.27] to calculate the surface-peaked component of the real potential for the elastic scattering of protons by ^{40}Ca. The results provided a natural explanation of the Fermi surface effect.

Since the pioneer work of Ahmad and Haider many dispersive optical model analyses have been made. The aim is to find the parametrization of the optical potential as a function of a radial distance and also of energy that will give the best overall fit to the experimental data. The main difficulties are that at negative energies the only data are the bound state energies and sometimes the widths. At positive energies extensive total and differential cross-sections are available, but at the lower energies the interpretation is complicated by the presence of compound elastic scattering, and for protons the Coulomb barrier limits the range of available energies.

There are two ways of carrying out a dispersive optical model analysis: the first parametrises the potential and seeks the optimum fit to selected data, while the second uses as data the phenomenological potentials obtained from standard optical model analyses.

In the first dispersive optical model method (DOMA), the Hartree–Fock potential is parametrised by a Saxon–Woods potential with a fixed radius and diffuseness and a depth that varies linearly or exponentially with energy. The imaginary potential is taken as the sum of volume and surface-peaked terms. In these analyses, the dispersive term is calculated from the imaginary potential and added to the Hartree–Fock term to give the real potential as in Eq. (2.30). The energy dependence of the volume integral of the real potential that fits the experimental values is shown in Fig. 2.2.

A detailed analysis of the interaction of neutrons with ^{208}Pb from -20 to $+165$ MeV is made by *Johnson* et al. [2.28]. The results of this work for the

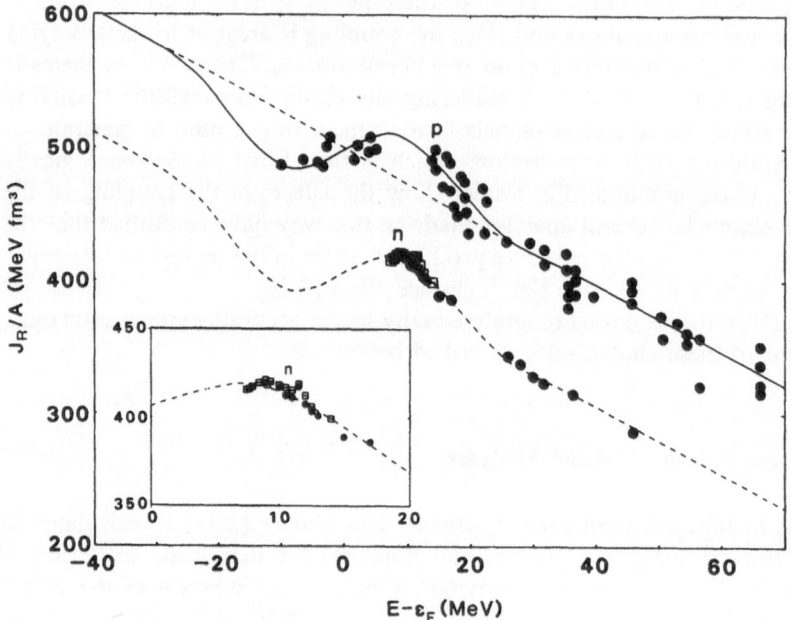

Fig. 2.2. Energy dependence of the volume integral of the real potential for neutrons and protons on ^{208}Pb showing the Fermi surface anomaly [2.16]

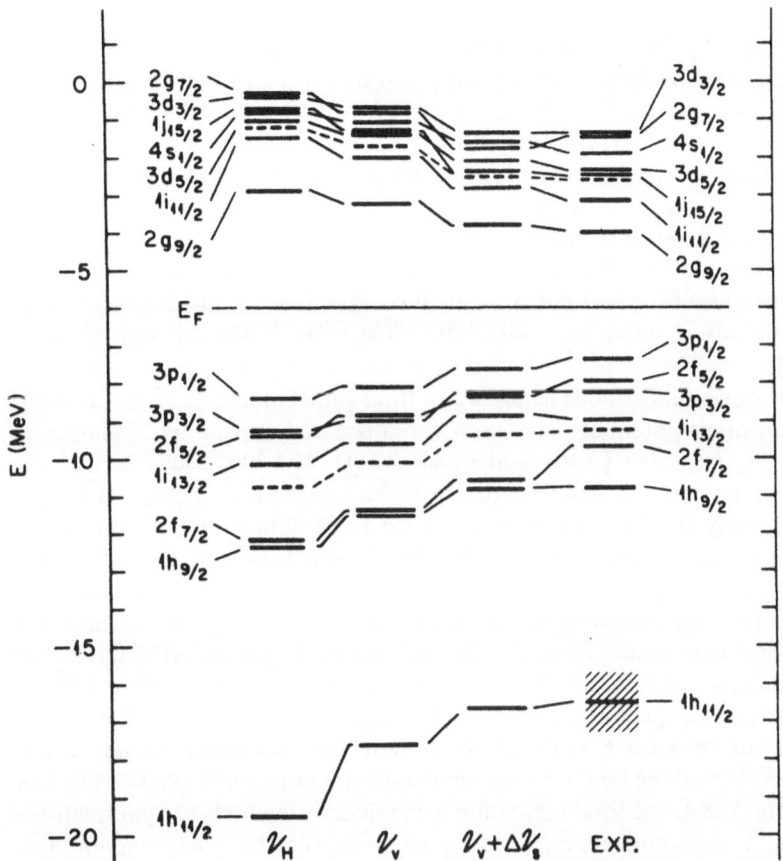

Fig. 2.3. Energies of neutron states in ^{208}Pb, compared with Hartree–Fock calculations with and without the dispersion corrections [2.28]. v_H is the Hartree–Fock field and v_v and Δv_s the additions due to the volume and surface dispersive corrections

energies of bound neutron states in ^{208}Pb using the addition of the dispersion terms are given in Fig. 2.3. At low energies ($0 < E < 10$ MeV), *Jeukenne* et al. [2.29] showed that the fits can be improved by allowing the diffuseness of the surface imaginary potential to decrease with decreasing energies, and its depth to depend on the orbital angular momentum. Further evidence of angular momentum dependence has been found by *Johnson* and *Winters* [2.30]. A similar analysis of the interaction of neutrons with ^{40}Ca from -80 to $+80$ MeV was made by *Johnson* and *Mahaux* [2.31]. The potentials were similar to those found for ^{208}Pb. The interactions of protons with ^{40}Ca from -60 to $+200$ MeV were analysed by *Tornow* et al. [2.32], and of neutrons with ^{86}Kr by *Johnson* et al. [2.33]. An analysis of the interactions of neutrons and protons with ^{208}Pb by *Finlay* et al. [2.34] showed dispersive effects for neutrons but not for protons. The interaction of neutrons with ^{208}Bi has been analysed by *Das* and *Finlay* [2.35]. The dispersive optical model has been extended to include analysing powers by *Roberts* et al. [2.36], and by *Chen* and *Tornow* [2.37], and the Coulomb correction term has been studied by *Tornow* and *Delaroche* [2.38] and by *Mahaux* and *Sartor* [2.39].

The dispersive model analyses that use potentials obtained by standard optical model fits can be made in several different ways. One of them is to calculate the volume integrals given by (2.36) with $q = 2$, to parametrize the energy variation of the imaginary part and then to use the DR to calculate volume integrals of the real potential at all energies (see, e.g. the applications to the interactions of neutrons with ^{51}V, ^{59}Co, ^{89}Y, ^{93}Nb, In and ^{209}Bi in [2.40–42]).

Several powerful methods using two or three radial moments of the phenomenological optical potentials have been developed and applied by *Mahaux* and *Sartor* [2.43] (see also [2.40] and [2.44, 2.45]). The basic idea is that the moments of the potential suffice to define its three parameters V_u, r_u and a_u. The moments satisfy the DR and so can be used to fit data over a wide range of energies, treating the parameters of the Hartree–Fock field as variable. A few iterations suffice to give a good overall fit to both scattering and bound state data. The optimum moments are then used to obtain the energy variations of the optical model parameters. Since this iterative moment approach (IMA) does not give the imaginary potentials themselves, it cannot be used to calculate differential cross-sections, spectral functions and occupation probabilities. It was worthwhile trying to remove its defects by a more sophisticated parametrization, and this has been done by the variational moment approach (VMA) [2.43f, 2.46, 2.47]. In the VMA, the total and volume imaginary potentials are parametrised separately. The parameters are fixed by requiring that the corresponding moments fit those of the phenomenological potentials. From the moments the dispersive correction is obtained, and subtraction from the phenomenological moments of the real potential gives those of the Hartree–Fock potential. The VMA has been successfully applied to the interactions of neutrons and protons with ^{40}Ca [2.47] and ^{208}Pb [2.48]. It is particularly notable that the VMA fixes the potential from the moments of the scattering data only, and its extrapolation to lower energies fits the bound state data.

The dispersion relation method has been applied to a well-deformed nucleus by *Merchant* et al. [2.49].

iii) The Energy Variation of the Imaginary Potential

Evaluation of the dispersion integral requires a knowledge of the imaginary potential over the whole range of energies; for positive energies it can be obtained from optical model analyses of elastic scattering but for negative energies it is only obtainable with difficulty and with less accuracy from the widths of single-particle states. It is therefore usual in dispersive optical model analyses to assume that the imaginary potential is symmetric about the Fermi energy, and this enables the dispersion correction to be written in the form (2.32).

The reliability of this assumption may be studied by examining the energy dependence of the bound state widths. It is shown in [2.46] that the widths are given by:

$$\Gamma_{nlj} = \alpha(E - E_F)^2,\tag{2.38}$$

where $\alpha \simeq 0.04$ in the vicinity of the Fermi surface and $\alpha \approx 0.009$ for the deeper states. This is illustrated in Figs. 2.4 and 2.5.

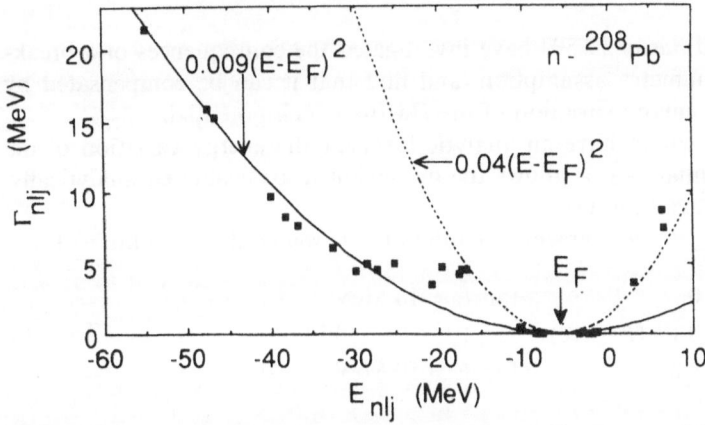

Fig. 2.4. Calculated quasi-particle bound-state widths as a function of energy for neutrons on ^{208}Pb [2.50]

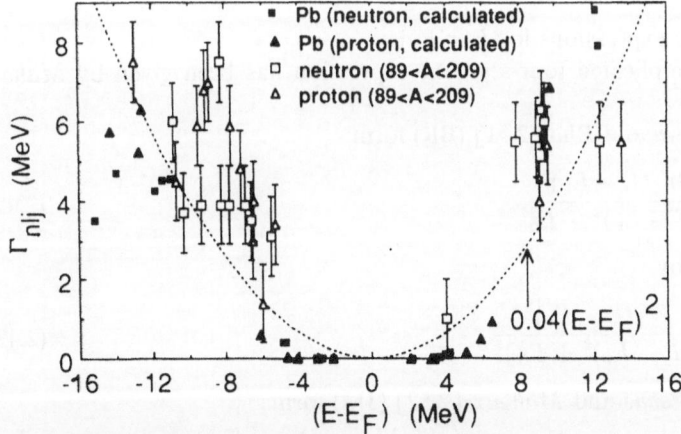

Fig. 2.5. Empirical and calculated quasi-particle bound-state widths as a function of energy for neutrons and protons on ^{208}Pb [2.50]

The widths of the bound states in a complex potential are related to the imaginary part of the potential by

$$\Gamma = 2 \int_0^\infty W(r)|\psi(r)|^2 \, dr. \tag{2.39}$$

If the form factor of the imaginary potential is suitably parametrized, its depth can be obtained using the experimental values of Γ. In the limit of infinite nuclear matter,

$$\Gamma = 2W. \tag{2.40}$$

If this relation is applied to finite nuclei, the symmetry assumption is approximately confirmed.

Mahaux and *Sartor* [2.50] have investigated the consequences of a breakdown in the symmetry assumption, and find that it can be compensated by a change in the energy variation of the Hartree–Fock potential.

It is convenient to have an analytic form for the energy variation of the imaginary potential as this allows the dispersion to be evaluated analytically. Several forms have been used:

1. The straight-line segment potential of *Johnson* et al. [2.28] for ^{208}Pb:

$$W_v(E) \begin{cases} = 0, & \text{for } E < 10 \text{ MeV} \\ = 0.17.(E - 10), & \text{for } 10 < E < 50 \text{ MeV} \\ = 6.8, & \text{for } E > 50 \text{ MeV} \end{cases} \tag{2.41a}$$

and

$$W_s(E) \begin{cases} = 0.4.(E - E_F), & \text{for } -6 < E < 10 \text{ MeV} \\ = -0.103.(E - 27), & \text{for } 10 < E < 72 \text{ MeV} \\ = 0, & \text{for } E > 72 \text{ MeV} \end{cases} \tag{2.41b}$$

(and symmetric expressions for $E < E_F$).

A more complicated four-segment expression has been given by *Mahaux* et al. [2.23].

2. The *Brown* and *Rho* [2.51] (BR) form:

$$W_v(E) = \frac{W_1(E - E_F)^2}{(E - E_F)^2 + E_0^2} \tag{2.42a}$$

supplemented by

$$W_s(E) = \frac{W_1(E - E_F)^2}{[(E - E_F)^2 + E_0^2]^2} \tag{2.42b}$$

3. The *Jeukenne* and *Mahaux* [2.52] (JM) form:

$$W(E) = W_1 \frac{(E - E_F)^4}{(E - E_F)^4 + E_0^4}. \tag{2.43a}$$

This has been modified by *Delaroche* et al. [2.53] to give the exponential form

$$W(E) = W_1 \frac{(E - E_F)^4}{(E - E_F)^4 + E_0^4} \exp[-C(E - E_F)].$$ (2.43b)

4. The potentials used by *Hicks* and *McEllistrem* [2.54] for osmium and platinum

$$W_v(E) \begin{cases} = 0, & \text{for } E < 8 \text{ MeV} \\ = 2.33 \ (E^{1/2} - 8^{1/2}), & \text{for } 8 < E < 40 \text{ MeV} \\ = 2.33 \ (40^{1/2} - 8^{1/2}) & \text{for } E > 40 \text{ MeV} \\ \approx 8.1 \text{ MeV} \end{cases}$$ (2.44a)

and

$$W_s(E) = \frac{a_0 + a_2(E - E_F)^2}{[E_0^2 + (E - E_F)^2]^2}$$ (2.44b)

(and symmetric expressions for $E < E_F$).

5. The BR–BR parametrisation of *Mahaux* and *Sartor* [2.47, 2.50]

$$W_v(E) = \gamma_v \frac{(E - E_p)^2}{(E - E_p)^2 + \mu_v^2},$$ (2.45a)

$$W_s(E) = \beta_s \left[\frac{(E - E_p)^2}{(E - E_p)^2 + \rho_s^2} - \frac{(E - E_p)^2}{(E - E_p)^2 + \mu_v^2} \right].$$ (2.45b)

6. The BR–JM parametrisation of *Mahaux* and *Sartor* [2.43f]

$$W_v(E) = \gamma_v \frac{(E - E_p)^4}{(E - E_p)^4 + \mu_v^4},$$ (2.46a)

$$W_s(E) = \beta_s \left[\frac{(E - E_p)^2}{(E - E_p)^2 + \rho_s^2} - \frac{(E - E_p)^4}{(E - E_p)^4 + \mu_v^4} \right].$$ (2.46b)

The analyses summarized in this section show that the dispersive optical model is able to give more accurate fits to the neutron data than the simple optical model, and also shows how a better global potential may be constructed.

3 Basic Correlation Methods

A critical analysis of the mean-field approximation is presented in Sect. 3.1 based on the natural orbital method in nuclear theory. It is shown that the nucleon density and momentum distributions cannot be simultaneously described within the Hartree–Fock method. Correlation methods in nuclear matter theory, such as the perturbation expansion and the hole-line or low-density expansion of the mass operator and related to them methods for study of finite nuclei (Brueckner theory, $\exp(S)$-method) are described in Sect. 3.2. Various correlation methods (that of *Brueckner, Eden, Francis* [3.68], Jastrow-type methods, the two-body correlation operator method) are considered in Sect. 3.3.

3.1 An Analysis of Mean-field Theories

The successful applications of the phenomenologically introduced shell model lead to the conclusion that the physical picture in terms of the individual states of the nucleons plays an important role in nuclear models. The average potential in which the nucleons move independently can be postulated on the basis of empirical data (that is the case of the shell model) or derived from a variational principle on the basis of a realistic nucleon–nucleon interaction in vacuum (the method of Hartree–Fock) [3.1–3]. The Hartree–Fock method with effective interactions [3.3] occupies an intermediate position between the microscopical methods based on vacuum nucleon–nucleon interactions and the various versions of the shell model with different phenomenological potentials.

Concerning the physical assumptions for the nucleons which are moving independently in a mean field the following remark has to be made regarding the "independence" of the motions of the nucleons: the average mean field is created by the nucleons themselves and hence it depends on their states. This means that the Hartree–Fock approximation contains a part of the nucleon–nucleon correlations. The coherent movement of a group of nucleons will lead to a change of the self-consistent field and therefore of the motions of the other nucleons.

In the Hartree–Fock approximation the nucleons are in stable single-particle states. The successful application of this approximation requires the particle interaction does not lead to strong local correlations and a condensation of particles [3.4]. For systems with an infinite hard core interaction between two particles the matrix elements of this interaction which determine the mean field become infinitely large. This makes it necessary to take account for the correlation effects more exactly. It is known that in the case of nuclear matter with strong repulsion at small distances between nucleons the Hartree–Fock method is not able to give the binding energy of the system [3.5].

The problem of correctly accounting for the correlation effects in the many-body theory arises also from the existence of nuclear experimental data which cannot be correctly described by mean-field methods. For instance [3.6], the Hartree–Fock calculations do not reproduce the central part of the density distributions in nuclei (e.g. ^{208}Pb [3.7]) even when the long-range correlation effects are included. The experimental data for the single-particle occupation probabilities show deviations from the Hartree–Fock predictions (unity below and zero above the Fermi level), see [3.8–13] and references and discussion in Chap. 6. The ground state nucleon momentum distributions show high-momentum components with a large excess over the predictions of the shell model (see [3.14–17] as well as [3.18] and references therein; this behaviour of the momentum distribution is discussed in Chap. 7). The knowledge of these momentum distributions is of great importance for the correct description of various interactions of particles (electrons, protons, heavy ions etc.) with nuclei (see Chaps. 7, 9–12). The experimental data which show the deviation of the deep-hole nuclear spectral function and widths from the mean-field predictions are discussed in Chap. 5 of this book. These data show the important role of the nucleon–nucleon correlations in nuclei which are related to the characteristic features of the nucleon–nucleon interaction at small distances such as the repulsive core and tensor part of the forces.

We shall discuss the deficiences of the Hartree–Fock approximation taking as an example the nucleon momentum and density distributions in nuclei. This problem has been considered in detail by *Jaminon* et al. [3.19, 3.6] on the basis of the one-body density matrix $\rho(r, r')$. It is known that the density distribution can be correctly described by a single Slater determinant total wave function using suitably chosen single-particle wave functions [3.20, 3.6]. Generally, this is true for any isolated nuclear ground state characteristic. It has been emphasized, however, that the nucleon–nucleon correlation effects can be observed when more than one quantity are investigated. This is related to the analysis of both the diagonal and non-diagonal elements of the one-body density matrix. It has been shown by *Bohigas* and *Stringari* [3.21] for the example of the ^{4}He nucleus that it is impossible to describe correctly both the density and momentum distributions.

An appropriate criterion for the proximity of the one-body density matrix $(\rho(r, r'))$ of the true (correlated) ground state to that of an uncorrelated system $(\rho_0(r, r'))$ is that the "mean-square deviation per particle"

$$\sigma = A^{-1} \cdot \mathrm{Tr}[(\rho - \rho_0)^2] \tag{3.1}$$

(where A is the mass number) should be minimal [3.22, 3.23, 3.19, 3.6]. The quantity σ reflects the deviation of the one-body density matrix ρ corresponding to correlated many-body wave function from ρ_0 which corresponds to single Slater determinant many-body wave function. It has been shown by *Kobe* [3.23] that $\sigma = \sigma_{min}$ when ρ_0 corresponds to Slater determinant constructed with the natural orbitals, i.e. with the single-particle wave functions $\varphi_\alpha(r)$ for which $\rho(r, r')$ has the form ([3.24] and see Sect. 4.3):

$$\rho(r, r') = \sum_\alpha n_\alpha \varphi_\alpha^*(r) \varphi_\alpha(r'). \qquad (3.2)$$

In (3.2) n_α are the occupation probabilities (or natural occupation numbers) of the state α ($0 \leqslant n_\alpha \leqslant 1$, $\Sigma_\alpha n_\alpha = A$).

The value of σ_{min} is determined by the properties of the N–N interaction [3.19]. The one-body density matrix $\rho_0^{HF}(r, r')$ obtained in the Hartree–Fock approximation is generally different from the matrix $\rho_0(r, r')$ leading to σ_{min}. As was shown by *Jaminon* et al. [3.19] if the Hartree–Fock matrix diagonal elements $\rho_0^{HF}(r, r)$ are fitted to reproduce correctly the exact density distribution $\rho(r, r)$, i.e. $\rho_0^{HF}(r, r) \simeq \rho(r, r)$ the extremum property $\sigma = \sigma_{min}$ leads inevitably to an increasing deviation between the non-diagonal elements of these two matrices ($\rho(r, r')$ and $\rho_0^{HF}(r, r')$ at $r \neq r'$). The non-diagonal matrix elements are related, however, to the nucleon momentum distribution:

$$n(k) = \int \rho(r, r') \exp[ik(r - r')] \, dr \, dr'. \qquad (3.3)$$

Therefore, it can be concluded [3.19] that generally the Hartree–Fock method is unable to give simultaneously a correct description of the two basic nuclear ground state characteristics, namely density and momentum distributions.

In the case of nuclear matter (NM) the natural orbitals are plane waves and the minimum value of σ^{NM} (σ_{min}^{NM}) corresponds to the uncorrelated one-body density matrix ρ_0^{NM} from the independent-particle model in which the single-particle wave functions are plane waves. It is shown in [3.6] that σ_{min}^{NM} can be estimated using the momentum probability distribution $n^{NM}(k)$ and that the available estimates of $n^{NM}(k)$ (disregarding the long-range correlations) leads to values of $\sigma_{min}^{NM} = 0.01$ to 0.04, where the upper limit corresponds to the hard core nucleon–nucleon interaction. The effects of short- and medium-range correlations are supposed to be almost the same in finite nuclear systems and in nuclear matter. This is not the case with the long-range correlations. For nuclear matter they increase σ_{min}^{NM} by 25–50%. In finite nuclei their effect can be evaluated taking account of the nuclear collective excitations.

The study of σ_{min} for finite nuclei [3.6] has been carried out in terms of the occupation probabilities of the natural orbitals (n_q)

$$\sigma_{min} = \sigma_< + \sigma_>, \qquad (3.4)$$

where

$$\sigma_< = A^{-1} \sum_{q \in F} (2j + 1)(1 - n_q)^2, \qquad (3.5)$$

$$\sigma_> = A^{-1} \sum_{q \notin F} (2j + 1) n_q^2 \tag{3.6}$$

("F" refers to the "Fermi sea" and $q = n, l, j$). $\sigma_<$ accounts for the depletion of the Fermi sea and $\sigma_>$ contains the occupation probabilities of those natural orbitals which are empty in the uncorrelated system.

The analyses in [3.6] lead to the observation that, while in the Hartree–Fock theory σ vanishes, the RPA contribution to σ is about 0.002, which shows that RPA includes long-range correlation effects only. For realistic nucleon–nucleon interactions the short- and medium-range correlation effects lead to a value of $\sigma_{min} \simeq 0.02$–0.03 that implies a Fermi sea depletion of about 10–15%. This estimate is confirmed by the experimental data [3.13, 3.25]. It has been emphasized in [3.6] that if a given model for the correlated ground state yields a value of σ_{min} which is significantly smaller than 0.01, then this model does not contain short- and medium-range correlations implied by the nucleon–nucleon interaction. As already shown, the fact that the short- and medium-range and tensor properties of the nucleon–nucleon forces are not included in the Hartree–Fock methods limits their applicability.

3.2 Correlation Methods Related to Mass Operator Expansions

A convenient way to consider the nucleon–nucleon correlations in infinite nuclear matter is given within the Green-function method. The one-particle Green function for A interacting particles is defined by means of the exact ground state wave function Ψ_0 and the field operators $\psi(r, t)$ and $\psi^+(r, t)$:

$$G(r, t; r', t') = -i \langle \Psi_0 | T\{\psi(r, t)\psi^+(r', t')\} | \Psi_0 \rangle, \tag{3.7}$$

where the time-ordering operator T acts in the case of a fermion system as follows [3.26]:

$$T\{\psi(r, t)\psi^+(r', t')\} = \begin{cases} \psi(r, t)\psi^+(r', t'), & \text{for } t > t' \\ -\psi^+(r', t')\psi(r, t), & \text{for } t < t'. \end{cases} \tag{3.8}$$

Due to the space- and time-translational invariance of infinite systems the Green function for nuclear matter has the following form in the momentum and energy representation:

$$G(k, \omega) = (\omega - k^2/2m - \Sigma(k, \omega))^{-1}, \tag{3.9}$$

where $\Sigma(k, \omega)$ is a complex operator (the so-called "mass operator"). In the vicinity of the Fermi energy ω_F its imaginary part has an asymptotic form [3.27, 3.28]:

$$\text{Im}\,\Sigma \approx C(\omega - \omega_F)^2. \tag{3.10}$$

The real and imaginary parts of the mass operator Σ are related by the dispersion relation:

$$\text{Re}\,\Sigma(\boldsymbol{k}, \omega) = \int\limits_{-\infty}^{\infty} \frac{d\omega'}{2\pi} \frac{\text{Im}\,\Sigma(\boldsymbol{k}, \omega')}{\omega - \omega'}. \tag{3.11}$$

The relation of the optical potential (Chap. 2) to the mass operator in the cases of infinite and finite nuclear systems is discussed in the reviews of *Jeukenne* et al. [3.29a] and *Mahaux* and *Sartor* [3.29b].

If the effective nucleon–nucleon forces are assumed to be sufficiently weak, the mass operator can be expanded in powers of the strength of the interaction $\hat{v}$. If realistic nucleon–nucleon forces are used (e.g. Reid's hard core force) the perturbation series diverges and must be rearranged. Following *Jeukenne* et al. [3.29a] we consider first the perturbation expansion of the mass operator:

$$\Sigma(k, \omega) = \Sigma_{1a}(k) + \Sigma_{1b}(k, \omega) + \Sigma_{1c}(k, \omega)$$
$$+ \Sigma_2(k, \omega) + \Sigma_3(k, \omega) + \cdots \tag{3.12}$$

In this series the first-order term of the nucleon–nucleon interaction corresponds to the Hartree–Fock approximation:

$$\Sigma_{1a}(k) = V_{HF}(k) = \sum_j \Theta(k_F - |j|)\{\langle k, j|\hat{v}|k, j\rangle - \langle k, j|\hat{v}|j, k\rangle\}$$

$$\equiv \sum_j \Theta(k_F - |j|)\langle k, j|\hat{v}|k, j - j, k\rangle, \tag{3.13}$$

where $|k, j\rangle = \exp\{i(kr_1 + jr_2)\}$. \tag{3.14}

This is static (i.e. is independent of the energy) and so local in time, as well as momentum dependent and therefore non-local in space.

The Green function in the Hartree–Fock approximation is:

$$G_{1a}(k, \omega) = \frac{\theta(k_F - k)}{\omega - \hat{e}(k) - i\eta} + \frac{\theta(k - k_F)}{\omega - \hat{e}(k) + i\eta} \tag{3.15}$$

with

$$\hat{e}(k) = k^2/2m + \Sigma_{1a}(k). \tag{3.16}$$

The momentum distribution $n(k)$ can be related to the Green function:

$$n(k) = -\frac{i}{2\pi} \int\limits_C d\omega\, G(k, \omega), \tag{3.17}$$

where the integration contour C is formed of the real axis and is closed by a semi-circle in the upper half plane. Using (3.15) one can obtain from (3.17):

$$n_{1a}(k) \equiv n_{HF}(k) = \theta(k_F - k). \tag{3.18}$$

Equation (3.18) shows that n_{HF} coincides with the free Fermi-gas momentum distribution. Hence, the ground-state motion is not correlated in the Hartree–Fock approximation.

The terms $\Sigma_{1b}(k, \omega)$ and $\Sigma_2(k, \omega)$ (which are of second order in $\hat{v}$) give the first corrections to the Hartree–Fock approximation

$$\Sigma_{1b}(k, \omega) = \frac{1}{2} \sum_{j,a,b} \Theta(k_F - |j|)\Theta(|a| - k_F)\Theta(|b| - k_F)$$

$$\times \frac{|\langle k, j|\hat{v}|a, b - b, a\rangle|^2}{\omega + \hat{e}(j) - \hat{e}(a) - \hat{e}(b) + i\eta}, \tag{3.19}$$

$$\Sigma_2(k, \omega) = \frac{1}{2} \sum_{l,j,a} \Theta(k_F - |j|)\Theta(k_F - |l|)\Theta(|a| - k_F)$$

$$\times \frac{|\langle j, l|\hat{v}|k, a - a, k\rangle|^2}{\omega + \hat{e}(a) - \hat{e}(j) - \hat{e}(l) - i\eta}. \tag{3.20}$$

In (3.19) and (3.20):

$$\hat{e}(d) = d^2/2m + \hat{U}(d) \tag{3.21}$$

and $\hat{U}(d)$ is the auxiliary potential chosen in such a way as to improve the convergence of the expansion.

The graphs corresponding to Σ_{1a}, Σ_{1b}, Σ_{1c} and Σ_2 are given in Fig. 3.1 in the case of particle states (the upward- and downward-pointing arrows represent particles and holes respectively).

The corrections to the Hartree–Fock nuclear matter momentum distributions have the form:

$$n_{1b}(k) = -\frac{1}{2} \sum_{l,c,d} \theta(k_F - |l|)\theta(|c| - k_F)\theta(|d| - k_F)$$

$$\times \frac{|\langle k, l|\hat{v}|c, d - d, c\rangle|^2}{[\hat{e}(k) + \hat{e}(l) - \hat{e}(c) - \hat{e}(d)]^2}, \quad k < k_F \tag{3.22}$$

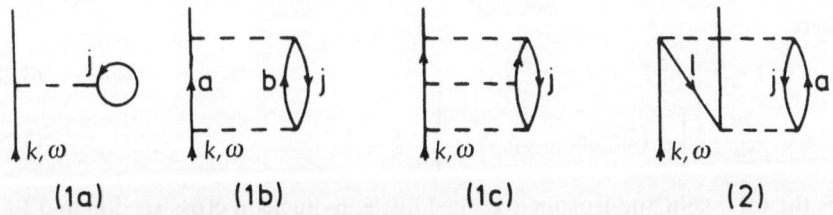

(1a)   (1b)   (1c)   (2)

Fig. 3.1. The graphs for Σ_{1a}, Σ_{1b}, Σ_{1c} and Σ_2 from the perturbation expansion in the case of particle states [3.29a]. The exchange terms are not given

and

$$n_2(k) = \frac{1}{2} \sum_{a,j,l} \theta(|a| - k_F)\theta(k_F - |j|)\theta(k_F - |l|)$$

$$\times \frac{|\langle k, a|\hat{v}|j, l - l, j\rangle|^2}{[\hat{e}(k) + \hat{e}(a) - \hat{e}(j) - \hat{e}(l)]^2}, \quad k > k_F. \tag{3.23}$$

The resulting momentum distribution for $k < k_F$ up to second order is

$$n_1(k) = n_{HF}(k) + n_{1b}(k). \tag{3.24}$$

The depletion of the Fermi sea to this order is $n_{1b}(k)$ and there exist components of $n(k)$ at $k > k_F$ $(n_2(k))$.

The total energy E of the system is determined using the Green function:

$$E = -\frac{i}{4\pi} \sum_k \int_C d\omega (k^2/2m + \omega) G(k, \omega), \tag{3.25}$$

where the integration contour is the same as in (3.17).

The Hartree–Fock binding energy is then

$$E_{HF} = \sum_k \theta(k_F - |k|)k^2/2m + \frac{1}{2} \sum_{k,j} \theta(k_F - |k|)\theta(k_F - |j|)$$

$$\times \langle k, j|\hat{v}|k, j - j, k\rangle. \tag{3.26}$$

A phenomenological model was suggested by *Orland* and *Schaeffer* [3.30] in which the imaginary part of the mass operator is parametrised using the nucleon–nucleon scattering data. The use of the dispersion relation (3.11) leads to the determination of the real part of the first two terms of the mass operator beyond the Hartree–Fock term, namely Σ_1 (the "core polarization" term) and Σ_2 (the "correlation graph"). As a result the nucleon momentum distribution is obtained in the form:

$$n(k) = \theta(k_F - k) - \text{sgn}(x)\frac{\alpha\beta}{2\pi}\frac{k_F^2}{2m}\left[1 + \frac{|x|}{\beta}\right.$$

$$\left. + \left(2\frac{|x|}{\beta} + \frac{x^2}{\beta^2}\right)\exp\left(\frac{|x|}{\beta}\right)\text{Ei}\left(-\frac{|x|}{\beta}\right)\right], \tag{3.27}$$

where

$$x = 1 - k^2/k_F^2, \tag{3.28}$$

$$\alpha = \frac{m}{(2\pi)^2}\sigma\left[\frac{1}{2}(k^2/2m + k_F^2/2m)\right], \quad \beta \simeq 3, \tag{3.29}$$

σ is the total spin and isospin averaged nucleon–nucleon cross-section and Ei is the integral exponential function. The result is applied to finite systems and it is shown in the case of the ^{40}Ca nucleus that the depletion of the Fermi sea is about 15%, i.e. a large number of nucleons (about 6) are above the Fermi level. This depletion value is close to the result of the calculations using the random-phase approximation and in the Brueckner–Hartree–Fock method presented in [3.30].

The role of the dynamical correlations produced by 2-particle (hole) – 1-hole (particle) intermediate states on various nuclear quantitites, such as momentum distributions, mean-free path of nucleons, single-particle level densities etc. can be studied within the semiclassical model of *Hasse* and *Schuck* [3.31, 3.32] by calculating the second-order terms of the mass-operator. The polarization and correlation contributions to the real part of the mass operator lead to the smearing out of the step-function momentum distribution with a remaining finite discontinuity at $k = k_F$. The value obtained for the gap (0.5) is in agreement with the results of *Orland* and *Schaeffer* [3.30], *Flynn* et al. [3.33] and *Fantoni* and *Pandharipande* [3.34]. The semiclassical approximation consists in writing all the expressions in Wigner space and then replacing the operators by their classical counterparts and their traces by phase-space integration. The *Gogny* D1 effective interaction [3.35, 3.36] is used for the determination of the semiclassical Hartree–Fock potential [3.1]. In this respect the model differs from the use of the bare nucleon–nucleon forces in perturbation theory.

In the case of realistic strong nucleon–nucleon interactions the perturbation expansion (3.12) is not appropriate. For nuclear matter densities close to the saturation value ($\rho_0 = 0.17\,\mathrm{N/fm^3}$), using the fact that the strong part range of the nuclear force is smaller than the average nucleon–nucleon distance, the perturbation series is reordered grouping together all graphs containing the same number of hole lines. Since all graphs which contain a single intermediate hole are proportional to k_F^3 (i.e. to the density $\rho = 2k_F^3/3\pi^2$), as well as all graphs with two intermediate hole momenta are roughly proportional to k_F^6 (i.e. to ρ^2) etc., the new expansion is called hole-line or low-density expansion (*Brueckner, Bethe* [3.37], *Sprung* [3.38], *Köhler* [3.39]).

Summing all the one-hole line graphs (i.e. the graphs (1a), (1b), (1c), . . . in Fig. 3.1) leads to the Brueckner–Hartree–Fock approximation for the mass operator:

$$\Sigma_{\mathrm{BHF}}(k, \omega) = \sum_j \theta(k_F - |j|)\langle k, j|g[\omega + e(j)]|k, j - j, k\rangle. \tag{3.30}$$

The expression (3.30) is formally analogous to the Hartree–Fock term (3.13), where the interaction $\hat{v}$ is substituted by the reaction matrix g which is the solution of the Bethe–Goldstone integral equation:

$$g[w] = v + v\sum_{c,d}\theta(|c| - k_F)\theta(|d| - k_F)\frac{|c, d\rangle\langle c, d|}{w - e(c) - e(d) + i\delta}g[w], \tag{3.31}$$

where w is a parameter (the "starting energy") and $e(d) = d^2/2m + U(d)$. The interaction v in (3.31) may contain a hard-core repulsion. The graph in Fig. 3.2

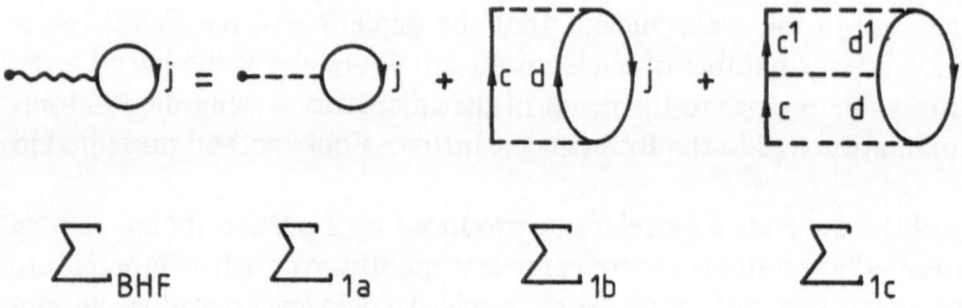

$$\sum_{\text{BHF}} \qquad \sum_{1a} \qquad \sum_{1b} \qquad \sum_{1c}$$

Fig. 3.2. The graph corresponding to Σ_{BHF} (3.30)

(with g represented by a wiggly line) corresponds to Σ_{BHF} (3.30). The BHF-approximation can be applied using arbitrary N–N interactions.

It should be noted that there are no particles with momentum $k > k_{\text{F}}$ in the BHF-approximation:

$$n_{\text{BHF}}(k) = 0 \quad \text{for} \quad k > k_{\text{F}}, \tag{3.32}$$

while at $k < k_{\text{F}}$:

$$n_{\text{BHF}}(k) = 1 - \frac{1}{2} \sum_{l,c,d} \theta(k_{\text{F}} - |l|)\theta(|c| - k_{\text{F}})\theta(|d| - k_{\text{F}})$$

$$\times \frac{|\langle k, l|g[e(k) + e(l)]|c, d - d, c\rangle|^2}{[e(k) + e(l) - e(c) - e(d)]^2}. \tag{3.33}$$

The first correction to $n(k)$ at $k > k_{\text{F}}$ originates from the second term Σ_2 in the hole-line expansion (Fig. 3.3):

$$\Sigma_2(k, \omega) = \frac{1}{2} \sum_{j,l,a} \theta(k_{\text{F}} - |j|)\theta(k_{\text{F}} - |l|)\theta(|a| - k_{\text{F}})$$

$$\times \frac{|\langle j, l|g[e(j) + e(l)]|k, a - a, k\rangle|^2}{\omega + e(a) - e(j) - e(l) - i\eta}. \tag{3.34}$$

The corresponding contribution to $n(k)$ is

$$n_2(k) = \frac{1}{2} \sum_{j,l,a} \theta(k_{\text{F}} - |j|)\theta(k_{\text{F}} - |l|)\theta(|a| - k_{\text{F}})$$

$$\times \frac{|\langle k, a|g[e(j) + e(l)]|j, l - l, j\rangle|^2}{[e(k) + e(a) - e(j) - e(l)]^2}, \quad \text{for } k > k_{\text{F}}. \tag{3.35}$$

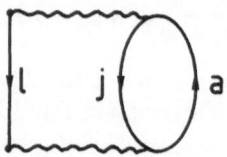

Fig. 3.3. The graph corresponding to Σ_2 (3.34)

Taking into account the higher order terms in applications of the BHF-approximation to finite nuclei leads to the replacement of Eq. (3.26) by [3.1, 3.40]

$$E_{BHF} = \sum_i \langle i|T|i\rangle + \frac{1}{2}\sum_{ij} \langle ij|g|i,j - j,i\rangle, \tag{3.36}$$

where the Brueckner reaction matrix g can be defined by the equation:

$$g = v + v\frac{Q}{e}g \tag{3.37}$$

with

$$e = E(i) + E(j) - E(p) - E(q), \tag{3.38}$$

where i, j are the initial nucleon states, p and q are intermediate states and the Pauli operator Q is equal to 1 if the p and q states are not occupied and zero otherwise.

The Brueckner–Hartree–Fock solution consists of a complicated doubly self-consistent procedure [3.40]. Self-consistency in the Brueckner sense is reached if the single-particle energies in (3.38) are determined from:

$$E(i) = \langle i|T|i\rangle + \sum_j \langle ij|g(E_i + E_j)|ij - ji\rangle. \tag{3.39}$$

The single-particle energies and potentials have to be consistent in the Hartree–Fock sense as well. This means that the following system of equations must be satisfied:

$$T\psi_i(r) + \sum_j \int \psi_j^*(r')g(r, r'; r_1, r_1')\{\psi_i(r_1)\psi_j(r_1')$$

$$- \psi_i(r_1')\psi_j(r_1)\}\,dr'\,dr_1\,dr_1' = \varepsilon_i\psi_i(r). \tag{3.40}$$

Due to the complications related to this double self-consistency, various approximate methods have been developed [3.40] and applied to studies of finite nuclei characteristics such as binding energies, density distributions and radii. The results of Brueckner–Hartree–Fock calculations using the Reid soft-core potential [3.40, 3.41] do not give even half the experimental binding energies for ^{16}O, ^{40}Ca and ^{208}Pb and the rms radii are about 10–20% smaller than the experimental ones. They also do not reproduce the single-particle spectrum for light and heavy nuclei.

Van Orden et al. [3.42] have used the Brueckner finite nuclei theory for calculations of nucleon momentum distribution in ^{16}O using realistic nucleon–nucleon potentials (Reid soft-core and De Tourreil–Sprung super-soft-core). The results presented in Fig. 3.4 show strong high-momentum components of $n(k)$ contrary to the shell-model predictions.

An substantial improvement of the BHF results for binding energies and rms radii has been achieved with the density-dependent Hartree–Fock (DDHF)

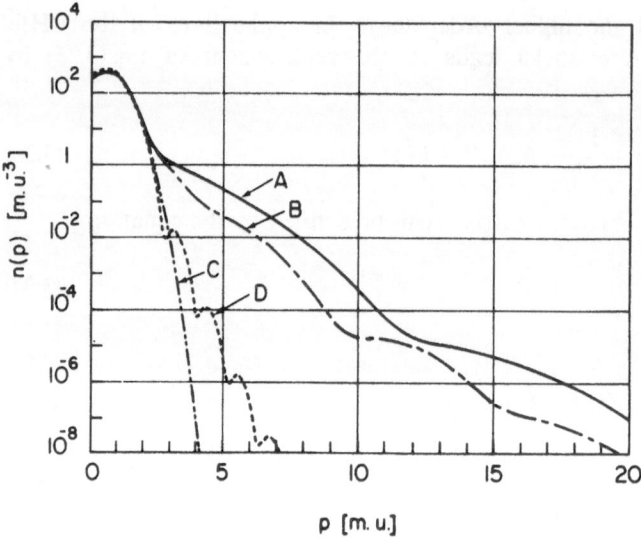

Fig. 3.4. Nucleon momentum distribution for ^{16}O [3.42] (m.u. $= m_\pi c = 139.57$ MeV/c). Curves A) and B): correlated distributions using the Reid soft-core and De Tourreil–Sprung super-soft-core potentials, respectively. Curves C) and D): shell model distributions using harmonic-oscillator and Saxon–Woods potentials

method of *Negele* [3.43]. In this method the starting energy w in (3.31) is replaced by an average value defined by twice the average hole energy in nuclear matter. In addition, the g-matrix is represented by an effective local operator $v^{\mathrm{eff}}(r)$. This effective interaction is renormalized phenomenologically to give the correct volume characteristics of nuclear matter. The use of the local density approximation in the determination of g leads [3.1] to very important new terms in the Hartree–Fock Hamiltonian (so-called "rearrangement" or "saturation potential") resulting from the density dependence in $v^{\mathrm{eff}}(\rho)$. It turns out that this term takes into account three-body scattering terms [3.44]. The DDHF method leads to a big improvement over the BHF approximation resulting in the correct description of the binding energies and rms radii in ^{16}O, ^{40}Ca and ^{208}Pb [3.43, 3.45]. We should also mention the DDHF results for charge distributions of *Campi* and *Sprung* [3.46] and of *Dechargé* et al. [3.47] using G0 and D1 interactions, respectively.

The main remark concerning the DDHF-methods is that the density-dependent forces cannot easily be used in the variational principle (1.45) from which the Hartree–Fock method originated and which is based on a linear Hamiltonian [3.1]. This difficulty is overcome by using the effective Skyrme force [3.48, 3.1, 3.3] considering it formally as containing a three-particle force. The use of the density-dependent Skyrme force with various sets of parameters in the DDHF approximation [3.49, 3.50] leads to good results for the bulk nuclear characteristics, such as binding energies and rms radii of spherical nuclei. The calculated charge distributions are also in agreement with the experimental data obtained from the analyses of electron and proton scattering on nuclei, although some oscillations in the nuclear central region are rather more prominent than those found experimentally. The correct ordering of the single-particle levels has also been obtained (e.g. for ^{208}Pb in [3.50]).

The situation with the nucleon momentum distribution, however, is quite different. The results for $n(k)$ in ^{4}He, ^{16}O, ^{40}Ca, ^{90}Zr and ^{208}Pb nuclei obtained in DDHF calculations with Skyrme-type forces [3.51] at $k > 2\,\mathrm{fm}^{-1}$ lie between those obtained from the shell model (or by the Hartree–Fock method using bare nucleon–nucleon forces) and those obtained by the more sophisticated correlation methods, such as the $\exp(S)$-method [3.52], which show prominent high-momentum tails. The increase in the high-momentum components of $n(k)$ compared with those in the Hartree–Fock method with bare nucleon–nucleon forces is due to the inclusion of a part of the nucleon–nucleon dynamical correlations when density-dependent nucleon–nucleon forces are used. The use, however, of a single Slater determinant as a total nuclear wave function still leads, as was shown in Sect. 3.1, to the inability to describe correctly the nucleon momentum distributions at high momenta.

The effects of the hard-core repulsion between two-particles on the nucleon momentum distribution is studied in the case of hard-sphere dilute Fermi gas by *Sartor* and *Mahaux* [3.53] using the low-density expansion of the mass operator up to terms of order $(k_F c)^2$, where k_F is the Fermi momentum and c is the hard-core radius. It is found that the behavior of $n(k)$ at large k is $k^{-(4+m)}$, where the value of m depends on the asymptotic behaviour of the imaginary part of the mass operator. The results for $n(k)$ are in agreement with those of *Czyż and Gottfried* [3.54] showing a small but rather long high-momentum tail and a discontinuity at $k = k_F$. They do not depend on the choice of $k_F c$ or on the spin–isospin degeneracy. A disagreement of the results of [3.53] with those of *Belyakov* [3.55] has been observed for $n(k)$ at $1 < k/k_F < 3$.

A formalism related to hole-line expansion of the mass operator has been developed in the $\exp(S)$-method (or the coupled-cluster many-body theory) [3.56–67, 3.52]. The total many-body wave function for doubly-closed shell nuclei is written in the form:

$$|\Psi\rangle = \exp(S)|\Phi\rangle, \tag{3.41}$$

where $|\Phi\rangle$ is a determinant built up from single-particle orbitals and S is the sum of operators S_n creating n-particle–n-hole excitations:

$$S = \sum_{n=1}^{A} S_n, \tag{3.42}$$

where

$$S_n = \frac{1}{(n!)^2} \sum_{\nu_i,\rho_i} \langle \rho_1,\ldots,\rho_n|S_n|\nu_1,\ldots,\nu_n\rangle_A\, a_{\rho_1}^+ \ldots a_{\rho_n}^+ a_{\nu_n} \ldots a_{\nu_1}. \tag{3.43}$$

The indices ν_i and ρ_i are used in (3.43) for unoccupied and occupied states respectively and

$$|v_1 v_2\rangle_A = |v_1 v_2\rangle - |v_2 v_1\rangle. \tag{3.44}$$

The Schrödinger equation $\hat{H}\Psi = E\Psi$ is used to obtain a system of non-linear coupled equations for E, $\langle \rho|S_1|v\rangle$ and other n-particle–n-hole amplitudes $\langle \ldots|S_n|\ldots\rangle$. The system can be truncated in the same way as is done for the hierarchy of equations for the density matrices or Green functions. The truncation of the system of equations neglecting respectively the terms S_n ($n \geqslant 2$), S_n ($n \geqslant 3$), S_n ($n \geqslant 4$) etc. reduces it to generalized Hartree–Fock equations, generalized BHF-equations, generalized three-particle Bethe–Faddeev equations, respectively, and so on.

The $\exp(S)$-method makes it possible to study the effects of nucleon–nucleon correlations on basic ground state nuclear characteristics. Unfortunately, the method is applicable only to light nuclei. The use of the method in the form of the generalized Brueckner–Hartree–Fock approximation [3.52] leads to the important conclusion that the nucleon–nucleon correlations are the main reason for the existence of high-momentum components in the nucleon momentum distributions in ^{4}He and ^{16}O at $k > 2\,\text{fm}^{-1}$, while the charge form factor is not so sensitive to the nucleon–nucleon correlations. The reason for this is that the nucleon momentum distribution is proportional at large momenta to the square of the amplitude S_2 in the momentum representation

$$n(k) \sim |S_2(k)|^2, \quad (S_2(k) = \langle k, -k|S_2|0, 0\rangle), \tag{3.45}$$

while the form factor is expressed by the integral

$$F(q) \sim \int d\mathbf{r}_1 \exp(i\mathbf{q}\mathbf{r}_1) \int d\mathbf{r}_2 |S_2(|\mathbf{r}_1 - \mathbf{r}_2|)|^2, \tag{3.46}$$

which smears the main features of the amplitude S_2.

The nucleon momentum distribution can be decomposed as follows:

$$n(q) = n_1(q) + n_2(q) + n_3(q), \tag{3.47}$$

where $n_1(q)$ describes the nucleon momentum distribution due to the uncorrelated motion of particles in occupied orbitals. The latter are Brueckner or maximum-overlap orbitals determined by the condition

$$\langle \Psi|\Phi\rangle = \max. \tag{3.48}$$

The terms $n_2(q)$ and $n_3(q)$ are related to the population of normally unoccupied orbitals and to the corresponding depletion of normally occupied orbitals, respectively. They have their origin from those parts of the total wave function Ψ which cannot be expressed in terms of a single Slater determinant.

The results of $\exp(S)$-method for the nucleon momentum distributions in ^{4}He and ^{16}O with significant high-momentum components [3.52] are discussed in Chap. 7.

The calculations of the local density distribution (or the form factor) by *Gari* et al. [3.65] have shown that the form factor is not strongly influenced by the nucleon–nucleon correlations.

3.3 Beyond Single Slater Determinant Methods

We shall consider in this section selected correlation methods in which the relation of the N–N correlations at short distances to the peculiarities of the nucleon–nucleon forces is investigated. Usually, the correlated many-body wave function is built up on the basis of the wave function for the uncorrelated system with an operator (or operators) to introduce correlations.

i) Brueckner, Eden and Francis Method [3.68]

The ground-state function of the correlated system $\Psi(\{r_i\})$ $(i = 1, 2, \ldots, A)$ is constructed by the method of *Brueckner* et al. [3.68] using the action of the correlation operator F on the shell-model many-body wave function $\Phi(\{r_i\})$:

$$\Psi(\{r_i\}) = F\Phi(\{r_i\}). \tag{3.49}$$

The operator F is related to the two-body scattering operators in the following way: the explicit form of the transformation (3.49) can be represented by the set of coupled equations:

$$F = 1 + \frac{1}{e} \sum_{ij} I_{ij} F_{ij}, \tag{3.50}$$

$$F_{ij} = 1 + \frac{1}{e} \sum_{lm \neq ij} I_{lm} F_{lm}, \tag{3.51}$$

$$e = E_0 - \sum_i T_i - V_c, \tag{3.52}$$

where the energy can be determined from the equation:

$$(E_0 - \sum_i T_i - V_c)\Phi = 0. \tag{3.53}$$

If V_{ij} is the potential between two nucleons (i and j) then the two-nucleon scattering operators t_{ij} are defined by

$$t_{ij} = V_{ij} + V_{ij}(1/e)t_{ij}. \tag{3.54}$$

The operators t_{ij} determine the operators I_{ij} as parts of t_{ij} which are non-diagonal with respect to the nuclear states as well as to the uniform potential V_c:

$$V_c = \frac{1}{2} \sum_{ij} t_{cij}, \tag{3.55}$$

where t_{cij} is the diagonal part of t_{ij}.

The effect of the non-diagonal operators I_{ij} which cause transitions from the uncorrelated state $\Phi(\{r_i\})$ is similar to an inelastic scattering of particles, a pair at a time, out of the Fermi-gas to excited states. Thus the deviations of Ψ from Φ can be related to the strength and range of the two-nucleon potentials.

ii) Jastrow-type Methods

A method to account for the short-range repulsion in the nucleon–nucleon force has been developed by *Jastrow* [3.69]. In it the total many-body wave function is written in the form

$$\Psi(r_1,\ldots,r_A) = C_A^{-1/2} \prod_{1 \le i < j \le A} f(r_{ij})\, \Phi(r_1,\ldots,r_A), \qquad (3.56)$$

where Φ is a Slater determinant built up from single-particle wave functions $\varphi_\alpha(r)$ which correspond to the occupied states, and C_A is the normalization constant. The correlation function $f(r_{ij}) \equiv f(|r_i - r_j|)$ satisfies the conditions:

$$\begin{aligned} f(r_{ij}) &= 0, \quad \text{for } |r_i - r_j| \le r_{\text{c}} \\ f(r_{ij}) &= 1, \quad \text{for } |r_i - r_j| \to \infty, \end{aligned} \qquad (3.57)$$

where r_{c} is the radius of the nucleon–nucleon repulsive core. These conditions ensure the vanishing of the wave function Ψ, if $|r_i - r_j|$ becomes smaller than the repulsive core radius. The wave function Ψ is used as a trial function in variational calculations of the energy for a system with a given Hamiltonian. The variations of the expectation value of the Hamiltonian with respect to the single-particle functions $\varphi_\alpha(r)$ and the correlation function $f(r_{ij})$ lead to corresponding equations of Euler–Lagrange. Various approximations and appropriate techniques have been developed to treat the problems of solving these equations.

Gaudin et al. [3.20] suggested a perturbation expansion method for calculating the one- and two-body density matrices. These quantities are written as an expansion in terms of the functions

$$g(r) = |f(r)|^2 - 1 \qquad (3.58)$$

and

$$h(r) = f(r) - 1. \qquad (3.59)$$

Using the lowest-order-cluster (LOC) approximation in the function $g(r)$ for the expansion of the one-body density matrix the density distribution ρ and the charge form factor of ^{40}Ca have been calculated in [3.20]. It should be noted that the Jastrow result for $\rho(r)$ is very close to the calculations using a Slater determinant function built up from natural orbitals, which diagonalize the one-body Jastrow density matrix.

The LOC approximation has been used [3.33] to calculate the momentum distribution in symmetrical nuclear matter. The same quantity has been calculated using the leading approximation to $n(k)$ within the irreducible cluster formalism [3.70]. The Fermi-hypernetted-chain (FHNC) technique (e.g. [3.71, 3.72]) is used for calculations of $n(k)$. The result shows a finite discontinuity of $n(k)$ at $k = k_{\text{F}}$ and a tail at $k > k_{\text{F}}$.

Calculations of form factors $F(q)$ and nucleon momentum distributions $n(k)$ have been carried out for finite nuclei using various forms of correlation functions $f(r_{ij})$ and single-particle functions φ corresponding to harmonic oscillator or Saxon–Woods-type potentials as well as to solutions of self-consistent Hartree–Fock system of equations. The calculations of $F(q)$ and $n(k)$ for ⁴He in LOC-approximation (*Bohigas* and *Stringari* [3.21]) have been compared with the exact Jastrow calculations by *Dal Rì* et al. [3.73]. The comparison shows that the LOC-approximation works rather well for quantitative analyses in ⁴He. The results for $n(k)$ are very close to each other at $k \geqslant 2.5 \, \mathrm{fm}^{-1}$, while the differences for $F(q)$ in both cases are larger. As was mentioned in Sect. 3.1 an important conclusion has been made in [3.21] concerning the inability of a single Slater determinant wave function to describe simultaneously both the density and the nucleon momentum distribution of a correlated system.

The Jastrow correlation method has been extended [3.74–76, 3.34] by representing the trial wave function in the form:

$$\Psi(x_1, \ldots, x_A) = \left(S \prod_{i>j=1}^{A} \mathscr{F}_{ij} \right) \Phi(x_1, \ldots, x_A), \tag{3.60}$$

where the pair-correlation operators:

$$\mathscr{F}_{ij} = \sum_p f^p(r_{ij}) O_{ij}^p \tag{3.61}$$

with

$$O_{ij}^{p=1,\ldots,8} = 1, (\tau_i \cdot \tau_j), (\sigma_i \cdot \sigma_j), (\sigma_i \cdot \sigma_j)(\tau_i \cdot \tau_j), S_{ij},$$
$$S_{ij}(\tau_i \cdot \tau_j), L \cdot S, L \cdot S(\tau_i \cdot \tau_j) \tag{3.62}$$

account for central, isospin, spin, tensor and spin-orbit pair correlations, S is the symmetrizer, x_i denotes the radius-vector r_i and the spin and isospin of the ith nucleon. The correlation functions are suitably parametrized. The values of their parameters are determined by minimizing the energy expectation value. In the case of nuclear matter [3.34] $\Phi(x_1, \ldots, x_A)$ is a Slater determinant built up from plane-wave single-particle functions. Both variational calculations and calculations with a wave function generalized by using perturbation theory in a correlated basis (CBF) for the nucleon momentum distribution in equilibrium nuclear matter have been carried out and compared in [3.34]. It is established that the tensor correlations play an important role. The second order CBF corrections are reduced by almost a factor of 4 at $k/k_F > 1$ when the tensor components (S_{ij}) in Ψ are not included in the calculations.

Improved variational wave functions have been used in Monte Carlo calculations within the extended Jastrow-type method (3.60–62) for the nucleon and cluster momentum distributions in nuclei with $A = 3$ and $A = 4$ and the nucleon momentum distribution in nuclear matter [3.77a] (see Chap. 7). A realistic Hamiltonian including three-nucleon interactions is used. The nucleon momentum distributions contain high-momentum components comparable with the results of other correlation methods (e.g. for ⁴He from [3.52]). It is shown that the nucleon momentum distributions for $A = 2, 3, 4$ and for nuclear matter are proportional to each other.

Two-nucleon (Argonne v_{14}) and three-nucleon (Urbana VII) potentials are used in the extended Jastrow method calculations of the ground state characteristics of ^{16}O (binding energy, charge density distribution and longitudinal structure function) [3.77b]. The expectation values are calculated using a cluster expansion for the noncentral correlations in the wave functions. The central correlations and exchanges are treated to all orders by Monte Carlo integration.

The ansatz (3.60–62) within the cluster expansion technique [3.78] and using the Reid soft-core interaction (the V6 version) has been used by *Benhar* et al. [3.79] for calculations of the nucleon momentum distribution in ^{12}C, ^{16}O and ^{40}Ca nuclei. The results for ^{16}O agree with those obtained in [3.42] using the Brueckner finite nucleus theory. The nucleon momentum distributions have a much larger content of high-momentum components with respect to the Hartree–Fock method calculations. The results from [3.79] are compared with other correlation methods in Chap. 7.

iii) Two-body Correlation Operator Method

The short-range and tensor correlation effects on the nucleon momentum distributions and form factors have been studied with the phenomenological model of *Dellagiacoma, Orlandini* and *Traini* [3.80] and *Traini* and *Orlandini* [3.81]. The two-body correlation operator $u(1, 2)$ acting on the pair wave function is introduced in the two-body density matrix of the correlated system:

$$\rho(v_1, v_2; v_1', v_2') = \sum_{a,b} [\langle v_1 v_2 | u(1, 2) | ab \rangle \langle ab | u^+(1, 2) | v_1' v_2' \rangle$$
$$- \langle v_1 v_2 | u(1, 2) | ab \rangle \langle ab | u^+(1, 2) | v_2' v_1' \rangle], \qquad (3.63)$$

where $v_i \equiv (r_i, s_i^z, t_i^z)$. In the case of the single-particle wave functions of the harmonic oscillator the two-particle state function $|a(1), b(2)\rangle$ is expanded on the basis of the relative and c.m. coordinates, the total angular momentum, and spin and isospin of the pair:

$$|a(1)b(2)\rangle = \sum C_{ab} |nlm\rangle |NLM\rangle |SS^z\rangle |TT^z\rangle, \qquad (3.64)$$

where $N, L, M; n, l, m$ are the radial and angular c.m. and relative motion quantum numbers, S, T the spin and isospin of the pair, and S^z and T^z their third components.

The short-range correlation effects are included by means of the operator $u(1, 2)_{s.r.}$ acting on the radial part of the pair wave function

$$[u(1, 2)]_{s.r.} |nlm\rangle = N_{nl}^{-1/2} f(r) |(nlm\rangle, \qquad (3.65)$$

with

$$f(r) \to 0, \quad f(r) \to 1.$$
$$r \to 0 \qquad r \to \infty \qquad\qquad (3.66)$$

The tensor correlations are included by using the two-body operator $u(1, 2)_{\text{tens.}}$ that acts both on the angular and the radial parts of the relative motion of the pair. In practical applications the tensor operator is restricted to deuteron-like states only:

$$[u(1, 2)]_{\text{tens.}} |n, {}^3S_1, J^z, T = 0\rangle = (1 - \eta^2)^{1/2} \varphi_{n0}(r)|n, {}^3S_1, J^z, T = 0\rangle$$
$$+ \eta\varphi_{n2}(r)|n, {}^3D_1, J^z, T = 0\rangle, \tag{3.67}$$

where $\varphi_{nl}(r)$ are the radial wave functions, chosen to be the harmonic oscillator functions.

Explicit expressions for the nucleon momentum distributions and form factors have been obtained for ^{4}He, ^{16}O and ^{40}Ca nuclei in [3.81]. It is shown that the tensor correlation effects are stronger for light nuclei (^{4}He and ^{16}O) than for ^{40}Ca. The tensor correlations cause an additional minimum in the form factors of ^{4}He and ^{40}Ca. The calculations of the nucleon momentum distributions in this model are compared with results from other methods in Chap. 7. The predictions of the model for the high-momentum behaviour of $n(k)$ are compatible with the results of other correlation methods.

Short-range correlation effects have been studied using a similar correlating operator $u(1, 2)$ in [3.82–83].

In this chapter we considered some basic beyond Hartree–Fock methods taking into account short-range and tensor nucleon–nucleon correlations. They have been included to describe the available experimental data on nuclear characteristics sensitive to these correlations, such as nucleon momentum distributions, spectral functions of deep-hole nuclear states and occupation numbers, as well as the cross-sections of different reactions which will be considered in Part 2 and Part 3 of this book.

Here we briefly mention some correlations which are not considered in this book. It is known [3.1] that the mean-field approximation is justified if a sufficient gap between the last occupied and the first unoccupied levels exists. Otherwise, it is possible for virtual particle-hole pairs to be excited and the ground state many-body function will then be more complicated than a Slater determinant $|\Phi_0\rangle$, taking for instance the form:

$$|\Psi_0\rangle = C_0^0|\Phi_0\rangle + \sum_{mi} C_{mi}^0 a_m^+ a_i|\Phi_0\rangle$$
$$+ \frac{1}{4} \sum_{\substack{mi \\ nj}} C_{minj}^0 a_m^+ a_n^+ a_i a_j|\Phi_0\rangle + \ldots, \tag{3.68}$$

where the operators a_i, a_j create holes below the Fermi surface and a_m^+, a_i^+ create particles above it. Thus the Hartree–Fock method will describe better the ground states of magic nuclei than those of nuclei with a few particles away from the closed shell configuration. If there exist more nucleons in the next open shell a different type of correlations may be important. A theory of s–d shell nuclei between doubly magic nuclei ^{16}O and ^{40}Ca (the so-called shell-model configuration mixing theory) has been developed (e.g. see [3.84–86]) and calculations with this theory have been carried out [3.87–89].

As is well-known [3.1] there are no matrix elements in the Hartree–Fock method between the ground state and the particle–hole excitations. Therefore, the particle–hole part of the N–N interaction, i.e. the long-range part of the N–N force is partially taken into account in the Hartree–Fock method.

The random-phase approximation (RPA) theory includes two-particle–two hole (2p–2h) correlations in the ground state:

$$|RPA\rangle_{g.s.} = C_0^0|HF\rangle + \frac{1}{4}\sum_{\substack{mi\\nj}} C_{minj}^0 a_m^+ a_n^+ a_i a_j|HF\rangle, \tag{3.69}$$

defined by the condition:

$$Q_\nu|RPA\rangle_{g.s.} = 0, \tag{3.70}$$

where

$$Q_\nu = \sum_{mi} \chi_{mi}^\nu a_i^+ a_m - \sum_{mi} Y_{mi}^\nu a_m^+ a_i. \tag{3.71}$$

This theory is considered widely in the literature (e.g. [3.1, 3.2, 3.85]). Comparisons with the predictions of RPA on occupation numbers and the nucleon momentum distribution of some nuclei will be given in Chaps. 6 and 7 of this book.

Specific short-range pairing correlations of a superconductive type can be separated from the residual interaction. The theory of superconductivity in nuclei [3.90–92] makes it possible to describe a system of strongly interacting nucleons with small excitation energies as a gas of quasiparticles [3.28] moving in an effective mean field. The spectrum of the collective excitations can be determined by introducing an interaction between quasiparticles which contains the part of the interaction not included in the mean field. The quasiparticle interaction turns out to be quite different from the interaction between two free particles.

A variational procedure going beyond the quasiparticle mean-field approximation and including the most important additional correlations has been suggested by *Schmid* et al. [3.93]. Each state in this method is approximated by a linear combination of several symmetry projected quasiparticle determinants.

The pairing correlations of superconductive type are discussed widely in the literature (see e.g. [3.1, 3.2, 3.85]) and are beyond the scope of this book.

4 Further Correlation Methods

In this chapter we describe the main assumptions and formalism of the generator coordinate method (Sect. 4.1), the coherent density fluctuation model (Sect. 4.2) and the natural orbital method (Sect. 4.3). These methods are applied to the problems of correlation effects in nuclear structure and nuclear reactions discussed in Part 2 and Part 3 of this book.

4.1 The Generator Coordinate Method

The generator coordinate method (GCM) has been suggested by *Hill* and *Wheeler* [4.1] and by *Griffin* and *Wheeler* [4.2] as a variational method for studying the collective motions in atomic nuclei. Using the shell model as a basis, the GCM connects the collective motion to appropriate variational parameters (the so-called "generator coordinates") of the single-particle wave functions. Different characteristics of the single-particle potential (or "construction potential"), such as the orientation, the size, the shape etc. can play the role of generator coordinates.

In the GCM the trial many-particle wave function $\Psi(\{r_i\})$ of a system of A nucleons is written in a form of a linear combination:

$$\Psi(r_1, \ldots, r_A) = \int f(x_1, x_2, \ldots) \Phi(r_1, \ldots, r_A; x_1, x_2, \ldots) dx_1 dx_2 \ldots, \quad (4.1)$$

where the generating function $\Phi(\{r_i\}; x_1, x_2 \ldots)$ depends on the radius-vectors of the nucleons $\{r_i\}$ and on the generator coordinates $x_1, x_2, \ldots$. This function is usually chosen to be a Slater determinant built up from single-particle wave functions corresponding to a given construction potential parametrized by $x_1, x_2, \ldots$. It is obvious that in this case the wave function (4.1) of the system, being a superposition of Slater determinants, goes beyond the limits of the mean-field approximation. In this Section we consider the main points and relationships of the GCM. In Chaps. 6 and 7 we study those aspects of the method which are related to the N–N correlation effects in nuclei.

The function $f(x_1, x_2, \ldots)$ (the so-called "weight" or "generator" function) can be determined using the variational principle. From (4.1) the energy functional

$$E[\Psi] = \langle \Psi | \hat{H} | \Psi \rangle / \langle \Psi | \Psi \rangle, \quad (4.2)$$

where $\hat{H}$ is the Hamiltonian of the system, becomes a functional of the weight function f, which has to be chosen so that E has an extreme value, i.e.

$$\delta E = 0. \tag{4.3}$$

The Eq. (4.3) leads to the following integral equation for the weight function:

$$\int [\mathscr{H}(x, x') - EI(x, x')] f(x') dx' = 0, \tag{4.4}$$

where the overlap kernel $I(x, x')$ and the energy kernel $\mathscr{H}(x, x')$ have the following forms, respectively:

$$I(x, x') = \langle \Phi(\{r_i\}, x) | \Phi(\{r_i\}, x') \rangle, \tag{4.5}$$

$$\mathscr{H}(x, x') = \langle \Phi(\{r_i\}, x) | \hat{H} | \Phi(\{r_i\}, x') \rangle. \tag{4.6}$$

In (4.4–6) the generator coordinate x denotes a set $x_1, x_2, \ldots$.

In some cases the weight function f turns to be not square integrable and so has the meaning of a statistical distribution [4.3]. The solutions $f_1, f_2, \ldots$ of (4.4) corresponding to the eigenvalues of the energy $E_1, E_2, \ldots$ satisfy the following orthonormality condition:

$$\int f_i^*(x) I(x, x') f_j(x') dx\, dx' = \delta_{ij}. \tag{4.7}$$

The applications of the GCM to describe the characteristics of many-particle systems require the use of various approximations. These are based on the observation that: i) the overlap kernel $I(x, x')$ has a Gaussian behaviour with respect to the difference $x - x'$ [4.4, 4.5], ii) for many-fermion systems $I(x, x')$ is close to delta-function, iii) the integral kernel

$$\langle \Phi(\{r_i\}, x) | \hat{H} - E | \Phi(\{r_i\}, x') \rangle \tag{4.8}$$

has a maximum at $x' \sim x$ [4.6], iv) the both kernels I and $\mathscr{H}$ are differentiable functions of x.

For many-fermion systems the kernels I and $\mathscr{H}$ peak strongly at $x \sim x'$ and can be written in the form [4.6]:

$$I(x, x') \simeq I(x, x) \mathscr{F}(x - x'), \tag{4.9}$$

$$\mathscr{H}(x, x') \simeq \mathscr{H}(x, x) \mathscr{F}(x - x'), \tag{4.10}$$

where $\mathscr{F}$ is peaked at $x \sim x'$.

The behaviour of (4.8) at $x \sim x'$ gives the possibility of using the expansion of the weight function

$$f(x') = f(x) + \frac{df}{dx}(x' - x) + \frac{1}{2!}\frac{d^2 f}{dx^2}(x' - x)^2 + \ldots \tag{4.11}$$

in (4.4). The use of the first three terms in the right-hand side of (4.11) and also (4.9) and (4.10) leads to the Schrödinger-type of equation [4.6]:

$$-\frac{\hbar^2}{2m_{\text{eff}}(x, E)}\frac{d^2 f}{dx^2} + V(x) f(x) = E f(x), \tag{4.12}$$

where

$$V(x) = \langle \Phi(\{r_i\}, x) | \hat{H} | \Phi(\{r_i\}, x) \rangle \tag{4.13}$$

and

$$m_{\text{eff}}(x, E) = -\hbar^2 \left[\frac{1}{\int \mathscr{F}(x' - x) \, dx'} \int \langle \Phi(\{r_i\}, x) | \hat{H} - E | \Phi(\{r_i\}, x') \rangle \right.$$
$$\left. \times (x' - x)^2 \, dx \right]^{-1}. \tag{4.14}$$

As shown by *Dirac* [4.7] the difference between the integral equation (4.4) and the differential equation (4.12) is to a great extent a formal one. The Schrödinger-type equation (4.12) can be obtained by the substitutions:

$$I(x, x') \rightarrow \delta(x - x'), \tag{4.15}$$

$$\mathscr{H}(x, x') \rightarrow -\frac{\hbar^2}{2m_{\text{eff}}} \delta''(x - x') + \delta(x - x') V\left(\frac{x + x'}{2}\right), \tag{4.16}$$

where the delta-function form of the potential energy part of the kernel $\mathscr{H}(x, x')$ (4.16) is appropriate only in the case of velocity-independent forces. It is shown in [4.2] that the delta-function approximation (4.15) and (4.16) is valid in the GCM in the case of many-fermion systems and it leads to the Eq. (4.12) with an effective mass dependent on the generator coordinate.

A different way for obtaining of the collective mass and potential is to use the Gaussian approximation for the overlap kernel [4.1, 4.8]:

$$I(x, x') = \exp[-\gamma(q)(x - x')^2/2] \tag{4.17}$$

with

$$q = (x + x')/2. \tag{4.18}$$

The collective Hamiltonian H_c is then:

$$H_c = -\frac{1}{\gamma(q)^{1/2}} \frac{\partial}{\partial q} \gamma(q)^{1/2} \frac{\hbar^2}{2M(q)} \frac{\partial}{\partial q} + V(q) \tag{4.19}$$

with a parameter of inertia

$$M^{-1}(q) = \frac{1}{2\gamma^2(q)} \left[\frac{\partial^2}{\partial x \partial x'} - \frac{\partial^2}{\partial x^2} + \frac{\partial \ln \gamma}{\partial q} \frac{\partial}{\partial x} \right] \tilde{\mathscr{H}}(x, x'), \tag{4.20}$$

where

$$\tilde{\mathscr{H}}(x, x') = \mathscr{H}(x, x')/I(x, x'). \tag{4.21}$$

The collective potential is:

$$V(q) = \langle q | \hat{H} | q \rangle - \varepsilon_0(q), \tag{4.22}$$

where $\varepsilon_0(q)$ is the quantum energy correction:

$$\varepsilon_0(q) = \frac{1}{2\gamma(q)} \left[\frac{\partial^2 \tilde{\mathscr{H}}(x, x')}{\partial x \partial x'} \right]_{x = x' = q} + \left\langle q \left| \frac{\partial \hat{H}}{\partial q} \frac{\partial}{\partial q} \right| q \right\rangle. \tag{4.23}$$

The assumptions of differentiability and sharply-peaked overlap functions can also be used in the symmetric moment expansion [4.3, 4.8]. It is based on the relation:

$$\int ds\, A(q, s) F(s) = \int ds\, e^{is\hat{P}/2\hbar} A\left(q + \frac{s}{2}, q - \frac{s}{2}\right) e^{is\hat{P}/2\hbar} F(q), \qquad (4.24)$$

which is valid for each operator having the properties of $I(x, x')$ and $\mathscr{H}(x, x')$. The operator $\hat{P}$ is

$$\hat{P} = -\frac{\hbar}{i}\frac{\partial}{\partial q}. \qquad (4.25)$$

The expansion of the right-hand side of (4.24) including the terms of second order on s makes it possible to obtain the effective mass

$$\frac{1}{M(q)} = \frac{I_2}{I_0}\frac{H_0}{I_0} - \frac{H_2}{I_0} \qquad (4.26)$$

and the collective potential

$$V(q) = \frac{H_0}{I_0} + \frac{1}{8}\frac{H_2''}{I_0} - \cdots . \qquad (4.27)$$

The momenta of $\mathscr{H}$ and I: $H_0, H_2, I_0, I_2, \ldots$ are defined by

$$A_n(q) = \int ds\, s^n A\left(q + \frac{s}{2}, q - \frac{s}{2}\right), \quad A = H, I. \qquad (4.28)$$

In (4.27) the two most important terms of $V(q)$ are given and the primes indicate derivatives with respect to q.

Various applications of the GCM to the nuclear problems are simplified using the effective nucleon–nucleon interaction of $Skyrme$ [4.9]. The appearance of the effective interaction between two nucleons in a medium, which is different from that between nucleons in vacuum, is due to the effects of the other nucleons on the pair of particles considered. The approach of Skyrme gives a direct parametrization of the realistic effective density-dependent nucleon–nucleon interaction. It has the following form in coordinate space [4.11–13]:

$$\begin{aligned}
v(r_1, r_2) = {}& t_0(1 + x_0 P_\sigma)\delta(r_1 - r_2) + \tfrac{1}{2}t_1(1 + x_1 P_\sigma)[\delta(r_i - r_2)\hat{k}^2 \\
& + \hat{k}'^2\delta(r_1 - r_2)] + t_2(1 + x_2 P_\sigma)\hat{k}' \cdot \delta(r_1 - r_2)\hat{k} \\
& + \tfrac{1}{6}t_3\rho^\sigma(1 + x_3 P_\sigma)\delta(r_1 - r_2) \\
& + iW_0[\hat{k}' \times \delta(r_1 - r_2)\hat{k}] \cdot (\sigma_1 + \sigma_2),
\end{aligned} \qquad (4.29)$$

where $t_0, t_1, t_2, t_3, W_0, \sigma, x_{1,2,3}$ are parameters, $P_\sigma = 1/2(1 + \sigma_1 \cdot \sigma_2)$, the operator $\hat{k}$ acts on the wave functions in the initial state and $\hat{k}'$ acts on the final state wave functions ($\hat{k} = 1/2i(\nabla_{r_1} - \nabla_{r_2}), \hat{k}' = -1/2i(\nabla_{r_1} - \nabla_{r_2})$). The values of the parameters of the Skyrme interaction are determined so as to give correct values of the binding energies of nuclei, as well as the nuclear matter binding energy (E/A), the incompressibility K, the effective mass m^* and the equilibrium density ρ_0 by the relationships [4.11–13, 4.14–16]:

$$E/A = \frac{\hbar^2}{2m}\beta\rho^{2/3} + \frac{3}{8}t_0\rho + \frac{1}{16}(3t_1 + 5t_2)\beta\rho^{5/3} + \frac{1}{16}t_3\rho^{1+\sigma}, \qquad (4.30)$$

$$K = 2\frac{\hbar^2}{2m}\beta\rho^{2/3} + \frac{9}{4}t_0\rho + \frac{5}{4}(3t_1 + 5t_2)\beta\rho^{5/3} +$$

$$+ \frac{9}{16}t_3(1 + \sigma)(\sigma + 2/3)\rho^{1+\sigma}, \qquad (4.31)$$

$$m^* = m\left[1 + \frac{2m}{\hbar^2}\frac{1}{16}(3t_1 + 5t_2)\rho\right]^{-1}, \qquad (4.32)$$

$$\frac{\partial}{\partial\rho}(E/A) = \frac{2}{3}\frac{\hbar^2}{2m}\beta\rho^{-1/3} + \frac{3}{8}t_0 + \frac{5}{48}(3t_1 + 5t_2)\beta\rho^{2/3}$$

$$+ \frac{(1 + \sigma)}{16}t_3\rho^\sigma = 0, \qquad (4.33)$$

where $\beta \equiv \frac{3}{5}\left(\frac{3\pi^2}{2}\right)^{2/3}$.

The use of the Skyrme interaction leads to significant simplifications in the GCM applications to studies of nuclear structure. Monopole, dipole, and quadrupole isoscalar and isovector vibrations in light double magic nuclei are considered in [4.17, 4.18].

If the generating function Φ in (4.1) for a $Z = N$ nucleus is a Slater determinant built up from single-particle wave functions $\varphi_\lambda(r)$ corresponding to a given construction potential then the energy kernel (4.6) has the form [4.19]:

$$\mathscr{H}(x, x') = \langle \Phi(\{r_i\}, x)| \Phi(\{r_i\}, x')\rangle \int H(x, x', r)\,dr. \qquad (4.34)$$

In the case of Skyrme effective N–N forces (for $Z = N$ nucleus without Coulomb and spin-orbital interaction):

$$H(x, x', r) = \frac{\hbar^2}{2m}T + \frac{3}{8}t_0\rho^2 + \frac{1}{16}(3t_1 + 5t_2)(\rho T + j^2)$$

$$+ \frac{1}{64}(9t_1 - 5t_2)(\nabla\rho)^2 + \frac{1}{16}t_3\rho^{2+\sigma}, \qquad (4.35)$$

where the density $\rho(x, x', r)$, the kinetic energy density $T(x, x', r)$ and the current density $j(x, x', r)$ are given by the expressions:

$$\rho(x, x', r) = 4\sum_{\lambda,\mu=1}^{A/4}(N^{-1})_{\mu\lambda}\varphi_\lambda^*(r, x)\varphi_\mu(r, x'), \qquad (4.36)$$

$$T(x, x', r) = 4\sum_{\lambda,\mu}^{A/4}(N^{-1})_{\mu\lambda}\nabla\varphi_\lambda^*(r, x)\cdot\nabla\varphi_\mu(r, x'), \qquad (4.37)$$

$$j(x, x', r) = 2 \sum_{\lambda, \mu}^{A/4} (N^{-1})_{\mu\lambda} \{ \varphi_\lambda^*(r, x) \nabla \varphi_\mu(r, x')$$

$$- [\nabla \varphi_\lambda^*(r, x)] \varphi_\mu(r, x') \}. \tag{4.38}$$

In Eqs. (4.36)–(4.38) A is the mass number and

$$N_{\lambda\mu} = \int \varphi_\lambda^*(r, x) \varphi_\mu(r, x') \, dr. \tag{4.39}$$

The overlap kernel (4.5) has the form:

$$I(x, x') = [\det(N_{\lambda\mu})]^4. \tag{4.40}$$

The one-body density matrix is given by

$$\rho(r, r') = \int \int f_0(x) f_0(x') I(x, x') \rho(x, x', r, r') \, dx \, dx', \tag{4.41}$$

where

$$\rho(x, x', r, r') = 4 \sum_{\lambda, \mu = 1}^{A/4} (N^{-1})_{\mu\lambda} \varphi_\lambda^*(r, x) \varphi_\mu(r', x') \tag{4.42}$$

and $f_0(x)$ is the solution of the Eq. (4.4) corresponding to the lowest energy eigenvalue. It follows from (4.41) that the nuclear density distribution $\rho(r)$ and the nucleon momentum distribution $n(k)$ can be expressed as

$$\rho(r) = \int \int f_0(x) f_0(x') I(x, x') \rho(x, x', r) \, dx \, dx' \tag{4.43}$$

and

$$n(k) = \int \int f_0(x) f_0(x') I(x, x') \rho(x, x', k) \, dx \, dx', \tag{4.44}$$

where

$$\rho(x, x', k) = 4 \sum_{\lambda, \mu = 1}^{A/4} (N^{-1})_{\mu\lambda} \, \tilde{\varphi}_\lambda^*(k, x) \tilde{\varphi}_\mu(k, x') \tag{4.45}$$

and $\tilde{\varphi}(k, x)$ is the Fourier transform of $\varphi(r, x)$.

The study of the N–N correlation effects within the GCM depends on the choice of the construction potential. In particular, the use of the harmonic oscillator construction potential

$$V(r) = -V_0 + \tfrac{1}{2} m \omega^2 r^2 \tag{4.46}$$

affects strongly the behaviour of some physical quantities due to the specific asymptotic behaviour of the oscillator function.

The application of the GCM using the square-well potential with infinite walls as a construction potential:

$$V(r) = \begin{cases} -V_0 & r < x \\ \infty, & r > x, \end{cases} \tag{4.47}$$

(where the radius of the well x is a generator coordinate) can also be done. According to this choice, the generating function $\Phi(\{r_i\}, x)$ corresponds to a state of A nucleons confined in a finite spatial volume (a sphere with radius x).

The N–N correlation effects depending on the choice of the construction potential on the nucleon momentum and density distributions [4.20–22] as well as on two-nucleon centre-of-mass and relative motion momentum distribution [4.23] are considered in Chap. 7. The natural orbital representation (occupation numbers and natural single-particle functions) within the GCM [4.24] is considered in Chap 6. The energies and density distributions of collective monopole states calculated within the GCM are presented in Sect. 8.2.

4.2 The Coherent Density Fluctuation Model

The coherent density fluctuation model (CDFM) has been suggested in [4.25–27, 4.22] as a model for studying characteristics of nuclear structure and nuclear reactions based on the local density distribution as a variable of the theory and using the essential results of the infinite nuclear matter theory.

The CDFM is introduced using the main ansatz of the GCM for the many-body function (Eq. (4.1)) and the delta-function approximation for the overlap and energy kernels (4.5) and (4.6):

$$I(x, x') \to \delta(x - x'), \tag{4.48}$$

$$\mathcal{H}(x, x') \to -\frac{\hbar^2}{2m_{\text{eff}}} \delta''(x - x') + \delta(x - x') V\left(\frac{x + x'}{2}\right), \tag{4.49}$$

which has been discussed in Sect. 4.1. The delta-function approximation (4.48) and (4.49) reflects the fact that for many-fermion systems the kernels $I(x, x')$ and $\mathcal{H}(x, x')$ are strongly peaked at $x \simeq x'$ [4.3]. The delta-function approximation leads to more transparent relationships between the physical quantities. At the same time, it is shown in the case of the nucleon and cluster momentum distribution (Chap. 7) that the difference between the results in the CDFM and in the GCM with square-well construction potential (i.e. without the delta-function approximation) does not lead to a change of the physical conclusions.

If the trial wave function $\Psi(\{r_i\})$ in the GCM (4.1) is normalized to the mass number A and the weight function is determined under the condition

$$\int_0^\infty |f(x)|^2 \, dx = 1, \tag{4.50}$$

then the delta-function approximation (4.48) leads to the relationship:

$$\int \Phi^*(\{r_i\}, x')\Phi(\{r_i\}, x)\, dr_1 \ldots dr_A = A\delta(x - x'). \tag{4.51}$$

Taking into account Eqs. (4.9) and (4.51) one can suggest that the following approximation holds in the case of many-fermion systems [4.28]:

$$\int \Phi^*(r, r_2, \ldots, r_A, x')\Phi(r', r_2, \ldots, r_A, x)\, dr_2 \ldots dr_A$$

$$\cong \rho_{x, x}(r, r')\delta(x - x'), \tag{4.52}$$

where $\rho_{x,x}(r, r')$ is the one-body density matrix corresponding to the wave function $\Phi(\{r_i\}, x)$:

$$\rho_{x,x}(r, r') \equiv \rho_x(r, r')$$

$$= \int \Phi^*(r, r_2, \ldots, r_A, x') \Phi(r', r_2, \ldots, r_A, x) \, dr_2 \ldots dr_A. \qquad (4.53)$$

We note that the integration of (4.52) over r at $r' = r$ would lead to a relation of the type (4.51). In this way the main approximation of the CDFM (4.52) is related to the delta-function limit in the GCM.

If the generating function $\Phi(\{r_i\}, x)$ describes a system with a uniform density:

$$\rho_x(r) = \rho_x(r, r' = r) = \rho_0(x)\theta(x - |r|), \qquad (4.54)$$

where

$$\rho_0(x) = 3A/4\pi x^3 \qquad (4.55)$$

and the generator coordinate x is the radius of a sphere containing all A nucleons uniformly distributed in it (the so-called "flucton"), it is appropriate to use for such a system one-body density matrix of the form:

$$\rho_x(r, r') = 3\rho_0(x) \frac{j_1(k_F(x)|r - r'|)}{k_F(x)|r - r'|} \theta\left(x - \frac{|r + r'|}{2}\right). \qquad (4.56)$$

In this expression j_1 is the first-order spherical Bessel function.

The Eq. (4.56) (without the θ-function) is the one-body density matrix for a system of particles described by plane waves [4.29], where

$$k_F(x) = (3\pi^2 \rho_0(x)/2)^{1/3} \equiv \alpha/x \qquad (4.57)$$

and

$$\alpha = (9\pi A/8)^{1/3} \simeq 1.52 A^{1/3}. \qquad (4.58)$$

Due to the function $\theta\left(x - \dfrac{|r + r'|}{2}\right)$ the centre-of-mass of two nucleons with radius-vectors r and r' remains in the sphere of radius x. The function $3j_1(t)/t$ is essentially different from zero at $t \lesssim 4$. This leads to the relation $|r - r'| \lesssim 4x/\alpha$. For all nuclei with the exception of the lightest ones $4/\alpha \lesssim 1$. Therefore, the one-body density matrix (4.56) describes to a good approximation a system of particles confined in a sphere of radius x and density $\rho_0(x)$. Its diagonal elements $\rho_x(r, r)$ coincide with the density (4.54)

Using (4.1) and (4.52) the one-body density matrix can be written as a coherent superposition of one-body density matrices for spherical "pieces" of nuclear matter with densities $\rho_0(x)$ [4.26, 4.27]:

$$\rho(r, r') = \int_0^\infty |f(x)|^2 \rho_x(r, r') \, dx \qquad (4.59)$$

with $\rho_x(r, r')$ from Eq. (4.56).

In the CDFM the Wigner distribution function (e.g. [4.30]):

$$W(r, k) = \int \rho\left(r + \frac{\eta}{2}, r - \frac{\eta}{2}\right) \exp(-ik\eta)\,d\eta \tag{4.60}$$

can be written using (4.59) and (4.56) in the form:

$$W(r, k) = \int_0^\infty |f(x)|^2 \theta(x - |r|)\theta(k_F(x) - |k|)\,dx. \tag{4.61}$$

The relations of the density and momentum distribution with the Wigner function:

$$\rho(r) = 4 \int W(r, k) \frac{dk}{(2\pi)^3}, \tag{4.62}$$

$$n(k) = \int W(r, k)\,dr \tag{4.63}$$

lead to the following expressions in the CDFM:

$$\rho(r) = \int_0^\infty |f(x)|^2 \rho_0(x)\theta(x - |r|)\,dx \tag{4.64}$$

and

$$n(k) = \int_0^\infty |f(x)|^2 \tfrac{4}{3}\pi x^3 \theta(k_F(x) - |k|)\,dx. \tag{4.65}$$

In the case of monotonically-decreasing density distributions ($d\rho/dr \leqslant 0$) one can obtain from (4.64) the relation of the weight function $f(x)$ with the density distribution:

$$|f(x)|^2 = -\frac{1}{\rho_0(x)} \frac{d\rho(r)}{dr}\bigg|_{r=x}. \tag{4.66}$$

As a result the nucleon momentum distribution can be obtained as a functional of the density distribution ρ [4.26, 4.27, 4.22]:

$$n(k) = \left(\frac{4\pi}{3}\right)^2 \frac{4}{A}\left[6\int_0^{\alpha/k} \rho(x)x^5\,dx - (\alpha/k)^6 \rho(\alpha/k)\right] \tag{4.67}$$

with the normalization:

$$\int n(k) \frac{dk}{(2\pi)^3} = A. \tag{4.68}$$

The basic properties of the nucleon momentum distribution (4.67) are: the power-law fall-off

$$n(k) \sim k^{-8} \quad \text{as } k \to \infty \tag{4.69}$$

$$(\text{when } d\rho/dr|_{r=0} = 0 \quad \text{and} \quad d^2\rho/dr^2|_{r=0} \neq 0) \tag{4.70}$$

and its behaviour as $k \to 0$:

$$n(k \to 0)/A \simeq (128\pi^2/3A^2) \int\limits_0^\infty \rho(x)x^5 \, dx. \tag{4.71}$$

These properties are discussed in Chap. 7.

The CDFM expressions for $n(k)$ in the case of density distributions which are not monotonic decreasing functions are obtained in [4.31, 4.32].

The results of the calculations of nucleon momentum distributions for various nuclei as well as of cluster momentum distributions within the CDFM are presented in Chap. 7.

As was shown in Sect. 4.1, the delta-function approximation (Eqs. (4.15) and (4.16)) in the GCM leads to the Schrödinger-type equation (4.12) which is the basic equation of the CDFM [4.25, 4.26]. It describes coherent vibrations of the nuclear density. All nucleons take part in this motion in a coherent way keeping a homogeneous density for any size of the system. The function $V(x)$ in (4.12) corresponds to the potential energy of the breathing collective motion of all A nucleons. As the intermediate states in the CDFM (i.e. the states with different values of the generator coordinate x) are systems with uniform density $\rho_0(x)$ containing all A nucleons, one can use for $V(x)$ the corresponding expressions for the energy of the nuclear matter with density $\rho_0(x)$, for instance, from the work of *Brueckner* et al. [4.33], see Eqs. (7.26–28). It should be noted that if the generating functions $\Phi(\{r_i\}, x)$ from (4.1) correspond in the CDFM to the states of nuclear matter with density $\rho_0(x)$, an interval (x_1, x_2) in which $V(x) \leqslant E_0$ ($E_0 = \min \langle \Psi | \hat{H} | \Psi \rangle$ is the ground state energy of the finite nucleus) will exist. The effective mass m_{eff} depending in principle on x is considered in the CDFM to be a constant. It is the only parameter of the model and can be determined by the fit of the calculated ground state energy of various nuclei to the experimental data.

The density distributions of the ground ($n = 0$) and the excited collective states ($n \neq 0$) have the form

$$\rho_n(r) = \int\limits_0^\infty |f_n(x)|^2 \rho_0(x)\theta(x - |r|) \, dx, \tag{4.72}$$

$$(n = 0, 1, 2, \ldots),$$

where $f_n(x)$ are solutions of (4.12) corresponding to the energies E_n.

The study of the collective motions within the CDFM is presented in Chap. 8.

It is of interest to note that (4.12) obtained in the framework of the delta-function limit of GCM has been used by G. *Brown* et al. [4.34] to describe the breathing collective motion of the nucleon in the three-quark bag model. The radius of the bag is the dynamic variable and the quark energy is considered to be the potential energy of the bag surface motion. In this approach the nucleon Roper resonance at about 1440 MeV is interpreted as a first breathing excitation of the quark bag. Though essentially non-relativistic, this model can be a start-ing point in the account of the internal nucleon dynamics of nuclear systems. As shown in [4.35] the result of the calculations of the asymptotic D/S-ratio in the deuteron using this model is in agreement with the experimental one.

It has been shown in the CDFM [4.36] that the amplitude of the elastic scattering of point-like particles on nuclei $A_{00}(q)$ can be obtained approxim-ately in the form:

$$A_{00}(q) = \int_0^\infty dx \, |f_0(x)|^2 A_0(x, q),$$ (4.73)

where $A_0(x, q)$ is the amplitude of the elastic scattering of the initial particle from a flucton with radius x. This amplitude is averaged by the CDFM weight function $f_0(x)$ corresponding to the ground state, and q is the transfer momentum.

Equation (4.73) has been used to calculate the elastic electron, proton, alpha-particle and heavy ion scattering by nuclei [4.22b]. It gives the possibility of accounting for the zero-motion nuclear density vibrations in the interaction processes. In other words, it is possible to study the effects of the nuc-leon–nucleon correlations included in the CDFM on the scattering cross-sections. At the same time (4.73) leads to significant simplifications of the calculations in the cases when the amplitude of the elastic scattering on a system with uniform density is known. It can be noted that the elastic electron scatter-ing form factor $F(q)$ is expressed exactly using (4.64) and the flucton form factor $F_\theta(q, x)$:

$$F(q) = \int_0^\infty dx \, |f_0(x)|^2 F_\theta(q, x),$$ (4.74)

where

$$F_\theta(q, x) = j_1(qx)/(qx).$$ (4.75)

The results of the CDFM calculations of elastic electron, proton as well as alpha-particle and heavy ion scattering on nuclei are presented in Chaps. 9, 11 and 12, respectively.

4.3 The Natural Orbital Method

The wave function $|\Psi\rangle$ of a system of A fermions can be written using a com-plete set of Slater determinants containing all possible excitations and construc-ted from an arbitrary complete set of orthonormal single-particle wave functions $\psi_\alpha(r)$. The one-body density matrix of the system:

$$\rho(r, r') = A \int \Psi^*(r, r_2, \ldots) \Psi(r', r_2, \ldots) dr_2 \ldots dr_A$$ (4.76)

can be expressed by:

$$\rho(\mathbf{r}, \mathbf{r}') = \sum_{\alpha, \beta} n^1_{\alpha\beta} \psi^*_\alpha(\mathbf{r}) \psi_\beta(\mathbf{r}'), \tag{4.77}$$

where

$$n^1_{\alpha\beta} = \langle \Psi | a^+_\alpha a_\beta | \Psi \rangle = n^{1*}_{\beta\alpha}. \tag{4.78}$$

In (4.78) a^+_α and a_β are operators that create or annihilate particles in states α and β, respectively. The Hermitian matrix $n^1_{\alpha\beta}$ is also called the one-body density matrix. The diagonal elements

$$n^1_{\alpha\alpha} = \langle \Psi | a^+_\alpha a_\alpha | \Psi \rangle = \langle n_\alpha \rangle \tag{4.79}$$

are the single-particle occupation numbers. They satisfy the conditions:

$$0 \leqslant \langle n_\alpha \rangle \leqslant 1 \quad \text{and} \quad \text{Tr}(n^1) = \sum_\alpha \langle n_\alpha \rangle = A. \tag{4.80}$$

The expectation value of any one-body operator $\hat{Q} = \sum_{i=1}^{A} \hat{q}_i$ can be expressed by the matrix n^1:

$$\langle \hat{Q} \rangle = \langle \Psi | \hat{Q} | \Psi \rangle = \sum_{\alpha, \beta} q_{\alpha\beta} n^1_{\beta\alpha} = \text{Tr}(qn^1), \tag{4.81}$$

where

$$q_{\alpha\beta} = \langle \psi_\alpha | \hat{q} | \psi_\beta \rangle. \tag{4.82}$$

It is seen from (4.81) that the one-body density matrix n^1 contains all the information on the single-particle properties of the fermion system.

Using the unitary transformation U one can diagonalize the matrix n^1 (4.78). The transformed matrix is

$$n^1_{\alpha\beta} \equiv (U^+ n^1 U)_{\alpha\beta} = n_\alpha \delta_{\alpha\beta}. \tag{4.83}$$

The unitary transformation U leads to a new set of single-particle functions:

$$\varphi_\alpha(\mathbf{r}) = \sum_\beta U_{\alpha\beta} \psi_\beta(\mathbf{r}). \tag{4.84}$$

The functions $\varphi_\alpha(\mathbf{r})$ are called "natural orbitals" and n_α are the occupation numbers of the natural orbitals (or "natural occupation numbers"). In this new basis the expectation value of any one-body operator $\hat{Q}$ has the form:

$$\langle \hat{Q} \rangle = \sum_\alpha \tilde{q}_{\alpha\alpha} n_\alpha, \qquad \tilde{q}_{\alpha\alpha} = \langle \varphi_\alpha | \hat{q} | \varphi_\alpha \rangle. \tag{4.85}$$

In the natural orbital representation the one-body density matrix $\rho(\mathbf{r}, \mathbf{r}')$ is

$$\rho(\mathbf{r}, \mathbf{r}') = \sum_\alpha n_\alpha \varphi^*_\alpha(\mathbf{r}) \varphi_\alpha(\mathbf{r}'). \tag{4.86}$$

This form of $\rho(\mathbf{r}, \mathbf{r}')$ was introduced by *Löwdin* [4.37]. Eq. (4.86) leads to the following expressions for the density and momentum distributions in the new basis:

$$\rho(r) = \sum_\alpha n_\alpha |\varphi_\alpha(r)|^2 \tag{4.87}$$

and

$$n(k) = \sum_\alpha n_\alpha |\varphi_\alpha(k)|^2, \tag{4.88}$$

where $\varphi_\alpha(k)$ are the Fourier transform of $\varphi(r)$.

In the Hartree–Fock approximation:

$$n_\alpha^{HF} = \begin{cases} 1, & \alpha \leqslant \alpha_F \\ 0, & \alpha > \alpha_F, \end{cases} \tag{4.89}$$

where α_F denotes the Fermi level.

If the wave function of the system is chosen as a single Slater determinant built up from natural orbitals $\varphi_\alpha(r)$ the one-body density matrix has the form

$$\rho_0(r, r') = \sum_\alpha^{\alpha_F} \varphi_\alpha^*(r)\varphi_\alpha(r'). \tag{4.90}$$

It has been shown by *Kobe* [4.38] that if $\rho(r, r')$ is the exact one-body density matrix, the following relation holds:

$$\mathrm{Tr}[(\rho - \rho_0)^2] = \text{minimum}. \tag{4.91}$$

Considering $\rho(r, r')$ for an arbitrary basis of single-particle wave function (4.77–80) it is seen that diagonal elements (4.79) of n^1 correspond to the chosen ψ representation. The sum of the single-particle occupation numbers up to the Fermi level $\sum_\alpha^{\alpha_F} \langle n_\alpha \rangle$ is different for the various sets of single-particle basis functions. Changing the sets of functions ψ one can rearrange the particles below and above the Fermi level conserving the condition $\sum_\alpha \langle n_\alpha \rangle = A$. There is a relation between the optimal arrangement of the particles and the type of the chosen basis. It is shown by *Kobe* [4.38] that the basis of the natural orbitals gives an optimal arrangement of the particles. This means that the number of particles in the Fermi sea

$$A_< = \sum_{\alpha, \beta}^{\alpha_F} \int dr\, n_{\alpha\beta}^1 \psi_\alpha^*(r)\psi_\beta(r) \tag{4.92}$$

will be a maximum (corresponding to the minimal depletion of the Fermi sea) in the natural orbital basis. Consequently, the number of particles above the Fermi level $A_>$ will be minimal. This is demonstrated by varying the Eq. (4.92) with respect to an arbitrary change of $\psi_\alpha(r)$:

$$\delta\psi_\beta = \sum_{\gamma(\neq\beta)} \xi_{\beta\gamma}\psi_\gamma, \tag{4.93}$$

$$\delta\psi_\alpha^* = \sum_{\gamma(\neq\alpha)} \eta_{\alpha\gamma}\psi_\gamma^*, \tag{4.94}$$

including the orthonormality condition:

$$\int \psi_\alpha^*(r)\psi_\beta(r)\,\mathrm{d}r = \delta_{\alpha\beta}. \tag{4.95}$$

Then

$$\delta A_< = \sum_{\substack{\gamma,\beta \\ (\gamma \neq \beta)}}^{\alpha_F} \xi_{\beta\gamma}\langle \Psi|a_\gamma^+ a_\beta|\Psi\rangle + \text{c.c.} \tag{4.96}$$

vanishes for arbitrary variations of $\xi_{\beta\gamma}$ only if

$$\langle \Psi|a_\gamma^+ a_\beta|\Psi\rangle = 0, \quad \gamma \neq \beta, \quad \gamma \leqslant \alpha_F, \quad \beta \leqslant \alpha_F. \tag{4.97}$$

But (4.97) is just the condition for the natural orbital representation. It can be shown that $\delta^2 A_<$ is negative, so the quantity $A_<$ is maximal. It follows from the conservation of the total number of particles

$$A = A_< + A_> \tag{4.98}$$

that the particles above the Fermi sea $A_>$ is minimal in the natural orbital representation.

Considering the one-body density matrix (4.77) and the one-body density matrix $\rho_0(r, r')$ corresponding to a Slater determinant wave function in an arbitrary single-particle basis:

$$\rho_0(r, r') = \sum_\alpha^{\alpha_F} \psi_\alpha^*(r)\psi_\alpha(r'), \tag{4.99}$$

Eq. (4.91) can be presented in the form (after performing the trace):

$$\sum_{\alpha,\beta} n_{\alpha\beta}^1 n_{\beta\alpha}^1 - 2\sum_\alpha^{\alpha_F} n_{\alpha\alpha} + A = \text{minimum}. \tag{4.100}$$

Since the first and third terms of (4.100) are constants, the minimum can be reached if

$$\sum_\alpha^{\alpha_F} n_\alpha \equiv A_< = \text{maximum}. \tag{4.101}$$

It has been already shown that (4.101) is satisfied if the single-particle functions ψ are natural orbitals.

The density distribution $\rho(r)$ in an arbitrary single-particle basis

$$\rho(r) = \sum_{\alpha\beta} n_{\beta\alpha}^1 \psi_\alpha^*(r)\psi_\beta(r) \tag{4.102}$$

and in the natural orbital basis (Eq. (4.87)) has a simple form if the system has total angular momentum $J = 0$. In this case, any choice of the single-particle wave functions having good quantum numbers l, j, m and good parity $(-1)^l$ leads to the partial diagonalization of n^1 [4.39]:

$$J = 0 \to n_{\alpha\beta}^1 \sim \delta_{l_\alpha l_\beta}\delta_{j_\alpha j_\beta}\delta_{m_\alpha m_\beta}. \tag{4.103}$$

This means that the diagonalization of the one-body density matrix for a system with $J = 0$ has to be performed only within the $\{ljm\}$-subspaces of the total model space $\{\psi_\alpha\}$. In this case the transformation (4.84) "mixes" only single-particle wave functions with the same l, j, m and with different quantum numbers n. Such a system has spherical symmetry. Using the explicit form:

$$\psi_\alpha(r) = \sum_{\mu\sigma} (l_\alpha s\mu\sigma | j_\alpha m_\alpha) \frac{1}{r} y_{n_\alpha l_\alpha m_\alpha}(r) Y_{l_\alpha}^\mu(\Omega) \chi_s^\sigma, \tag{4.104}$$

as well as accounting for (4.103), one can obtain for the density distribution [4.40]:

$$\rho(r) \equiv \frac{1}{4\pi} \int d\Omega \, \rho(r) = \sum_{lj} \rho_{lj}(r), \tag{4.105}$$

where

$$4\pi r^2 \rho_{lj}(r) = \sum_{nn'} (2j + 1) n_{n'lj,\,nlj}^1 y_{nlj}^*(r) y_{n'lj}(r) \tag{4.106}$$

in an arbitrary single-particle basis $(\psi_\alpha(r))$ and

$$4\pi r^2 \rho_{lj}(r) = \sum_n (2j + 1) n_{nlj} |\tilde{y}_{nlj}(r)|^2 \tag{4.107}$$

in the natural orbital basis $(\varphi_\alpha(r))$. The function $\tilde{y}(r)$ is the radial part of the natural orbital $\varphi(r)$.

Using the notation

$$\tilde{R}_{nlj}(r) = \tilde{y}_{nlj}/r, \tag{4.108}$$

the nuclear density distribution $\rho(r)$ and the nucleon momentum distribution $n(k)$ for nuclei with $J = 0$ can be written in the following form in the natural orbital representation:

$$\rho(r) = \frac{1}{4\pi} \sum_{lj} (2j + 1) \sum_n n_{nlj} |\tilde{R}_{nlj}(r)|^2, \tag{4.109}$$

$$n(k) = \frac{1}{4\pi} \sum_{lj} (2j + 1) \sum_n n_{nlj} |\tilde{R}_{nlj}(k)|^2, \tag{4.110}$$

where

$$\tilde{R}_{nlj}(k) = (2/\pi)^{1/2} (-i)^l \int_0^\infty dr \, r^2 j_l(kr) \tilde{R}_{nlj}(r). \tag{4.111}$$

$j_l(kr)$ are spherical Bessel functions of order l.

The applications of the natural orbital representation to the problems of correlation effects in nuclei are considered in Chaps. 6 and 7. This representation gives the possibility of using the single-particle description (which is well known from the shell model) in the complicated correlated methods going beyond the limits of the mean-field approximation.

Part 2
Nucleon Correlations and Nuclear Structure

5 Spectral Functions

In this chapter the correlation effects on the spectral function of deep-hole nuclear states and related quantities are discussed. The definitions are given in the Introduction (Sect. 5.1) and the description of the spectral function in the framework of the Green-function method in Sect. 5.2. The coherent density fluctuation model result for the hole nuclear state spectral function is presented in Sect. 5.3. In Sect. 5.4 the sum rules for the spectral function are considered. The role of the short-range two-body interactions (not taken into account in the Hartree–Fock method and in the shell model) in the sum-rule analyses, as well as the momentum and energy range conditions necessary for the obtaining of experimental information are discussed in the same Section.

5.1 Introduction

Systematic experimental and theoretical studies of one-nucleon transfer and knockout reactions give information about the applicability of the shell model, or more generally, of the quasiparticle approach [5.1–3] to the description of single-particle states in nuclei. The investigation of single-particle characteristics, such as occupation numbers, deep-hole state energies and widths is closely related to this problem. It is of interest to compare the values of these quantities for deep-hole states with those for states near the Fermi level. Important experimental information in this respect can be obtained in (e, e′ p) (e.g. [5.4, 5.5]) and (p, 2p) reactions (e.g. [5.6, 5.7]). In the Plane-Wave Impulse Approximation (PWIA) the cross-section of the (e, e′ p) reaction can be written in the form [5.8]:

$$d^4 \sigma / d\Omega_{e'} \, d\varepsilon_{e'} \, d\Omega_p \, dT_p = C(d\sigma / d\Omega)_{ep} \, S(E, k), \tag{5.1}$$

where C is a kinematical factor, $(d\sigma / d\Omega)_{ep}$ is the cross-section of the elementary e + p interaction and $S(E, k)$ is the diagonal element of the hole spectral density (or spectral function) in the momentum representation. The latter contains the information on the nuclear structure which can be extracted from the (e, e′ p) reaction. It is defined as [5.9, 5.10]:

$$S(E, k) = \langle \Psi_0 | a^+(k) \delta(E - \hat{H}) a(k) | \Psi_0 \rangle, \tag{5.2}$$

where $|\Psi_0\rangle$ is the ground state wave function of the target nucleus with

A nucleons, $a^+(k)$ and $a(k)$ are creation and annihilation operators for a nucleon with momentum k, E is the energy of the residual nucleus with $A-1$ nucleons with respect to the ground state energy of the target nucleus. $\hat{H}$ is the Hamiltonian of the system with $A-1$ nucleons. The function $S(E,k)$ can be interpreted as the probability of finding a nucleon with momentum k in the target nucleus leaving the residual nucleus with an energy E [5.11]. Obviously $S(E,k)$ is different from zero only if the final state wave function of the residual nucleus overlaps with $a(k)|\Psi_0\rangle$, i.e. it can be partially described in terms of a hole created in $|\Psi_0\rangle$ in the process of a particle knockout.

The spectral function can also be expressed in the form [5.10, 5.12, 5.13]:

$$S(E,k) = \frac{1}{2J_A + 1} \sum_f |\langle \Psi_f | a(k) | \Psi_A \rangle|^2 \, \delta(E - E_f), \qquad (5.3)$$

where J_A and Ψ_A are the spin and the wave function of the target nucleus ground state, Ψ_f and E_f are the wave function and the energy of the state f of the residual nucleus, respectively.

In the shell model (i.e. with no account taken of the residual interaction) the spectral function can be written as [5.10]:

$$S(E,k) = \sum_\alpha N_\alpha |\varphi_\alpha(k)|^2 \, \delta(E - E_\alpha), \qquad (5.4)$$

where $\varphi_\alpha(k)$ is the single-particle wave function of the state α in the momentum representation, and N_α is the corresponding occupation number.

Thus, the contributions of different nuclear shells to $S(E,k)$ can in principle be obtained and the momentum distribution for each state α can be studied in the framework of PWIA using the shell model.

Actually, however, due to the residual interactions, the hole state is not an eigenstate of the residual nucleus but is a mixture of several single-particle states [5.4a]. The decay widths, when the excitation energy of the residual nucleus is above the threshold of particle emission, also lead to the spreading of the shell structure. This turns the narrow-peaked structure of $S(E,k)$ for $-20 < E < -8$ MeV (corresponding to the bound states of the final nucleus) into broad maxima for $E < -20$ MeV and only a careful study of the momentum dependence of $S(E,k)$ can separate the contributions from different shells.

To generalize (5.4) we consider the matrix element from (5.3):

$$\Phi_{f_A}(k) \equiv \langle \Psi_f | a(k) | \Psi_A \rangle = \int d\mathbf{r} \, \Phi_{f_A}(\mathbf{r}) \exp(-i k \cdot \mathbf{r}), \qquad (5.5)$$

which is expressed by the overlap integral

$$\Phi_{f_A}(\mathbf{r}) = \int \Psi_f^*(\mathbf{r}'_2, \ldots \mathbf{r}'_{A-1}) \, \Psi_A(\mathbf{r}, \mathbf{r}'_2, \ldots, \mathbf{r}'_{A-1}) \, d\mathbf{r}'_2 \ldots d\mathbf{r}'_{A-1}. \qquad (5.6)$$

In (5.6) relative coordinates are introduced:

$$r = r_1 - R_B, \quad r'_i = r_i - R_B \quad (i = 2, \ldots, A), \qquad (5.7)$$

where

$$R_B = \sum_{i=2}^{A} r_i/(A-1).$$

Using a complete orthonormal system of single-particle functions (orbitals) $\{\varphi\}$ generated by an effective potential in the shell model, the overlap integral $\Phi_{f_A}(r)$ can be presented in the form [5.10, 5.12]:

$$\Phi_{f_A}(r) = \sum_{\alpha} \Theta_{\alpha, f}\, \varphi_\alpha(r), \tag{5.8}$$

where $\Theta_{\alpha, f}$ is the amplitude of the contribution of the orbital α to the overlap integral for the eigenstate Ψ_f. In the case of the "pure" shell model the function $\Phi_{f_A}(r)$ is reduced to the single term $\Phi_{f_A}(r) = \varphi_\alpha(r)$. Equation (5.8) expresses the fact that the state Ψ_f of the final nucleus can be obtained by the knockout of a proton from more than one orbital in the target nucleus. Substituting (5.8) in (5.5) and (5.3) the spectral function can be obtained in the form [5.10, 5.11]:

$$S(E, k) = \sum_{\alpha, \beta} \varphi_\alpha^*(k)\, \varphi_\beta(k)\, \mathscr{S}(\alpha, \beta, E)$$

$$\approx \sum_{\alpha} |\varphi_\alpha(k)|^2\, \mathscr{S}(\alpha, E), \tag{5.9}$$

where

$$\mathscr{S}(\alpha, E) = \frac{1}{2J_A + 1} \sum_f |\Theta_{\alpha, f}(E_f)|^2\, \delta(E - E_f) \tag{5.10}$$

is the energy distribution of the strength of the hole state α. Equation (5.9) is a generalization of (5.4). The essential experimental information for $\mathscr{S}(\alpha, E)$ is obtained in [5.4a] for hole states in ^{28}Si, ^{40}Ca and ^{58}Ni nuclei by studying quasifree (e, e′ p) scattering on these nuclei. The most important results are the large widths of $\mathscr{S}(\alpha, E)$ in 1s-hole states which are more than 40 MeV.

The PWIA analysis of the ^{27}Al, ^{40}Ca and ^{51}V (e, e′ p) reactions in [5.5] shows the following widths of 1s-hole states: 31 ± 9 MeV for ^{27}Al, 34 ± 10 MeV for ^{40}Ca and 36 ± 11 MeV for ^{51}V.

Important information on the 1s-hole state widths is obtained also from (p, 2p) and other proton-induced reactions for ^{12}C, ^{16}O and ^{27}Al nuclei; these widths are 9, ~ 14 and ~ 20 MeV, respectively [5.14].

These results show the limited validity of the shell picture for deep-hole nuclear states. Actually, the initial and the final nucleons interact with other nucleons in the nucleus and this leads to their deviation and partial absorption. This can be taken into account in the framework of the Distorted-Wave-Impulse-Approximation (DWIA). The technical complications in this case lead to an effective spectral function [5.10, 5.11] in the expressions of the type of (5.1). The main physical conclusions on the spectral function properties, however, remain the same as in the PWIA case.

5.2 Green Function Description of the Spectral Function

The Green function method is appropriate for the consideration of the spectral function in nuclear theory. The "nondiagonal" spectral function

$$S(E, k, k') = \langle \Psi_0 | a^+(k') \delta(E - \hat{H}) a(k) | \Psi_0 \rangle \tag{5.11}$$

can be obtained using the advanced part of the time-ordered Green function $G_<(k, k', z)$ in momentum space [5.8, 5.10], which has poles and a cut as a function of z along the real axis representing bound and continuum hole states of the target-minus-proton system. The spectral function can be expressed in the form:

$$S(E, k, k') = \lim_{\eta \to 0_+} \frac{1}{2\pi i} \{ G_<(k, k', -E - i\eta)$$

$$- G_<(k, k', -E + i\eta) \}. \tag{5.12}$$

Many difficulties arising in the case of finite nuclei can be avoided using the Green function method in the case of nuclear matter. The results are applied to finite systems by introducing appropriate variables. Due to the fact that the properties of the hole distributions do not depend essentially on the details of nuclear structure, this procedure turns out to be quite convenient [5.10].

The spectral function for infinite systems can be expressed by the one-body Green function in the momentum and energy representation $G(k, \omega)$ [5.15]:

$$S(k, \omega) = \frac{i}{2\pi} [G(k, \omega + i\eta) - G(k, \omega - i\eta)]. \tag{5.13}$$

In infinite systems the function $G(k, \omega)$ has the form (see Sect. 3.2):

$$G(k, \omega) = \frac{1}{\omega - k^2/2m - \Sigma(k, \omega)}, \tag{5.14}$$

where $\Sigma(k, \omega)$ is the mass operator. Then the spectral function (5.13) can be represented by the real and imaginary parts of the mass operator:

$$S(k, \omega) = \frac{1}{\pi} \frac{\operatorname{Im} \Sigma(k, \omega)}{[\omega - k^2/2m - \operatorname{Re} \Sigma(k, \omega)]^2 + [\operatorname{Im} \Sigma(k, \omega)]^2}. \tag{5.15}$$

The following relations can be introduced:

$$G(k, \omega) = G_h(k, \omega) + G_p(k, \omega)$$

$$= \int_{-\infty}^{\varepsilon_F} d\omega' \frac{S_h(k, \omega')}{\omega - \omega' - i\eta} + \int_{\varepsilon_F}^{\infty} d\omega' \frac{S_p(k, \omega')}{\omega - \omega' + i\eta} \tag{5.16}$$

and

$$S(k, \omega) = S_p(k, \omega) + S_h(k, \omega), \tag{5.17}$$

with

$$\int\limits_{-\infty}^{\infty} d\omega \, S(k, \omega) = 1. \tag{5.18}$$

The functions $S_h(k, \omega \leqslant \varepsilon_F)$ and $S_p(k, \omega > \varepsilon_F)$ are the hole- and particle-spectral functions, respectively. They can be used for the evaluation of the widths of hole- and particle-states in nuclei. The occupation numbers (often identified with the momentum distribution) can be obtained by means of the hole spectral function:

$$n(k) = \int\limits_{-\infty}^{\varepsilon_F} d\omega \, S_h(k, \omega). \tag{5.19}$$

The 1s-hole state widths Γ in finite nuclei are calculated in [5.16] using the relation (see (2.40)):

$$\Gamma = 2 \mathrm{Im} \, \Sigma(\langle k \rangle, E(\langle k \rangle)), \tag{5.20}$$

where the imaginary part of the mass operator is calculated up to the second-order terms in the Brueckner theory of nuclear matter (Sect. 3.2) with a density equal to the mean density $\langle \rho \rangle$ felt by the hole in the finite nucleus. The quantity $\langle \rho \rangle$ and the average momentum of the hole $\langle k \rangle$ are calculated in the independent-particle model using a harmonic oscillator potential. The following 1s-hole state widths for various nuclei are obtained: ^{12}C (11.5 MeV), ^{16}O (13.7 MeV), ^{27}Al (19.6 MeV), ^{40}Ca (21.6 MeV) and ^{208}Pb (29.8 MeV).

In [5.17a] the spectral function is calculated using Eqs. (5.14) and (5.15). A simple phenomenological model is suggested to provide an explicit form of the imaginary part of the mass operator which is suitably parametrized using the nucleon–nucleon scattering data. By means of the dispersion relation the real part of the first two terms of the mass operator after the Hartree–Fock term, namely Σ_1 and Σ_2 (corresponding to "core polarization" graph and the "correlation" graph, respectively) is obtained. The calculations of the occupation numbers $n(k)$ (5.19) in ^{40}Ca show a substantial deviation from the Hartree–Fock predictions.

The occupation numbers for all hole orbitals in ^{40}Ca are reduced by about 15%. This means that about six nucleons in this case are in the shells above the Fermi level. This result is in agreement with the calculations in the Brueckner theory and random-phase approximation (RPA) methods presented in [5.17a].

In [5.17b] the influence of short-range correlations on the nuclear spectral functions is studied in the framework of a self-consistent Green function method by summing the so-called ladder diagrams in the perturbative expansion of the effective interaction. The hole–hole propagation is treated on the same footing as the particle–particle propagation. The use of [5.19] leads to an average depletion of 13% in nuclear matter with normal density due to the short-range nucleon–nucleon correlations. It is concluded that the tensor and long-range correlations produce a further reduction but a substantial part of the total depletion is due to short-range correlations.

The spectral functions and 1s-hole state widths in ^{8}Be, ^{12}C, ^{16}O and ^{27}Al nuclei are obtained in [5.18] by calculating the first two terms in the low-density expansion of the mass operator in nuclear matter (Sect. 3.2) and using the separable N–N interaction from [5.19]. This nuclear matter approach is applied for calculations of hole-state spectral functions and widths in finite nuclei by means of a local density approximation. The spectral function is calculated for various values of the density ρ, as for each value of ρ one computes $S_h(k, \omega)$ for k equal to the mean value $\langle k \rangle$ in the finite nucleus. The peak width corresponding to $\rho \to \langle \rho \rangle$ is then obtained by interpolation between the widths evaluated from these spectral functions.

In [5.20] the real and imaginary parts of the Hartree–Fock, core polarization and correlation graph contributions to the complex single-particle field in nuclear matter using the N–N interaction from [5.19] are obtained. The calculations are carried out using Brueckner's reaction matrix theory. The approach is applied to finite nuclei using the experimental difference $E_F - E(k)$ between the Fermi energy and the quasiparticle energy. In the vicinity of $E(k)$, the hole-state spectral function is written as [5.20, 5.10]:

$$S_h(k, E) \approx \pi^{-1} Z^2(k) \frac{\tilde{W}(k) + [E - E(k)] R(k)}{[E - E(k)]^2 + \frac{1}{4} [2Z(k) \tilde{W}(k)]^2}, \tag{5.21}$$

where

$$Z(k) = \left[1 - \frac{\partial}{\partial E} V(k, E) \right]_{E = E(k)}^{-1}, \tag{5.22}$$

$$R(k) = \left[\frac{\partial}{\partial E} W(k, E) \right]_{E = E(k)}^{-1}, \tag{5.23}$$

and $V(k, E)$ and $W(k, E)$ are the real and imaginary parts of the mass operator, respectively. $\tilde{W}$ is the on-shell $(E = E(k))$ expression of W. The spectral functions and widths obtained in [5.20] depend essentially on the choice of the nuclear matter density ρ in the calculations of the mass operator. It is shown that the calculated peaks of $S_h(k, E)$ are too narrow for the outer shells if the normal nuclear matter density $(\rho = 0.17 \, \text{fm}^{-3})$ is used and are in better agreement with the experimental data for ^{28}Si, ^{40}Ca and ^{58}Ni nuclei if the lower value of the density $(\rho = 0.11 \, \text{fm}^{-3})$ is used.

The Green function method is applied for determination of the overlap integrals (5.6) (the spectroscopic amplitudes) in [5.21, 5.22]. It is shown that they can be determined by a Schrödinger-type equation in the discrete [5.21, 5.22] and continuous [5.22] spectrum of the Hamiltonian for the $A - 1$ – body residual nucleus. The hole decay widths are related to the antihermitian part of the hole mass operator, while the Hermitian part of this operator acts as an effective potential.

Proton spectral functions for the 2p, 1f, 2s and 1d hole states in ^{89}Y have been calculated by the Hartree–Fock method using Skyrme forces in [5.23].

5.3 Spectral Functions within the Coherent Density Fluctuation Model

A method to calculate the spectral functions, widths and energies of single-particle states in finite nuclei has been developed [5.24a] in the framework of the coherent density fluctuation model (CDFM) [5.24b–d]. This approach is based on the possibility of using nuclear matter theory in calculations of characteristics of finite nuclei. In the CDFM the spectral functions and widths do not depend on the matter densities felt by the particle in a particular state (as in [5.16]) but are functionals of the nuclear ground state density distribution.

The Wigner distribution function $W(r, k)$ corresponding to the CDFM one-body density matrix $\rho(r, r')$ (4.59) is:

$$W(r, k) = \int_0^\infty dx |f(x)|^2 \, \Theta(x - |r|) \Theta(k_F(x) - |k|)$$

$$= \int_{|r|}^\infty dx |f(x)|^2 \, \Theta(k_F(x) - |k|). \tag{5.24}$$

The function $W(r, k)$ determines the particle density in the phase space in the vicinity of the point (r, k). Since the interior of the nucleus has properties similar to those of infinite nuclear matter, the function $W(r, k)$ will be taken at the point $r = 0$. Then

$$n(k) \equiv W(r = 0, k) = \int_0^\infty dx |f(x)|^2 \, \Theta(k_F(x) - |k|) \tag{5.25}$$

is the function which determines the occupation numbers in single-particle states with a given momentum k. The occupation numbers in the Fermi gas with a Fermi momentum $k_F(x)$ are given by

$$n_x(k) = \Theta(k_F(x) - |k|). \tag{5.26}$$

Keeping in mind the relation between the occupation numbers $n_x(k)$ and the spectral function for uniform system $S_x(k, \omega)$:

$$n_x(k) = \int_{-\infty}^{\varepsilon_F(x)} \frac{d\omega}{2\pi} S_x(k, \omega) \tag{5.27}$$

the Eq. (5.25) can be written in the form (since $\hbar^2 k^2/2m$ is positive):

$$n(k) = \int\limits_0^\infty dx |f(x)|^2 \int\limits_0^{\frac{\hbar^2 k_F^2(x)}{2m}} d\omega\, \delta\left(\omega - \frac{\hbar^2 k^2}{2m}\right), \tag{5.28}$$

where m is the nucleon mass. The following approximation which is valid for free particles is used to obtain (5.28) [5.25]:

$$S_x(k,\omega) = 2\pi\delta\left(\omega - \frac{\hbar^2 k^2}{2m}\right), \quad k < k_F(x). \tag{5.29}$$

After the substitution:

$$\omega' = \frac{\mu}{\hbar^2 k_F^2(x)/2m}\,\omega + E_F \tag{5.30}$$

the Eq. (5.28) has the form:

$$n(k) = \int\limits_{\mu+E_F}^{E_F} \frac{d\omega'}{2\pi} \int\limits_0^\infty dx |f(x)|^2\, 2\pi\delta(\omega' - E_F - \mu k^2/k_F^2(x)), \quad k < k_F(x), \tag{5.31}$$

with

$$k_F(x) = (9\pi A/8)^{1/3}/x \equiv \alpha/x, \qquad \alpha \equiv (9\pi A/8)^{1/3}. \tag{5.32}$$

The parameter E_F in (5.31) is the Fermi energy and is identified with the separation energy of the last nucleon. The parameter μ is interpreted as the energy of the deepest level, so that $\mu + E_F$ is the depth of the potential well. The spectral function of the hole states in the nucleus can be determined by the integrand in the integral over ω' in (5.31):

$$S(k,\omega \leqslant E_F) = 2\pi \int\limits_0^\infty dx |f(x)|^2\, \delta(\omega - E_F - \mu k^2/k_F^2(x))\,\Theta(k_F(x) - k). \tag{5.33}$$

Finally, the integration over x in (5.33) leads to the explicit expression for the hole-state spectral function:

$$S(k,\omega \leqslant E_F) = \frac{\pi\alpha}{k[\mu(\omega - E_F)]^{1/2}}\left| f\left(\frac{\alpha}{k}\left(\frac{\omega - E_F}{\mu}\right)^{1/2}\right)\right|^2 \tag{5.34}$$

with the condition $\mu + E_F \leqslant \omega \leqslant E_F$ following from the function $\Theta(k_F(x) - k)$ in (5.33).

As can be seen from (5.34) $S(k, \omega \leqslant E_F)$ depends on the weight function $|f|^2$ in CDFM, which can be determined by the nucleon density distribution (see e.g. (4.66)). The use of the symmetrized Fermi-type density distribution $\rho_{SF}(r, R, b)$

[5.26] gives the following form of the spectral function (5.34):

$$S(k, \omega \leqslant E_F) = \frac{\pi}{R^3 b\left[1 + \left(\dfrac{\pi b}{R}\right)^2\right]} \frac{\alpha^4 (E_F - \omega) e^{\left[\frac{\alpha}{k}\left(\frac{\omega - E_F}{\mu}\right)^{1/2} - R\right]/b}}{k^4 \mu^2 \left[1 + e^{\left[\frac{\alpha}{k}\left(\frac{\omega - E_F}{\mu}\right)^{1/2} - R\right]/b}\right]^2}. \tag{5.35}$$

The analytical dependence of the spectral function (5.35) on the momentum k and the energy ω is individual for each nucleus and is determined by means of its geometrical characteristics (R and b in (5.35)) and the mass number.

The spectral functions for the hole-states in ^{28}Si, ^{40}Ca and ^{58}Ni nuclei obtained by Eq. (5.35) are shown in Figs. 5.1, 5.2 and 5.3, respectively. The values of the parameters R and b are obtained from the electron–nuclei scattering data [5.26]: ^{28}Si ($R = 3.085$ fm, $b = 0.563$ fm), ^{40}Ca ($R = 3.556$ fm, $b = 0.578$ fm), ^{58}Ni ($R = 4.153$ fm, $b = 0.566$ fm). The value of the parameter μ is chosen to be equal to -50 MeV for the three nuclei. The proton separation energies E_F are -11.585 MeV, -8.33 MeV and -8.17 MeV for ^{28}Si, ^{40}Ca and ^{58}Ni, respectively. The corresponding empirical strength functions from [5.4a] are also given in the figures for comparison.

Since the data do not yield directly the normalized spectral function, following [5.20] a normalization coefficient is attached to the theoretical curves, chosen in such a way that the height of the 1p-peak coincides with the experimental value.

The spectral functions are calculated [5.24a] at such values of the momentum k, for which the positions of the peaks of $S(k, \omega)$ coincide with the empirical ones. This procedure is close to the one used in [5.18a]. The corresponding values of k are as follows: for the states 1d, $1p_{1/2}$, $1p_{3/2}$, 1s in ^{28}Si $k = 0.15$ fm^{-1}, 0.35 fm^{-1}, 0.80 fm^{-1}, 1.15 fm^{-1}; for the states 1d, 1p, 1s in ^{40}Ca $k = 0.45$ fm^{-1}, 0.95 fm^{-1}, 1.30 fm^{-1}; and for the states 1f, 1d, 1p, 1s in ^{58}Ni $k = 0.25$ fm^{-1}, 0.55 fm^{-1}, 0.95 fm^{-1}, 1.30 fm^{-1}.

It should be noted that, while in [5.18a] the hole-state width in a finite nucleus is determined by an interpolation of widths obtained from $S(k, \omega)$ at $\rho = \langle \rho \rangle$, the spectral function in the CDFM is a superposition of spectral functions for nuclear matters with different densities $\rho_0(x)$ multiplied by the corresponding weight factor $|f(x)|^2$. Moreover, Eq. (5.33) gives the possibility of determining the interval Δx (or the interval of densities $\Delta \rho$) which takes part in the formation of the peak of $S(k, \omega)$ in the interval $\Delta \omega$ in the vicinity of $\omega = \omega_{max}$ (at a fixed momentum k).

It can be noted that the asymmetry of the curves in Figs. 5.1–3 and the increasing widths for deeply bound states are characteristic features and such behaviour of the spectral functions is in accordance with the experimental data. The evaluation of the hole-state widths Γ are given in Table 5.1. The results describe satisfactorily the observed widths (and particularly those of the 1s-hole

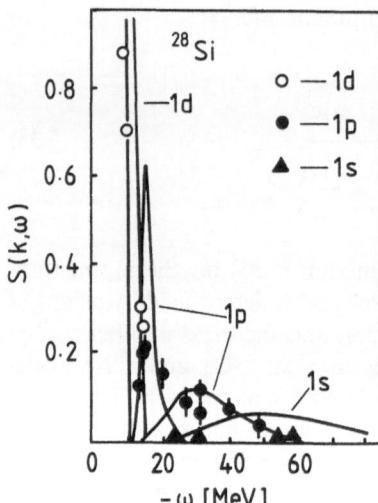

Fig. 5.1. Comparison between empirical strength functions [5.4a] and spectral functions calculated from Eq. (5.35) for ^{28}Si [5.24a]. The ordinate scale is in arbitrary units

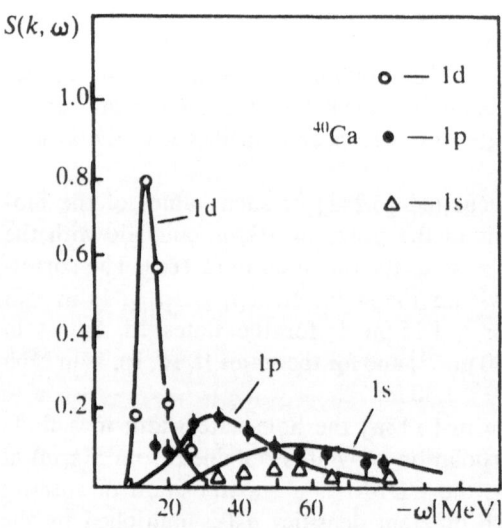

Fig. 5.2. Same as in Fig. 5.1. for ^{40}Ca

states) [5.4a] (see also for comparison the results for ^{208}Pb in Figs. 2.4 and 2.5 confirming Eq. (2.38)).

The centroid energy [5.17a]

$$E(k) = \int\limits_{-\infty}^{E_F} \frac{d\omega}{2\pi} \, \omega \, S(k,\omega) \bigg/ \int\limits_{-\infty}^{E_F} \frac{d\omega}{2\pi} \, S(k,\omega) \qquad (5.36)$$

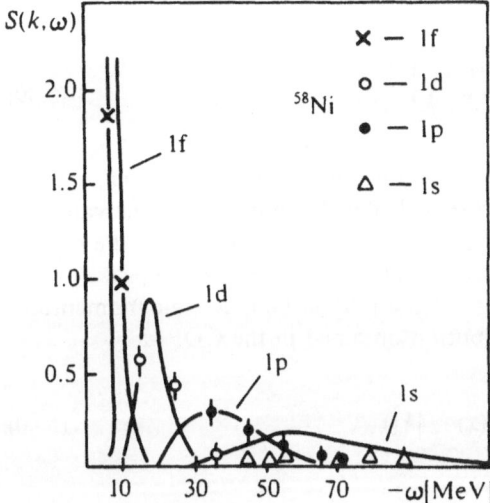

Fig. 5.3. Same as in Fig. 5.1. for ^{58}Ni

Table 5.1. Hole-state widths Γ (in MeV) obtained in the CDFM [5.24a]

Nuclei	State	Width Γ (MeV)
^{28}Si	1d	~ 1
	$1p_{1/2}$	~ 4
	$1p_{3/2}$	~ 21
	1s	~ 43
^{40}Ca	1d	~ 6
	1p	~ 26.5
	1s	~ 50
^{58}Ni	1f	$\lesssim 2$
	1d	~ 8
	1p	~ 24
	1s	~ 43.5

can be calculated in the CDFM. Substituting (5.33) in (5.36) one gets:

$$E(k) = \frac{\dfrac{\hbar^2 k^2}{2m} \mu \displaystyle\int\limits_0^{\alpha/k} dx\,|f(x)|^2 \Big/ \left(\dfrac{\hbar^2 k_{\mathrm{F}}^2(x)}{2m}\right)}{\displaystyle\int\limits_0^{\alpha/k} dx\,|f(x)|^2} + E_{\mathrm{F}}. \tag{5.37}$$

At small values of k:

$$E(k) \simeq \frac{\hbar^2 k^2}{2m^*} + E_{\mathrm{F}}, \tag{5.38}$$

where the quasiparticle effective mass m^* is given by

$$m^* = m\left[|\mu| \int_0^\infty dx |f(x)|^2 \Big/ \left(\frac{\hbar^2 k_F^2(x)}{2m}\right)\right]^{-1}.$$ (5.39)

The centroid energy $E(k)$ for ^{58}Ni is shown in Fig. 5.4. In this case $m^* \simeq 0.6m$. The values of the centroid energies $E(k)$ calculated using (5.37) at the values of k given above for various hole states are in good agreement with the experimentally observed positions (on the axis ω) of the spectral function peaks for the corresponding states. The parameter k has a role of the hole-state momentum.

If the nucleon momentum distribution obtained in the CDFM:

$$n(k) = \frac{4}{A} \int_0^\infty dx |f(x)|^2 \frac{4\pi x^3}{3} \Theta(k_F(x) - |k|)$$ (5.40)

with

$$\int n(k) \frac{dk}{(2\pi)^3} = 1$$

is used instead of the occupation numbers (5.25), a procedure similar to that of (5.25)–(5.34) allows us to obtain the spectral function $P(k, E)$ defined in [5.13, 5.27, 5.28]. It is related to the nucleon momentum distribution $n(k)$ by

$$n(k) = \int_{E_{min}}^\infty P(k, E) \, dE$$ (5.41)

and is normalized so that:

$$4\pi \int P(k, E) \frac{k^2 \, dk}{(2\pi)^3} \, dE = 1.$$ (5.42)

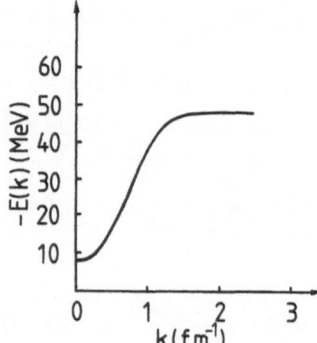

Fig. 5.4. The centroid energy $E(k)$ calculated from (5.37) for ^{58}Ni

The expression for the spectral function $P(k, E)$ obtained in the CDFM is:

$$P(k, E) = \frac{16\pi r_0^3}{3A} \frac{\alpha}{2|k|} \frac{|f(r_0)|^2}{[\mu(\omega - E_F)]^{1/2}}, \tag{5.43}$$

where

$$r_0 = \frac{\alpha}{|k|} ((\omega - E_F)/\mu)^{1/2}, \qquad \alpha = (9\pi A/8)^{1/3}. \tag{5.44}$$

Using the spectral function (5.43) the EMC experimental results for ^{12}C and ^{56}Fe are interpreted within the CDFM taking into account N–N correlations and binding effects [5.29] (see Chap. 9, Sect. 9.3).

In the CDFM the single-particle characteristics, such as the spectral functions, hole-state widths and centroid energies are functionals of the density in the ground state of the nucleus. The analysis of these quantities is made without fitted parameters. It should be emphasized that while in the nuclear matter methods the density ρ is specified for calculation of each single-particle spectral function, in the CDFM the main contributions to their peaks are related to certain intervals $\Delta\rho$ of the density distribution $\rho(r)$. In this way, the necessity of considering the density as a free parameter (as is done in nuclear matter calculations [5.16, 5.18]) is avoided in the CDFM.

Although the spectral functions $S_x(k, \omega)$ for different densities $\rho_0(x)$ are chosen in a maximum simplified form (Eq. (5.29)), it is seen that the decisive factor for the appearance of realistic properties of the finite nuclei spectral function $S(k, \omega)$ is the correct treatment of the weight $|f(x)|^2$ for each density $\rho_0(x)$ participating in the formation of the ground state nucleon density distribution $\rho(r)$.

5.4 Sum Rules for the Spectral Function

By means of the spectral function $S(E, k)$ (5.2) a model-independent sum rule has been established in [5.30, 5.31]. This relates the total binding energy per nucleon E_A/A to the mean kinetic $\langle T \rangle$ and mean removal $\langle E \rangle$ energies in the case of the Hamiltonian containing no more than two-body terms:

$$E_A/A = (\langle T \rangle - \langle E \rangle)/2. \tag{5.45}$$

In (5.45):

$$\langle T \rangle = n^{-1} \int dE \, dk \, \frac{k^2}{2m} S(E, k), \tag{5.46}$$

$$\langle E \rangle = n^{-1} \int dE \, dk \, E S(E, k), \tag{5.47}$$

$$n = \int dE \, dk \, S(E, k). \tag{5.48}$$

Taking account of the recoil energy of the centre-of-mass of the residual system in the reactions in which $S(E, \boldsymbol{k})$ is determined leads to the following sum rule for Z protons in a nucleus with A nucleons and a binding energy per proton E_Z/Z [5.10]:

$$\frac{E_Z}{Z} = \frac{1}{2}\left(\frac{A-2}{A-1}\langle T \rangle - \langle E \rangle\right). \tag{5.49}$$

The sum rule can be generalized for a Hamiltonian including n-body operators $\hat{H} = \sum_n \hat{H}_n$ [5.10]. For a Hamiltonian that includes three-body forces the sum rule has the form:

$$\frac{\langle H_3 \rangle}{Z} = \frac{(A-2)}{(A-1)}\langle T \rangle - \langle E \rangle - 2\frac{E_Z}{Z}. \tag{5.50}$$

The analysis of the (p, 2p) experiments of *James* et al. [5.6, 5.7] which provide data for the spectral function $S(E, \boldsymbol{k})$ showed reasonable agreement between the experimental binding energy of the proton system $(E_Z/Z)_{exp}$ and the sum-rule value [5.30] for ^{12}C, ^{40}Ca, ^{120}Sn and ^{208}Pb nuclei. In addition, this analysis shows that if the energy-momentum range used in the experiments to determine $\langle T \rangle$ and $\langle E \rangle$ is limited and therefore does not cover all contributions to the spectral function this can cause the breakdown of the sum rule.

The experimental data on the (p, 2p) reaction have been analyzed in [5.6, 5.7] by the Hartree–Fock method assuming that every proton has a shell-model orbit. The contributions to $S(E, \boldsymbol{k})$ from each filled orbit is considered. This orbital analysis is applied in [5.30] to calculate the orbital sum rule $(E_Z/Z)_{orb} = (T_{orb} + E_{orb})/2$ for the mean kinetic T_{orb} and removal energy E_{orb} analogous to (5.45–48). Systematically less binding per proton is obtained in this orbital analysis. The main conclusion from this is that in the orbital approach the number of protons in each orbit is fixed, while the spectral function shows more protons in high-energy orbits than are accounted for in the Hartree–Fock picture [5.30]. This raises the question of the role of the short-range two-body interactions not taken into account in the Hartree–Fock method in the sum-rule analyses. It turns out that these interactions shift some strength at both high energy E and high momentum k outside the energy and momentum range studied in the (e, e′ p) experiments for various nuclei (from ^{4}He to ^{208}Pb) [5.4, 5.5, 5.32]. Some attempts to take into account the short-range correlation effects have been made in the renormalized Brueckner–Hartree–Fock (RBHF) theory [5.33], where a partial depletion of the normally occupied orbitals ($\rho_\alpha < 1$ for α below the Fermi level) is introduced. The calculations in the density-dependent Hartree–Fock method, in which the "rearrangement energy" term in the Hartree–Fock sum rule appears to account for the short-range correlation effects can give also a reasonable explanation of the discrepancies in the sum-rule analyses obtained in the Hartree–Fock method.

An important problem in complete beyond-Hartree–Fock sum-rule analyses is the extent of the energy and momentum range of the experiments in which the spectral function is determined.

Table 5.2. Mean kinetic and removal energies [5.27a, 5.34a] (in MeV)

Nuclei		$\langle T \rangle$	$\langle E \rangle$
^{3}He	SRC	17.4	13.6
^{4}He	Shell model	17.1	19.8
	SRC	21.1	28.2
^{12}C	HF	17.0	23.0
	SRC	37.0	49.0
^{16}O	HF	15.0	24.0
	SRC	27.0	41.0
^{40}Ca	HF	16.5	26.6
	SRC	36.0	52.1
^{56}Fe	HF	17.0	25.0
	SRC	33.0	49.8

The studies of the effects of short-range nucleon–nucleon correlations in many-body calculations with realistic nucleon-nucleon interactions in [5.27, 5.28, 5.34] shows a substantial increase of the values of $\langle T \rangle$ and $\langle E \rangle$ with respect to their Hartree–Fock values, as well as a strong relation between the high-momentum components in the momentum distributions and the high values of the removal energies. In Table 5.2 the values of $\langle T \rangle$ and $\langle E \rangle$ calculated in the shell model, in the Hartree–Fock method and taking into account the short-range correlations are given for ^{3}He, ^{4}He (taken from [5.27a]) and for ^{12}C, ^{16}O, ^{40}Ca and ^{56}Fe (taken from [5.34a]). It is also shown in [5.34a] that the use of $\langle T \rangle_{HF}$ and $\langle E \rangle_{HF}$ in the Koltun sum rule cannot reproduce the binding energy per nucleon $|E_A/A| \approx 8$ MeV unless, e.g., the rearrangement energy which comes from the density-dependent effective interaction [5.35] is included. For nuclear matter $\langle T \rangle = 42$ MeV and 44 MeV at $k_F = 1.33$ fm^{-1} calculated in [5.36] using Urbana and Argonne interactions, respectively, with the account for three-nucleon forces, while for the Fermi gas $T_{FG} = 22$ MeV. It is established in [5.34a] that the use of various contemporary versions of realistic nucleon–nucleon interactions gives values of the mean kinetic and removal energies which differ only by a few MeV.

It is emphasized in [5.34] that the larger values of $\langle T \rangle$ and $\langle E \rangle$ obtained in correlation approaches can be related to the presence in the many-body spectral function $S(E, \langle k \rangle)$ of high-momentum and removal energy components. The latter cannot be measured in exclusive (e, e' p) reactions in the limited momentum-energy region ($k < 1.5$ fm^{-1}, $E \leqslant 60$ MeV). It is noted that (e, e' p) reactions on few-nucleon systems and ^{12}C at $k > 1.5$ fm^{-1} and $E > 60$ MeV show large contributions from scattering by correlated nucleon pairs and this strongly increases $\langle E \rangle$.

The problems of the sum rules for the spectral function are also discussed in the review of *Orlandini* and *Traini* [5.37].

The increased calculated values of $\langle T \rangle$ and $\langle E \rangle$ due to including short-range nucleon–nucleon correlations lead to reasonable agreement with the experimental results on the EMC (European Muon Collaboration) effect in the region $0.2 \leqslant x \leqslant 0.7$ in terms of nucleonic degrees of freedom [5.34] (see also Sect. 9.3).

6 Natural Orbitals and Occupation Numbers in Nuclei

In this chapter the effects of correlations on nuclear structure are discussed using the formalism of the occupation probabilities (or occupation numbers) and natural orbitals. The quenching of the single-particle probabilities with respect to the results of the mean-field approximation observed in various experiments is described in Sect. 6.1. The use of absolute, as well as of relative spectroscopic factors in the sum-rule analyses together with data from electron–nucleus scattering for the extraction of the occupation probabilities is considered in the same Section. The data for occupation numbers in various nuclei (^{206,208}Pb, ^{40,48}Ca, ^{51}V, ^{64}Ni, ^{64}Zn, ^{89}Y, ^{90}Zr) obtained mainly from (e, e'p)- and ($\vec{d}$, ^{3}He)-reactions are presented. The dependence of the validity of the orbital occupation experimental determination on the form of the single-particle wave functions which are used in the extraction from the data is emphasized.

Applications of the natural orbital method (*Malaguti* et al. [6.53], *Boffi* and *Pacati* [6.54], *Jaminon* et al. [6.45–47, 6.56], *Gaudin* et al. [6.61a]) are discussed in Sect. 6.2.

The occupation numbers and natural orbitals of various nuclei obtained by diagonalization of the one-body density matrix in the coherent density fluctuation model and the generator coordinate method are presented and compared in Sect. 6.3 and 6.4. The comparison of the results of both methods with available experimental data for occupation numbers is given in Sect. 6.5.

6.1 Nucleon Correlations and Single-particle Occupation Numbers

An important problem for the validity of the mean-field approximation (MFA) is related to the observation of the quenching of the single-particle occupation probabilities with respect to the MFA predictions. This is considered [6.1] as one of the cleanest signatures of correlation effects in nuclei. The depletion of the normally fully occupied states of the Fermi sea can be explained by the virtual scattering of nucleons to states above the Fermi level due to the strong nucleon–nucleon interaction at the Fermi surface. Here we shall consider some experimental evidence for such depletion obtained from particle–nucleus scattering data.

Electron elastic scattering on nuclei in the lead region and muon data enable the ratio of elastic cross-sections from ^{205}Tl to ^{206}Pb as a function of the transferred momentum [6.2–5] (Fig. 6.1) to be determined, and hence the charge density difference of ^{206}Pb and ^{205}Tl [6.3–6] (Fig. 6.2). It is pointed out that the MFA correctly predicts a peak of large amplitude in the ratio in Fig. 6.1. It can be seen, however that the description of the amplitude of the oscillation can only

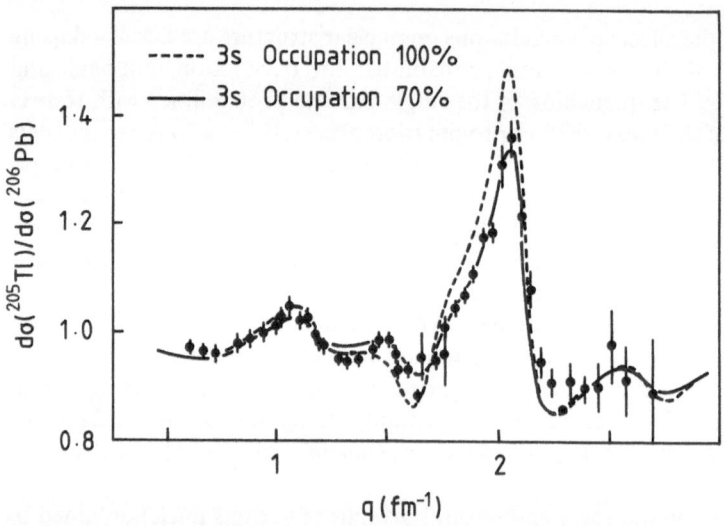

Fig. 6.1. The ratio of elastic cross sections from ^{205}Tl and ^{206}Pb [6.2–5] ([6.4])

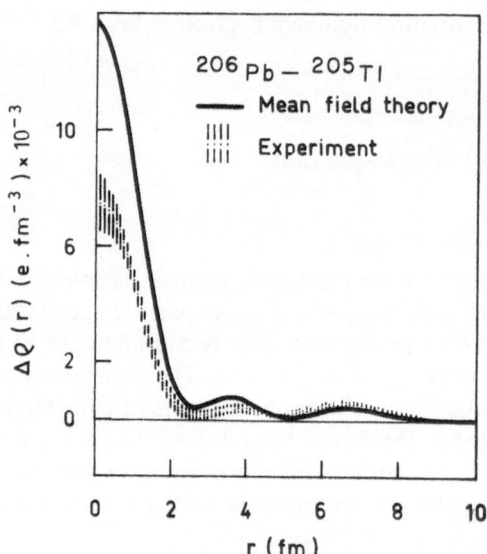

Fig. 6.2. The charge density difference of ^{206}Pb and ^{205}Tl [6.3–6] ([6.4])

be achieved with a reduction of 30 to 35% of the 3s-state occupation number. The charge density difference of ^{206}Pb and ^{205}Tl (Fig. 6.2) generally follows the prediction of the MFA but a reduction of this difference in the interior of the nucleus $(0 \lesssim r \lesssim 5$ fm) is observed. This means that some charge is removed from the centre of the nucleus and redistributed at the surface [6.4].

Electron elastic scattering also provides information on the ground-state magnetization densities of nuclei in the lead region [6.7, 6.4–5]. The experiments show that the magnetic form factors of ^{205}Tl and ^{207}Pb are strongly modified due to correlations (Fig. 6.3). The quenching of these form factors can be related to the partial occupancy in ^{208}Pb. The single-particle magnetic form factors of neighbouring nuclei (^{205}Tl, ^{207}Pb, ^{209}Bi) are quenched in comparison with MFA result [6.8] by a factor $Z = n_- - n_+$, where n_+ and n_- denote the occupation probabilities of orbitals lying immediately above and below the Fermi level. The value of $Z = 0.70 \pm 0.07$ is found for both the neutron and proton shells in ^{208}Pb. This value is comparable with the estimate of *Pandharipande, Papanicolas* and *Wambach* [6.11]: $Z = 0.6 \pm 0.1$ obtained adding RPA correlations to the nuclear matter result taking into account the short-range and tensor nucleon–nucleon correlations (Fig. 6.4). An estimate of Z between 0.7 and 0.8 is also found from low-energy neutron-scattering data using dispersion relation analysis [6.12]. The discontinuity in the occupation numbers at the Fermi level Z differ substantially from the value used in MFA analyses $(Z = 1.0)$ It is shown in [6.7] that taking into account the core polarization and

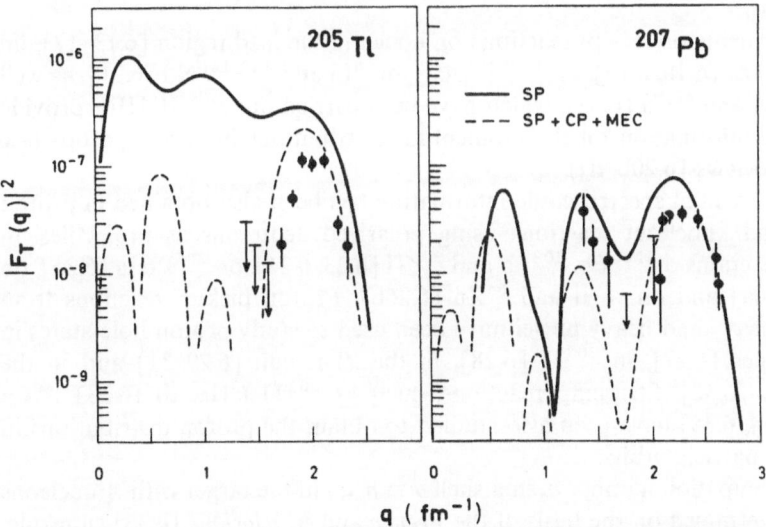

Fig. 6.3. The elastic magnetic form factors of ^{207}Pb and ^{205}Tl [6.7]. The solid line is the single-particle prediction [6.8]. The dashed curve shows the effects of core polarization and meson exchange [6.9, 6.10] ([6.4])

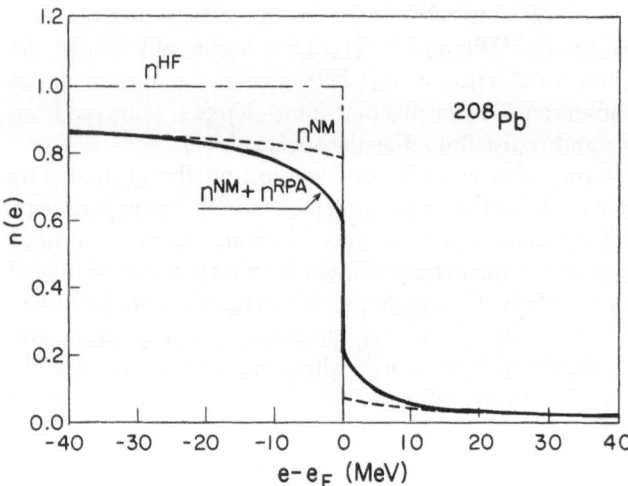

Fig. 6.4. An estimate of occupation numbers in ^{208}Pb (thick line) derived from nuclear matter calculations to which the effects of RPA correlations are added [6.11] ([6.4])

meson-exchange current effects [6.9–6.10] is not sufficient to explain the discrepancy between the magnetic form factors.

The results on the reduction of the 3s-orbit in the lead region modify the mean-field concept of magic nuclei as nuclear systems with closed shells. The electron experiments showed that this is an approximate picture and that the residual interaction strongly changes the distributions of nucleons in these shells [6.5].

Experiments on (e, e′p) reactions on nuclei in the lead region [6.13–17], on ^{51}V and ^{90}Zr [6.18, 6.19], on ^{12}C [6.20a], on ^{6}Li and $^{142, 146}$Nd [6.21], as well as on ^{40}Ca and ^{48}Ca [6.22], which have been carried out at NIKHEF provide additional information on the momentum distributions in various orbits (see also the reviews [6.20b–d]).

Some essential spectroscopic information has been also obtained in proton pickup and knockout reactions using polarized deuterons as projectiles: in $(\vec{d}, ^3\text{He})$ reactions on ^{206}Pb, ^{208}Pb and ^{205}Tl [6.23, 6.24], on ^{206}Pb in [6.25], on ^{58}Ni [6.26a] and on ^{64}Ni and ^{64}Zn [6.26b]. Proton pickup reactions from medium-heavy and heavy nuclei have been used to study proton hole-states in Pm isotopes [6.27], in ^{207}Tl [6.28], in the Zr-region [6.29–31] and in the Au-region [6.32]. Stripping reactions, such as ^{205}Tl (^{3}He, d) [6.33], ^{40}Ca $(\vec{d}, p)$ [6.34, 6.35] have been also studied to obtain the proton distributions in the single-particle orbits.

The occupation number n_α in a shell $\alpha \equiv n, l, j$ in the target with A nucleons can be determined on the basis of the *French* and *Macfarlane* [6.36] sum rule:

$$n_\alpha(A) = \sum_f C^2 S_\alpha^-(A) \equiv \sum_f G_\alpha^-(A), \tag{6.1}$$

where the spectroscopic factors $C^2 S_\alpha^- (A)$ for proton pickup is summed over all final states f [6.33], C being the isospin Clebsch-Gordon coefficient. The single-particle spectroscopic information is included essentially in the spectroscopic factor $C^2 S_\alpha^- (A)$ which is the overlap of the many-body wave function with the wave function of the state α in the shell model. The following sum rule may be used [6.36]:

$$\sum G_\alpha^+ (A) + \sum G_\alpha^- (A) = 2j + 1, \tag{6.2}$$

which means that for any target nucleus with A nucleons the number of particles $\sum G_\alpha^- (A)$ in a shell-model state α plus the number of holes $\sum G_\alpha^+ (A)$ in that state has to be equal to the shell-model limit of $(2j + 1)$. In (6.2) [6.33]:

$$\sum G_\alpha^+ (A) = \sum [(2J_f + 1)/(2J_i + 1)]\, C^2 S_\alpha^+ (A) \tag{6.3}$$

is the strength expressed by the spectroscopic factors for stripping on the same target nucleus.

The Eqs. (6.1–3) include absolute spectroscopic factors and summation over the full $(A - 1)$ and $(A + 1)$ spectra. The use of the sum rule (6.1) meets difficulties related to the absolute spectroscopic factors which cannot be obtained reliably in surface pickup reactions. The results obtained from the data are sensitive to the assumed form of the wave function of the transferred nucleon. In the $(e, e'p)$ reactions the accuracy of the absolute spectroscopic factors is estimated to be only to within 30% [6.33]. This is related to the difficulties of obtaining the shapes of the proton momentum distributions from $(e, e'p)$ reactions by DWIA analyses.

In addition, in the sum rules the sum has to be taken over the full $(A - 1)$ spectrum, while only a limited range of the spectrum can be studied with necessary accuracy in the experiments.

The difficulties mentioned above can be overcome using relative spectroscopic factors in a sum rule based on a method of combined evaluation of relative spectroscopic factors and electron scattering (CERES method) [6.23–6.24b]. The exact form of this sum rule in the case of the ^{208}Pb nucleus is [6.33]:

$$n(208) = Z \left[\frac{\sum G^- (206)}{\sum G^- (208)} - \frac{\sum G^- (205)}{\sum G^- (208)} \right]^{-1}. \tag{6.4}$$

In (6.4) Z represents the 3s-proton contribution to the accurately determined charge density difference $\Delta\rho = \rho(^{206}\text{Pb}) - \rho(^{205}\text{Tl})$ in [6.3]:

$$\Delta\rho \cong Z \rho_{3s}(r), \tag{6.5}$$

where $\rho_{3s}(r)$ is the single-particle charge density $|\Psi_{3s_{1/2}}|^2$ of $3s_{1/2}$-protons evaluated within a particular model. The relative spectroscopic factors used in (6.4) have to be accurate within 10% for proton removal from a given subshell. This is estimated for $(e, e'p)$ and $(d, {}^3\text{He})$ reactions at various energies and nuclei [6.33].

The final form of the CERES sum rule used for determination of the occupation numbers in ^{208}Pb is [6.24a, 6.33]:

$$n(208) \simeq Z \left[\frac{\sum' G^-(206)}{G_0^-(208)} - \frac{\sum' G^-(205)}{G_0^-(208)} \right]^{-1}, \tag{6.6}$$

where the truncated sums (denoted by primes) account for the experimentally accessible range of excitation energies and $\sum' G^-(208) \approx G_0^-(208)$.

As can be seen from (6.6) the correct determination of Z is very important for the estimation of the occupation probabilities n. It has been shown in [6.36, 6.17] that the deduced value of Z depends crucially on the core polarization ($\Delta \rho_{CP}$) contribution:

$$\Delta \rho = Z \rho_{3s}(r) + \Delta \rho_{CP}. \tag{6.7}$$

The value of Z has been determined by *Burghardt* [6.37a], who described the charge density difference $\Delta \rho (^{206}$Pb$-^{205}$Tl$)$ [6.3, 6.6] by the use of full occupations for ^{206}Pb and adjusted occupations for the $3s_{1/2}$ and $2d_{3/2}$ states in ^{205}Tl. The calculations of the core polarization using the G_σ nucleon–nucleon interaction lead to a good agreement of the experimental $\Delta \rho$ with the theoretical Hartree–Fock charge density difference. The derived value of Z is

$$Z = 0.66 \pm 0.08, \tag{6.8}$$

where the error is related to the statistical uncertainties of the data. The charge density difference between ^{206}Pb and ^{204}Hg [6.37a] leads to $Z/2 = 0.47$–0.66. Using the expression for the $3s_{1/2}$-occupancy in ^{208}Pb:

$$n(208) = (1.15 \pm 0.11) Z \tag{6.9}$$

and the value of Z from (6.8) it is shown in [6.17] that

$$n(208) = 0.76 \pm 0.12. \tag{6.10}$$

The CERES method has been critically analyzed by *Mahaux* and *Sartor* [6.37b]. They emphasized that the factor Z can be influenced by the difference between the radial shapes of the 3s-orbits in ^{205}Tl and in ^{206}Pb, as well as by the role of the 2s-orbit. These and other considerations lead to the conclusion in [6.37b] that, at present, no reliable empirical method exists for determining occupation probabilities. They point out that the difficulties are not mainly related to inaccuracies of the empirical data but to the intrinsic limitation of the concept of a mean-field orbit. This concept is valid only in the vicinity of the peaks of the spectral functions [6.37b] (see also Chap. 5).

The result of self-consistent Hartree–Fock calculations with G_σ interaction for the charge density $\rho(r)$ in ^{208}Pb [6.17] using the following occupation numbers for $3s_{1/2}$, $2d_{3/2}$, $1h_{11/2}$ and $2d_{5/2}$ proton orbitals: 0.76, 0.83, 0.86 and 0.90, respectively, is shown in Fig. 6.5 and compared with the experimental charge density. The use of partial occupations improves the agreement of the theoretical calculations with the empirical density.

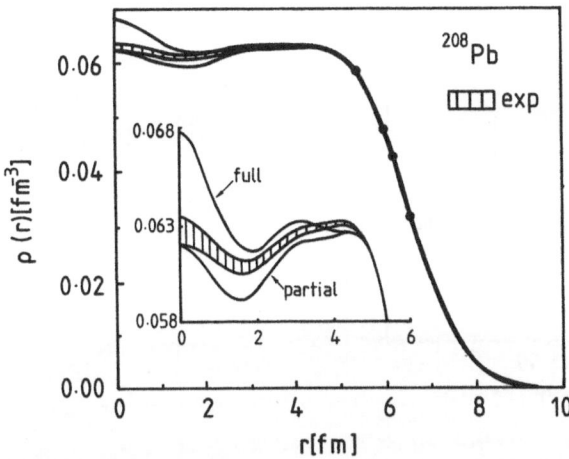

Fig. 6.5. Experimental charge density $\rho(r)$ of ^{208}Pb (hatched areas) compared to a Hartree-Fock result using G_σ interaction. The curve "full" shows calculations with 100% occupations for the states below the Fermi level, the curve "partial" shows the result with the partial occupations given in the text ([6.17])

Here we shall present other results for occupation numbers in various nuclei obtained from different reactions.

In Table 6.1 the proton occupation probabilities in ^{206}Pb and ^{208}Pb extracted from $(\vec{d}, {}^3\text{He})$ reaction are given [6.25].

Whereas a 3s-depletion of 20–30% is deduced in the Pb-region by means of nearly model-independent sum-rule analysis, a larger depletion of $\approx 50\%$ is obtained from the DWIA analyses of (e, e'p) reactions for 1f- and 2p-orbits in ^{51}V and ^{90}Zr [6.38, 6.19]. It is noted that the DWIA analysis of $(\vec{d}, {}^3\text{He})$ reactions leads to a similar (large) value of the depletion if bound state wave functions deduced from electron scattering are employed. The occupation probabilities for proton-hole states in ^{51}V and ^{90}Zr from [6.19] are shown in Fig. 6.6a. The proton occupation probabilities in ^{64}Zn and ^{64}Ni have been obtained from the ^{64}Zn $(\vec{d},{}^3\text{He})^{63}$Cu- and ^{64}Ni $(\vec{d}, {}^3\text{He})^{63}$Co-reactions by *Seeger* et al. [6.26b]. The results which are shown in Fig. 6.6b determine the Fermi surface of

Table 6.1. Proton occupation numbers in ^{206}Pb and ^{208}Pb [6.25]

Orbital (*nlj*)	$n(206)$	$n(208)$
$3s_{1/2}$	0.67	0.82
$2d_{3/2}$	0.69	0.83
$2d_{5/2}$	0.53	0.90
$1h_{11/2}$	0.70	0.90

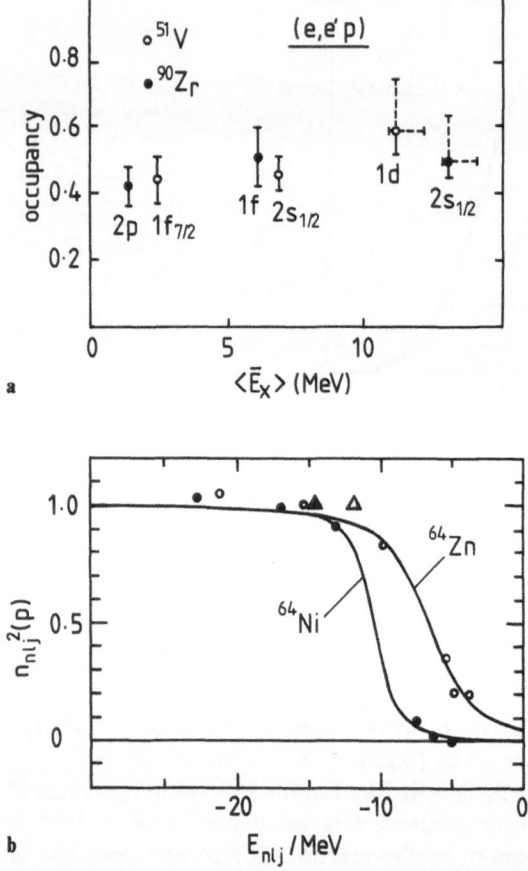

Fig. 6.6. (a) Occupation probabilities for proton-hole states in ^{51}V and ^{90}Zr deduced from (e, e'p)-reaction analysis [6.19] and plotted as a function of the mean excitation energy. (b) The proton occupation probabilities n_{nlj} versus single-particle energies E_{nlj} for ^{64}Zn (open circles) and ^{64}Ni (full circles) obtained from $(\vec{d}, {}^3$He)-reaction [6.26b]. The sequence of the subshells according to increasing binding energy is: $1f_{5/2}$, $2p_{1/2}$, $2p_{3/2}$, $1f_{7/2}$, $1d_{3/2}$ and $1d_{5/2}$. The solid curves shows the shapes of the Fermi surface as deduced from a fit with a BCS occupation probability function

both nuclei considered. They plausibly suggest that the more diffuse Fermi surface in ^{64}Zn is due to the additional $2p_{3/2}$ protons.

The data from the ^{40}Ca$(\vec{d}, p)$ reaction [6.34, 6.35] concerning the (2s, 1d) shell of ^{40}Ca show that the depletion factors are of the order of 5–10% for the deeper $1d_{5/2}$ and $2s_{1/2}$ hole states and increase to about 20% for the Fermi level ($1d_{3/2}$). This result is in accordance with the theoretical estimations [6.39a]. The account for 1p-1h-RPA and 2p-2h-RPA correlations in [6.39b] leads to the quenching of ^{41}Ca valence hole states by about 10% and 18%, respectively, in agreement with $(\vec{d}, p)$-reaction [6.34, 6.35] and ^{40}Ca$(\vec{p}, d)$-reaction [6.40] (with 79% $1d_{5/2}$ strength) results.

The occupation probabilities in ^{40}Ca, ^{48}Ca and ^{51}V [6.22] obtained from (e, e′p) and (d, ^{3}He) experiments are presented in Table 6.2. It is seen that the occupations of shell-model orbitals are about 90%. For ^{40}Ca the occupancies are consistent with theoretical model predictions from [6.41]. A check for the deduced occupancies is the requirement that the sum must be equal to the total number of protons. In [6.22] the proton sum for ^{40}Ca is 21.6 (5.9), for ^{48}Ca it is 18.1 (4.6) (without the $1s_{1/2}$ occupancy) and for ^{51}V it is 13.6 (2.6) (from knockout reactions only from 2s, 1d, 1f, 2p orbitals).

The experimentally-obtained occupation probabilities of proton states in ^{89}Y [6.42a] are presented in Table 6.3.

The theoretical estimates [6.42b] within the RPA using a density-dependent force of the Skyrme-type for the construction of the Hartree–Fock single-particle states give larger ($\gtrsim 15\%$) occupation probabilities.

The study of the (e, e′p)-reaction on $^{142, 146}$Nd and ^{6}Li [6.21] shows that the experimental spectroscopic factors of Nd isotopes are quenched by about 50%, whereas for the lightest systems the quenching amounts to typically 15%.

Table 6.2. The occupation probabilities (n) in ^{40}Ca, ^{48}Ca and ^{51}V obtained from (e, e′ p) and (d, ^{3}He) reactions (taken from [6.22]). The uncertainties are given in brackets

Orbital	^{40}Ca n %	Orbital	^{48}Ca n %	Orbital	^{51}V n %
$1s_{1/2}$	95 (20)				
1p	92 (37)	1p	92 (33).		
$1d_{5/2}$	90 (14)	$1d_{5/2}$	90 (14)	$1d_{5/2}$	92 (17)
$2s_{1/2}$	89 (19)	$2s_{1/2}$	90 (16)	$2s_{1/2}$	91 (25)
$1d_{3/2}$	88 (17)	$1d_{3/2}$	88 (17)	$1d_{3/2}$	90 (16)
$1f_{7/2}$	37 (14)	$1f_{7/2}$	24 (10)	$1f_{7/2}$	89 (17)
$2p_{3/2}$	12 (6)				

Table 6.3. Occupation probabilities (n) of proton–hole states in ^{89}Y (taken from [6.42b]). The uncertainties are given in brackets

Orbital	n
$1d_{5/2}$ $\Big\}$ $1d_{3/2}$	0.71 (10)
$2s_{1/2}$	0.64 (8)
$1f_{7/2}$	0.68 (9)
$1f_{5/2}$	0.60 (8)
$2p_{3/2}$	0.56 (6)
$2p_{1/2}$	0.36 (4)
$1g_{9/2}$	0.07 (2)

As shown in [6.43] the measured occupation probabilities of the 1s and 1p states in ^{12}C are 0.50 and 0.65.

The theoretical description of the depletion of the Fermi sea is one of the most complicated problems of nuclear theory. We shall consider some theoretical attempts on the basis of the natural orbital representation in the following sections. Here we would like to emphasize several important points of the theoretical studies in this field.

The difference $\Delta\rho$ between the experimental charge density in the interior of the nuclei in the lead region and the theoretical calculations in the mean-field approaches (e.g. Hartree–Fock, see Fig. 6.5) pose the question of the role of different types of nucleon–nucleon correlations in the description of nuclear characteristics. The account of long-range correlations using the *RPA* [6.44] turns out to be not sufficient to describe the difference $\Delta\rho$. Later this difference has been considered [6.11, 6.45–48] as evidence for the effects of short-range and tensor nucleon–nucleon correlations leading to the depletion of the Fermi sea in the nuclear ground state. As can be seen from Sect. 6.2 the theoretical analyses have been carried out in the framework of a particular mean-field approach and they depend on the quality of the theoretical basis. This question is discussed thoroughly by *Bennour* et al. [6.49]. It is shown that the experimental data for the charge distribution of nuclei in the lead region can be well described by the Hartree–Fock theory using SkM* interaction without an assumption of the depopulation of states below the Fermi level. The account for the pairing correlations slightly improves the agreement with the data. As noted in [6.49], however, the validity of the experimental determination of the orbital occupation depends strongly on how well the form of the single-particle wave function is known. Different equally reliable mean-field calculations lead to variations in the form of the wave functions which are of the same order of magnitude as the experimental error bars. It should be noted that the use of effective density-dependent nucleon–nucleon forces in the Hartree–Fock calculations means that some types of nucleon–nucleon correlations are already included in the approach. As will be shown in Chap. 7, the form of the single-particle function is decisive for the correct description of the nucleon momentum distributions at high momenta and nuclear density distributions near the centre, which are sensitive to the short-range nucleon–nucleon correlations. In other words, the correlated single-particle wave functions even at zero depletion of the Fermi sea can give an essential contribution for the description of the nuclear quantities sensitive to the correlations. Such an example is given in Sect. 6.4 within the generator coordinate method. Here we should emphasize the remark which has been made in [6.23] that the relation of the charge densities and spectroscopic factors has to be analyzed by means of more appropriate natural orbital functions which differ from the Hartree–Fock-type wave functions. The latter has been shown by *Jaminon* et al. in [6.45].

An alternative description of the charge density difference of ^{206}Pb and ^{205}Tl has been made by *Celenza* et al. [6.50] using modified electromagnetic form factors of nucleons in nuclear medium as well as occupation numbers of about 90 to 100% for the $3s_{1/2}$ orbital.

Mahaux and *Sartor* [6.51a] calculated characteristics of very deep, as well as of deep, weakly bound and quasibound single-particle states in a complex mean-field method, namely the extended and improved iterative moment approach (IMA, see Sect. 2.3). The real part of the mean field is written as a sum of a Hartree–Fock term and a dispersive contribution which is linked by a dispersion relation to the explicitly parametrized imaginary part. The theoretical results for the occupation numbers, spectroscopic factors, spectral functions, widths, energies and rms radii are in good agreement with the empirical data for the ^{208}Pb nucleus. The quasiparticle properties of protons in ^{208}Pb (occupation numbers, effective masses etc.) have been studied by *Ma* and *Wambach* [6.51b] using a quasiparticle Hamiltonian including correlations in a phenomenological way.

6.2 Applications of the Natural Orbital Method

The natural orbital method (NOM) (*Löwdin* [6.52]) and its application to nuclear structure studies have been considered in Chap. 4 (Sect. 4.3). It was shown that the use of the NOM in the approaches going beyond the framework of the method of Hartree–Fock can preserve the plausible features of the single–particle description taking account of the nucleon–nucleon correlation effects.

The one-body density matrix $\rho(r, r')$ has the following form in the basis of the natural oribtals $\varphi_\alpha(r)$:

$$\rho(r, r') = \sum_\alpha n_\alpha \varphi_\alpha^*(r) \varphi_\alpha(r'), \tag{6.11}$$

where n_α are the natural occupation numbers (or simply, occupation numbers),

$$0 \leqslant n_\alpha \leqslant 1 \quad \text{and} \quad \sum_\alpha n_\alpha = A, \tag{6.12}$$

A being the number of particles in the system. It has to be noted that in the Hartree–Fock approximation the occupation numbers are $n_\alpha = 1$ when the states α are below the Fermi level and $n_\alpha = 0$ above it.

The NOM has been applied to nuclear theory in two ways. In the first of them the natural orbitals are constructed using single-particle wave functions corresponding to some single-particle potential under the condition that certain nuclear characteristics, such as nucleon separation energies, density distributions, etc. are reproduced. The occupation numbers are taken from experimental data (e.g. in one-nucleon transfer reactions) or from theoretical calculations accounting for short-range and tensor nucleon–nucleon correlations. They can be determined also by the condition that some nuclear characteristics are correctly reproduced. It has to be emphasized that in such calculations the natural orbitals and occupation numbers are not determined on a common basis.

In the second method the one-body density matrix $\rho(r, r')$ obtained in particular correlation method is diagonalized directly and the natural orbitals and occupation numbers obtained on the same footing.

In this Section both methods of applying the NOM are presented:

i) Single-particle Potential (SPP) Method [6.53]

It was shown in Sect. 4.3 that for spherically symmetric systems (with total angular momentum $J = 0$) the density distribution has the form:

$$\rho(r) = \frac{1}{4\pi} \int d\Omega \, \rho(r) = \sum_{lj} \rho_{lj}(r), \qquad (6.13)$$

where

$$4\pi r^2 \rho_{lj}(r) = \sum_{n,n'} (2j + 1) \, n^1_{n'lj,nlj} \, y^*_{nlj}(r) \, y_{n'lj}(r) \qquad (6.14)$$

in an arbitrary single-particle basis and

$$4\pi r^2 \rho_{lj}(r) = \sum_{n} (2j + 1) \, n_{nlj} |\tilde{y}_{nlj}(r)|^2 \qquad (6.15)$$

in the natural orbital basis. The functions $\tilde{y}_{nlj}$ are the radial part of the natural orbitals. In the SPP method these functions are obtained by means of the radial eigenfunctions $y_{nlj}(r)$ of a local Saxon–Woods potential of standard form and depth V_L:

$$V(r) = -V_L f_1(r) + V_S \left(\frac{\hbar}{m_\pi c} \right)^2 \frac{1}{r} \frac{d f_2(r)}{dr} L \cdot \sigma + V_C(r), \qquad (6.16)$$

where

$$f_{1,2}(r) = \{1 + \exp[(r - R_{1,2})/a_{1,2}]\}^{-1} \qquad (6.17)$$

with $R_1 = RA^{1/3}$, $R_2 = 1.1 \, A^{1/3}$ and $a_2 = 0.65$ fm.

R and a_1 are treated as free parameters, V_C (only for protons) is the Coulomb potential of a uniformly charged sphere with the experimentally-measured rms charge radius. V_L and V_S are chosen so as to reproduce the single-particle energies of the $j = l \pm 1/2$ doublets, measured by one-nucleon transfer reactions.

It is shown in [6.53] that for a $J = 0$ nucleus the natural orbitals are simply admixtures of single-particle wave functions having the same l, j, m and different principal quantum numbers n. For nearly all cases of nuclei which are of practical interest there are at most two bound single-particle wave functions with the same l, j, m (the first exception in the periodic table are the $1, 2, 3s_{1/2}$ states in ^{208}Pb), so the unitary transformation (4.84) leading from the best Saxon–Woods radial functions $y(r)$ to the natural radial functions $\tilde{y}(r)$ can be considered as a rotation in the (1, 2) plane:

$$\tilde{y}_1(r) = \cos \chi \cdot y_1(r) + \sin \chi \cdot y_2(r),$$

$$\tilde{y}_2(r) = -\sin \chi \cdot y_1(r) + \cos \chi \cdot y_2(r). \tag{6.18}$$

In (6.18) χ is the appropriate mixing parameter. Then

$$\rho_{lj}(r) = \frac{(2j+1)}{4\pi r^2} [n_1 \tilde{y}_1^2(r) + n_2 \tilde{y}_2^2(r)]$$

$$= \frac{(2j+1)}{4\pi r^2} [n_1^{(1)} y_1^2(r) + n_2^{(1)} y_2^2(r) + \varepsilon y_1(r) y_2(r)], \tag{6.19}$$

where

$$n_1^{(1)} = n_1 \cos^2 \chi + n_2 \sin^2 \chi,$$

$$n_2^{(1)} = n_1 \sin^2 \chi + n_2 \cos^2 \chi, \tag{6.20}$$

$$\varepsilon = (n_1 - n_2) \sin 2\chi.$$

The expressions (6.18–20) are applied in [6.53] to the ^{40}Ca nucleus using the local Saxon–Woods single-particle wave functions $y_{1,2}(r)$ corrected for non-locality. Considering the bound single-particle states, the only $\{lj\}$ subspaces containing 2 n-values are the $\{s_{1/2}\}$ and the $\{p_{3/2}\}$ ones (forcing $2p_{3/2}$ which is slightly unbound to be bound).

The natural occupation numbers n_α are approximated by those extracted from one-nucleon transfer reactions (e.g. (p, 2p) and (e, e'p)) according to the *French-Macfarlane* [6.36] sum rules (Eqs. (6.1, 6.2)) for states near the Fermi level, and are taken to be unity for deep-lying orbits.

For proton sinlge-particle states the occupation numbers are determined to be:

$$n_{1s_{1/2}} = 1; \; n_{1p_{3/2}} = 1; \; n_{1p_{1/2}} = 1; \; n_{2s_{1/2}} = 0.850; \; n_{2p_{3/2}} = 0.0375;$$

$$n_{1d_{5/2}} = 1; \; n_{1d_{3/2}} = 0.8975; \; n_{1f_{7/2}} = 0.070.$$

The fit of (6.19) to the experimental data for the ^{40}Ca nucleus charge density is achieved using the following values of the parameters $\varepsilon(s_{1/2}) = 0.05$ and $\varepsilon(p_{3/2}) = -0.05$. The solution of (6.20) gives the values of the parameters $\chi(s_{1/2}) = 0.17$ and $\chi(p_{3/2}) = -2.6 \cdot 10^{-2}$. In this way the $1,2s_{1/2}$ and $1,2p_{3/2}$ natural orbitals in ^{40}Ca starting from Saxon–Woods model functions are constructed in the SPP model.

ii) The Method of Boffi and Pacati [6.54]

A phenomenological approach for the construction of natural orbitals and occupation numbers in ^{16}O and ^{40}Ca nuclei is developed in [6.54]. Its basic relations are analogous to the Eqs. (6.18–20). In this case, however, the natural orbitals $|\alpha\rangle$ are expanded in terms of harmonic oscillator eigenstates $|nljm\rangle$, which preserve all the usual symmetries for a spherical nucleus:

$$|\alpha\rangle = |v, ljm\rangle = \sum_{n=0}^{2} \chi_n^{\alpha}|nljm\rangle. \tag{6.21}$$

The expansion coefficients χ_n^{α} are fitting parameters satisfying the orthonormalization conditions. In this method the one-body density matrix ρ_{SD}, corresponding to a Slater determinant total wave function is compared with the one-body density matrix ρ_λ, depending on the occupation numbers λ_α (for which $\sum_\alpha \lambda_\alpha = A$). The parameters of ρ_λ are fixed by a best fit to the experimental radial density for ^{16}O and ^{40}Ca minimizing the rms percentage deviation between the experimental and trial radial density. The realistic one-body density matrix allows a single-nucleon Hamiltonian to be constructed and diagonalized. A satisfactory description of all ground state properties (ground state energy, rms radii, single-particle energies) is obtained, in contrast with the prediction of the Hartree–Fock approximation.

Harmonic oscillator wave functions and occupation numbers different from 1 and 0 are used in [6.55], in which the best fit of the calculated charge density distributions and differences as well as form factors to the experimental ones leads to the determination of the occupation numbers in s–p- and s–d–shell nuclei.

iii) The Method of Jaminon, Mahaux and Ngô [6.45–47, 6.56]

In the phenomenological model of *Jaminon* et al. [6.45–47] the natural orbitals of the first occupied states are chosen to be the single-particle wave functions corresponding to a Saxon–Woods potential of the type (6.16, 6.17) with a particular depth–energy dependence. The parameters of the potential are fitted so that the single-particle energies of bound and quasibound states are in good agreement with the Hartree–Fock calculations of *Decharg* and *Gogny* [6.8]. The wave functions of the positive-energy states up to tens of MeV are constructed using a cut-off parameter R_0 and the boundary condition for the radial part of the wave function

$$\left[\frac{du_\alpha(r)}{dr} \bigg/ u_\alpha(r) \right]_{r=R_0} = K, \quad K < 0. \tag{6.22}$$

In the region $r > R_0$ the natural orbitals constructed in this way are of the form

$$u_\alpha(r) = u_\alpha(R_0)\exp[K(r - R_0)]. \tag{6.23}$$

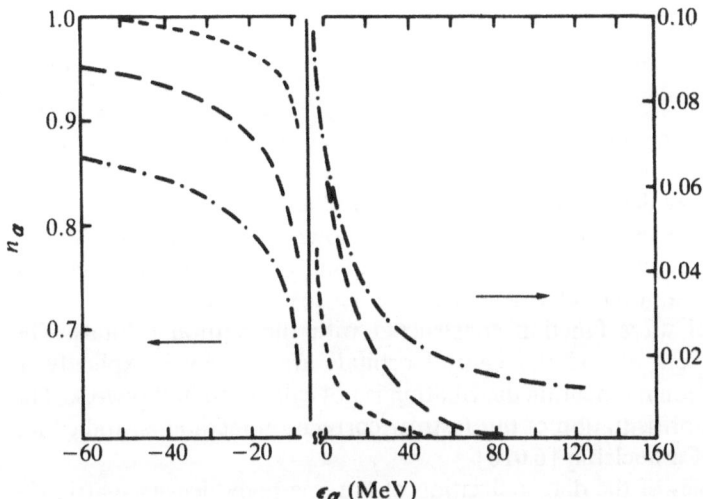

Fig. 6.7. Dependence of the proton occupation probabilities n_α (n,l,j) upon the single-particle energy ε_α in the case of RPA (short dashes) [6.44], in the case of PPW + RPA (dash dot) and PPWT + RPA (long dashes) [6.46, 6.47] for the ground state of ^{208}Pb

The occupation numbers in the model are taken from theoretical calculations in the framework of the RPA taking account of the long-range N–N correlations [6.44b] and from nuclear matter calculations adding the effects of the RPA to those produced by short-range and tensor nucleon–nucleon correlations [6.11]. The effect of the nuclear matter-type depletion can be seen in Fig. 6.7 where the proton occupation probabilities are given in the case of RPA [6.44b] (with 3.6% depletion of the Fermi-sea), in the case from [6.11] PPW + RPA (with 20.6% depletion) and in the intermediate case PPWT + RPA [6.46, 6.47] (with 11.6% depletion of the Fermi-sea).

The charge density distribution of ^{208}Pb calculated in the approach is in good agreement with the experimental one near the centre of the nucleus but is too small near $r \simeq 5$ fm. The effect of the depletion is that the charge distribution becomes flatter in the inner region.

The main effect of the depletion on the nucleon momentum distribution $n(k)$ is its rise in the domain $1.8 \text{ fm}^{-1} \leqslant k \leqslant 2.2 \text{ fm}^{-1}$ which is in accord with the predictions of more sophisticated correlation approaches, such as exp(S)-method [6.57], the Jastrow-type method [6.58], the Brueckner finite nuclei theory [6.59] for lighter nuclei and nuclear matter [6.60].

The methods i)–iii) are examples of the cases in which the natural orbitals are not consistent with the occupation numbers.

iv) The Method of Gaudin, Gillespie and Ripka [6.61a]

The more consistent way to obtain natural orbitals and occupation numbers on one and the same basis is to diagonalize the one-body density matrix $\rho(r, r')$

which is calculated including correlations. Thus the integral equation

$$\int \rho(r, r')\, \varphi_\alpha(r')\, dr' = n_\alpha \varphi_\alpha(r) \tag{6.24}$$

has to be solved.

In [6.61a] the one-body density matrix is calculated using a Jastrow many-body wave function in the framework of a perturbation expansion in the case of ^{40}Ca. The one-body density matrix is diagonalized and natural orbitals for the single-particle states in ^{40}Ca are obtained. It is shown that the density distribution obtained directly from the diagonal elements of the Jastrow one-body density matrix is in good agreement with the density calculated with a Slater determinant total wave function constructed with the natural orbitals. The occupation numbers n_α and the natural orbitals are not given explicitly in [6.61a]. The nucleon momentum distribution is not calculated in this work. The natural orbital representation of the Jastrow correlation method is studied for ^{4}He, ^{16}O and ^{40}Ca nuclei in [6.61b].

Other examples of the diagonalization of the one-body density matrix are given in the next Sections.

6.3 The Natural Orbital Representation within the Coherent Density Fluctuation Model (CDFM)

The one-body density matrix in the CDFM has the form (4.59, 4.56)

$$\rho(r, r') = 3 \int\limits_0^\infty dx\, |f(x)|^2\, \rho_0(x) \frac{j_1(k_F(x)|r - r'|)}{k_F(x)|r - r'|}\, \theta\left(x - \frac{|r + r'|}{2}\right), \tag{6.25}$$

where $k_F(x) = \alpha/x$, $\alpha \equiv (9\pi A/8)^{1/3}$, $\rho_0(x) = 3A/4\pi x^3$.

The weight function $f(x)$ can be determined either on the basis of the CDFM dynamical equation (4.12) or by means of the density distribution (4.66):

$$|f(x)|^2 = -\frac{1}{\rho_0(x)} \frac{d\rho(r)}{dr}\bigg|_{r=x} \tag{6.26}$$

in the case of monotonically-decreasing density distributions ($d\rho/dr \leqslant 0$ for all r).

The one-body density matrix (6.25) is independent of the spin and isospin variables. So that the summation over the spin and isospin has been already performed. For nuclei with $J = 0$ the one-body density matrix has to be diagonalized in the $\{ljm\}$ subspace. In the case of nuclei with spherical symmetry the natural orbitals have the form:

$$\varphi_{nlm}(r) = R_{nl}(r)\, Y_{lm}(\theta, \varphi) = \frac{u_{nl}(r)}{r}\, Y_{lm}(\theta, \varphi). \tag{6.27}$$

Substitution of (6.27) in (6.24) and integration over the angular variables gives the following equation for the radial part of the natural orbitals:

$$\int_0^\infty dr' \, K_l(r, r') \, u_{nl}(r') = 2n_{nl} \, u_{nl}(r),$$ (6.28)

where

$$K_l(r, r') = -6\pi\beta rr' \int_{-1}^{1} dy \, P_l(y)|r - r'| \int_0^{2\beta|r-r'|/|r+r'|} dt$$

$$\times \frac{(\sin t - t\cos t)}{t^5} \left.\frac{d\rho}{dx}\right|_{x=\beta\frac{|r-r'|}{t}},$$ (6.29)

$|r \pm r'| = (r^2 + r'^2 \pm 2rr'y)^{1/2}$ and $P_l(y)$ are the Legendre polynomials. The factor 2 in (6.28) accounts for the spin degree of freedom. Only one sort of particles (protons or neutrons) are considered, so $\beta = (9\pi N/4)^{1/3}$, where N is the number of the particles.

The natural orbitals and occupation numbers for protons are calculated in the CDFM solving Eqs. (6.28, 6.29) for ^{16}O, ^{40}Ca, ^{58}Ni and ^{208}Pb nuclei in [6.62]. The density distribution of the symmetrized Fermi form has been used:

$$\rho_{SF}(r) = \rho_0 \frac{\sinh(R/b)}{[\cosh(R/b) + \cosh(r/b)]}, \qquad \rho_0 = \frac{3Z}{4\pi R^3[1 + (\pi b/R)^2]}$$ (6.30)

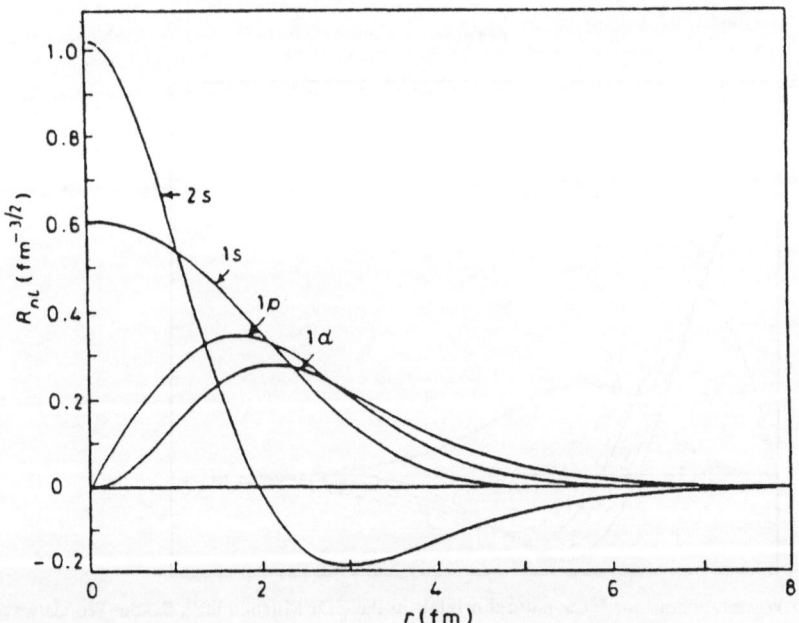

Fig. 6.8. Natural orbitals in the CDFM for 1s, 1p, 1d and 2s states in ^{16}O

with values of the parameters R and b taken from the analyses of electron–nuclei elastic scattering data [6.63]. The calculations for ^{40}Ca and ^{58}Ni nuclei have been repeated with more realistic density distributions [6.53, 6.64], extracted from different experimental data by means of model-independent analysis.

The natural orbitals (wave functions) for ^{16}O, ^{40}Ca, ^{58}Ni and ^{208}Pb are shown in Figs. 6.8–11. The corresponding natural occupation numbers are shown in Table 6.4. It was shown in the calculations that the differences of the wave functions and occupation numbers evaluated for ^{40}Ca and ^{58}Ni with different density distributions (from (6.30) and with the Sick densities [6.53, 6.64]) are rather small. For this reason only the values calculated with the symmetrized Fermi distributions are presented.

The variation of the occupation numbers follows the order of the states in the energy-level scheme of a square-well potential with infinite walls and agrees

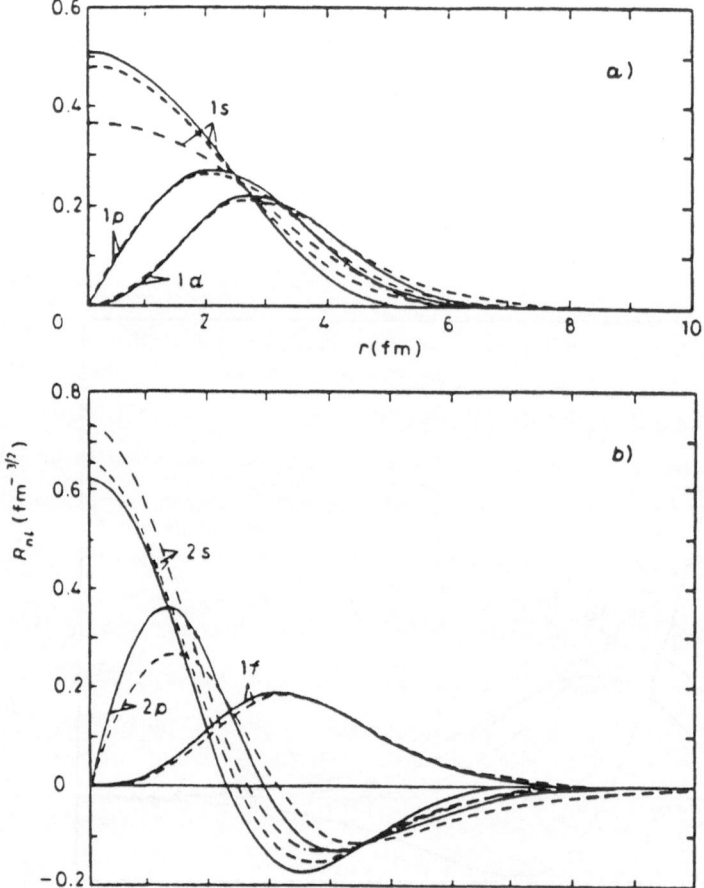

Fig. 6.9. Wave functions for ^{40}Ca: natural orbitals in the CDFM (solid line); Saxon–Woods wave functions in the SPP method [6.53] (dashed line); natural orbitals in the SPP method [6.53] (dash–dotted line). (**a**) For 1s, 1p and 1d states, (**b**) for 2s, 2p and 1f states

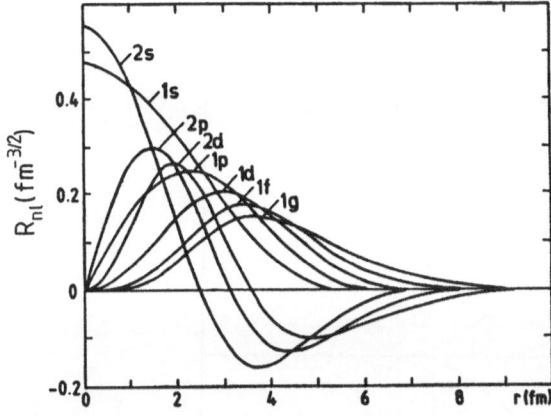

Fig. 6.10. Natural orbitals in the CDFM for ^{58}Ni 1s, 1p, 1d, 2s, 1f, 2p, 1g and 2d states

with the sequence of the empirical values of the single-particle energies available for some states and given in [6.53].

In the case of ^{40}Ca a comparison of the CDFM natural orbitals with the SPP functions (eigenfunctions of a Saxon–Woods potential with parameters adjusted to the rms charge radii and the single-particle energies) as well as with the natural orbitals constructed within the SPP method (6.18–20) is made in Fig. 6.9. It is seen that the CDFM natural orbitals are close to the SPP wave function for all states except for the 2p one. The natural orbitals constructed within the SPP method for the 1 s and 2 s states in ^{40}Ca differ significantly from the corresponding SPP wave functions, whereas the 1 p and 2 p SPP natural orbitals practically coincide with the initial SPP wave functions. Thus the SPP natural orbitals deviate from the CDFM ones. This can be expected because of the difference between the CDFM occupation numbers and those used in [6.53]. The latter give a depletion of the Fermi sea of about 3.5% which is, for instance, far smaller than the depletion of 17% suggested from nuclear matter calculations [6.11] including short-range and tensor correlations. This is also supported by the analysis in [6.65] where the nucleon momentum distribution for ^{40}Ca was evaluated using the natural orbitals of the SPP method and various sets of the occupation numbers with different depletion (see Fig. 7.10 in Chap. 7). It was shown in [6.65] that a depletion of 3–4% is not sufficient to reproduce the results of correlation calculations (e.g. those of [6.66–68]) of the nucleon momentum distribution.

The occupation numbers obtained in the CDFM are discussed and compared with available experimental data in Sect. 6.5.

6.4. The Natural Orbital Representation within the Generator Coordinate Method (GCM)

The natural orbitals and occupation numbers using the one-body density matrix obtained in the GCM approach [6.69, 6.70] (Sect. 4.1) are calculated in [6.71].

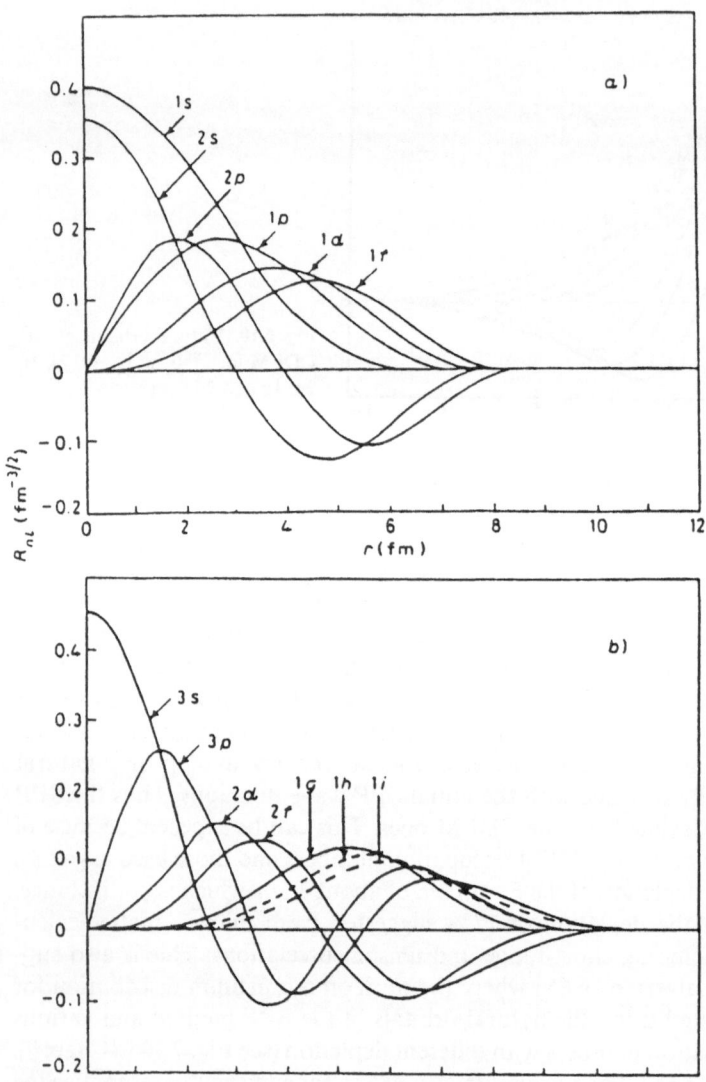

Fig. 6.11. Natural orbitals in the CDFM for ^{208}Pb. (a) For 1s, 1p, 1d, 2s, 1f and 2p states, (b) for 1g, 2d, 1h, 3s, 2f, 3p and 1i states

In this way the single-particle picture including correlations is restored in the GCM. The effective single-particle wave functions (natural orbitals) and occupation numbers containing the information about the N–N short-range correlations, whose effect on the physical quantities is quite substantial (e.g. concerning the existence of high-momentum components of the nucleon momentum distribution, see Sect. 7.1), are obtained.

Table 6.4. Occupation numbers and total depletions calculated in the generator coordinate method (GCM) with square-well construction potential with infinite walls (GCMSW), in GCM with harmonic oscillator construction potential (GCMHO) and in the coherent density fluctuation model (CDFM). The Fermi level is denoted by (*).

Nucleus	State	GCMSW	GCMHO	CDFM
^{4}He	3s	0.002	0.001	
	2s	0.014	0.005	
	1s*	0.984	0.995	
Total depletion		1.6%	0.5%	
^{16}O	2p	0.013	$< 10^{-3}$	0.044
	2s	0.008	$< 10^{-3}$	0.191
	1p*	0.986	1.000	0.536
	1s	0.992	1.000	0.804
Total depletion		1.2%	$< 0.1\%$	39.9%
^{40}Ca	3s	0.014	$< 10^{-3}$	0.061
	2d	0.010	$< 10^{-3}$	0.073
	2p	0.007	$< 10^{-3}$	0.211
	2s*	0.986	0.999	0.442
	1d	0.990	0.999	0.548
	1p	0.993	0.999	0.768
	1s	1.000	1.000	0.925
Total depletion		0.9%	$< 0.1\%$	36.6%
^{58}Ni	2d			0.14
	1g			0.22
	2p			0.31
	1f*			0.44
	2s			0.56
	1d			0.66
	1p			0.84
	1s			0.96
Total depletion				35.1%
^{208}Pb	3p			0.23
	1i			0.30
	2f			0.31
	3s*			0.46
	1h			0.51
	2d			0.53
	1g			0.71
	2p			0.75
	1f			0.85
	2s			0.91
	1d			0.94
	1p			0.99
	1s			1.00
Total depletion				25.9%

The ground state one-body density matrix in the GCM-scheme [6.69, 6.70] has the form (see also Sect. 4.1):

$$\rho(r, r') = \int \int f_0(x) f_0(x') I(x, x') \rho(x, x', r, r') \, dx \, dx'. \tag{6.31}$$

The function $f_0(x)$ is the solution of the Hill–Wheeler equation:

$$\int_0^\infty [\mathcal{H}(x, x') - EI(x, x')] f(x') \, dx' = 0 \tag{6.32}$$

corresponding to the lowest energy eigenvalue,

$$\rho(x, x', r, r') = 4 \sum_{\lambda, \mu = 1}^{A/4} (N^{-1}(x, x'))_{\mu\lambda} \, \varphi_\lambda^*(r, x) \, \varphi_\mu(r', x'), \tag{6.33}$$

where $\varphi_\lambda(r, x)$ are neutron and proton orbitals by means of which the generating wave function $\Phi(\{r_i\}, x)$ is built as a single Slater determinant. The expressions for the overlap $(I(x, x'))$ and energy $(\mathcal{H}(x, x'))$ kernels, as well as of the matrix $N_{\lambda\mu}(x, x')$ are given in Sect. 4.1.

In the case of nuclei with total spin $J = 0$ the natural orbitals $\varphi_{nlm}(r)$ have the form (6.27), where their radial parts $u_{nl}(r)$ and occupation numbers n_{nl} are solutions of the equation

$$\int_0^\infty K_l(r, r') u_{nl}(r') \, dr' = n_{nl} u_{nl}(r), \tag{6.34}$$

where

$$K_l(r, r') = rr' \int \int f_0(x) f_0(x') I(x, x') \sum_{nn'} (N^{-1}(x, x'))_{nl, n'l}$$

$$\times R_{n'l}^*(r, x) R_{nl}(r', x') \, dx \, dx'. \tag{6.35}$$

The summation in (6.35) is performed over all single-particle functions φ with given l forming the Slater determinant of the generating function and $R_{nl}(r, x)$ are the radial parts of these functions.

Eq. (6.34) has nonzero solutions for the occupation numbers only for values of $l \leqslant l_{max}$, where l_{max} is the maximum angular momentum of the single-particle states whose wave functions form the Slater determinant of the generating function $\Phi(\{r_i\}, x)$. This is a restriction of the GCM with Slater determinant generating functions applied to spherical nuclei. To overcome it more complicated generating functions may be used, but the simplicity of the method will then be lost.

The natural orbitals and occupation numbers in ^{4}He, ^{16}O and ^{40}Ca have been determined in [6.71] using both square-well and harmonic oscillator "construction" potentials (i.e. the single-particle wave functions in (6.33) and in $(N^{-1}(x, x'))_{\mu\lambda}$ are obtained from these potentials). The Skyrme force parameters in $\mathcal{H}(x, x')$ ((4.34), (4.35)) in the case of square-well construction potential, were obtained in [6.71] so as to give an optimal fit to the binding energies of the three nuclei studied by solving the Hill–Wheeler equation. In the case of harmonic oscillator construction potential, SkM* parameter sets [6.72] giving realistic binding energies have been used. The GCM natural orbitals in coordinate space are plotted in Figs. 6.12–6.14 and the corresponding occupation numbers are shown in Table 6.4. As can be seen the depletion of the Fermi sea varies from 0.9% for ^{40}Ca to 1.6% for ^{4}He when a square-well potential is used and is less than 0.5% for harmonic oscillator potential. This is an evidence that as a whole

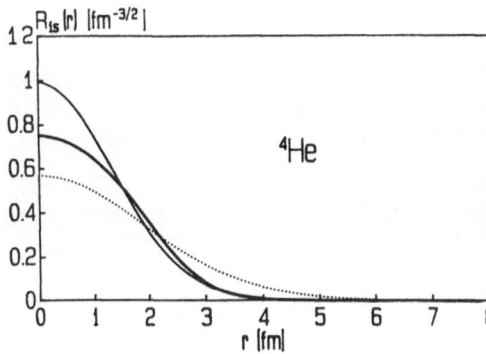

Fig. 6.12. Natural orbitals in coordinate space for ^{4}He, 1s state, calculated using the GCM with a square-well construction potential (thick solid line), the GCM with a harmonic oscillator construction potential (dotted line) and the pure harmonic oscillator wave function (thin solid line)

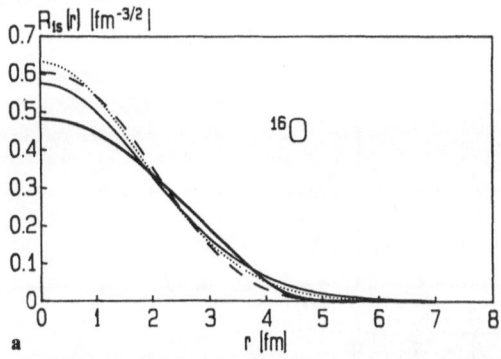

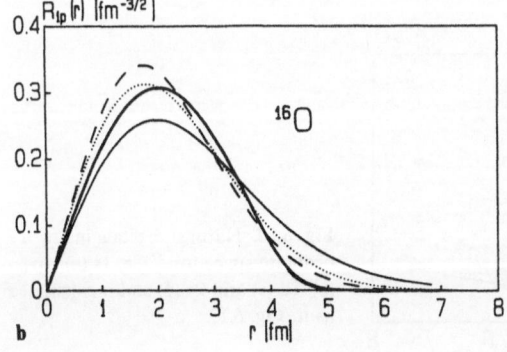

Fig. 6.13. Natural orbitals in coordinate space for ^{16}O, 1s (a) and 1p (b) states calculated using the GCM with a square-well construction potential (thick solid line), the GCM with a harmonic oscillator construction potential (dotted line), the CDFM [6.62] (dashed line) and the Hartree–Fock wave function [6.73] (thin solid line)

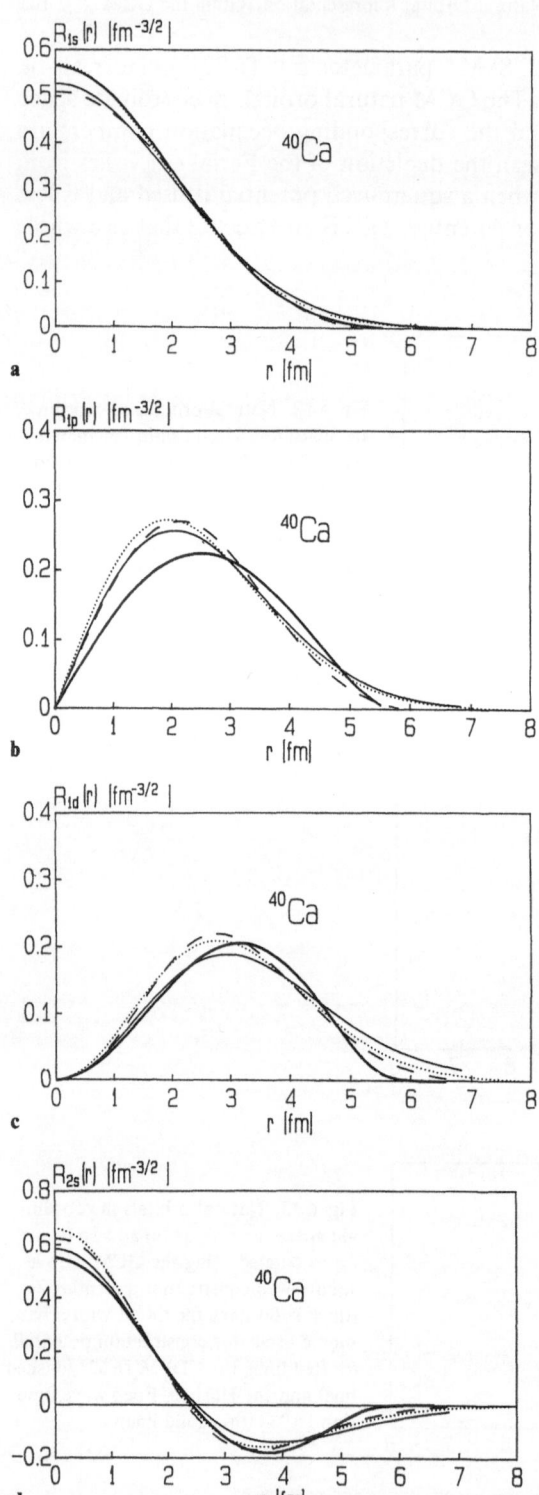

Fig. 6.14. Natural orbitals in coordinate space for ^{40}Ca, 1s (**a**), 1p (**b**), 1d (**c**) and 2s (**d**) states calculated as in Fig. 6.13

the short-range correlations have a small effect upon the occupation numbers. This effect is stronger in the case of square-well construction potential. For comparison, the total depletion for ^{40}Ca obtained in beyond Hartree–Fock model of *Orland* and *Schaeffer* [6.74a] (see Chap. 3) is 15%, while in the CDFM it is 37%.

The natural orbitals for both construction potentials can be compared with the CDFM natural orbitals (from Sect. 6.3) and with the Hartree–Fock single-particle functions [6.73] for ^{40}Ca and ^{16}O and with pure harmonic oscillator

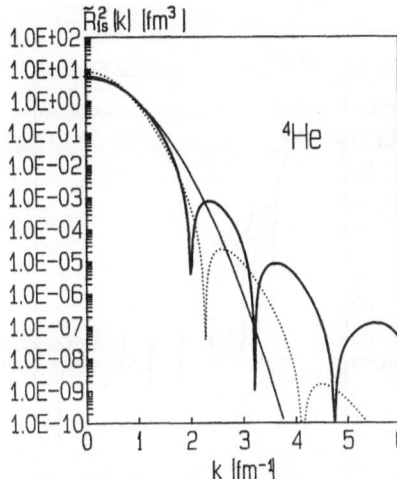

Fig. 6.15. Natural orbitals in momentum space for ^{4}He, 1s state calculated as in Fig. 6.12

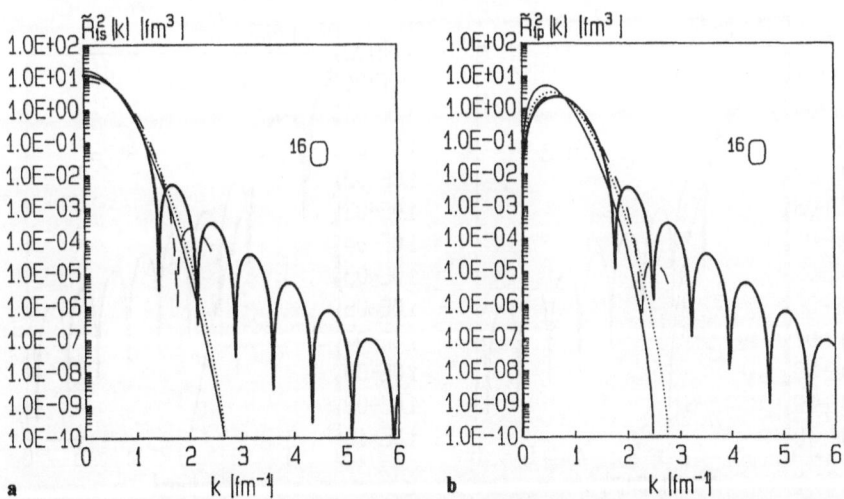

Fig. 6.16. Natural orbitals in momentum space for ^{16}O, 1s (a) and 1p (b) states calculated as in Fig. 6.13

wave functions for ^{4}He. It is found that all sets of wave functions differ significantly, mainly in the central region and this results in differences in the nuclear density distributions [6.69, 6.70].

The effect of N–N correlations becomes, however, more pronounced when the natural orbitals are compared in momentum space. The squares of their radial parts are shown in Figs. 6.15–17. The corresponding nucleon momentum distributions (see Figs. 7.1–3 in Chap. 7) calculated in GCM with a square-well construction potential have pronounced high-momentum components which is in general agreement with the predictions of other theoretical methods taking

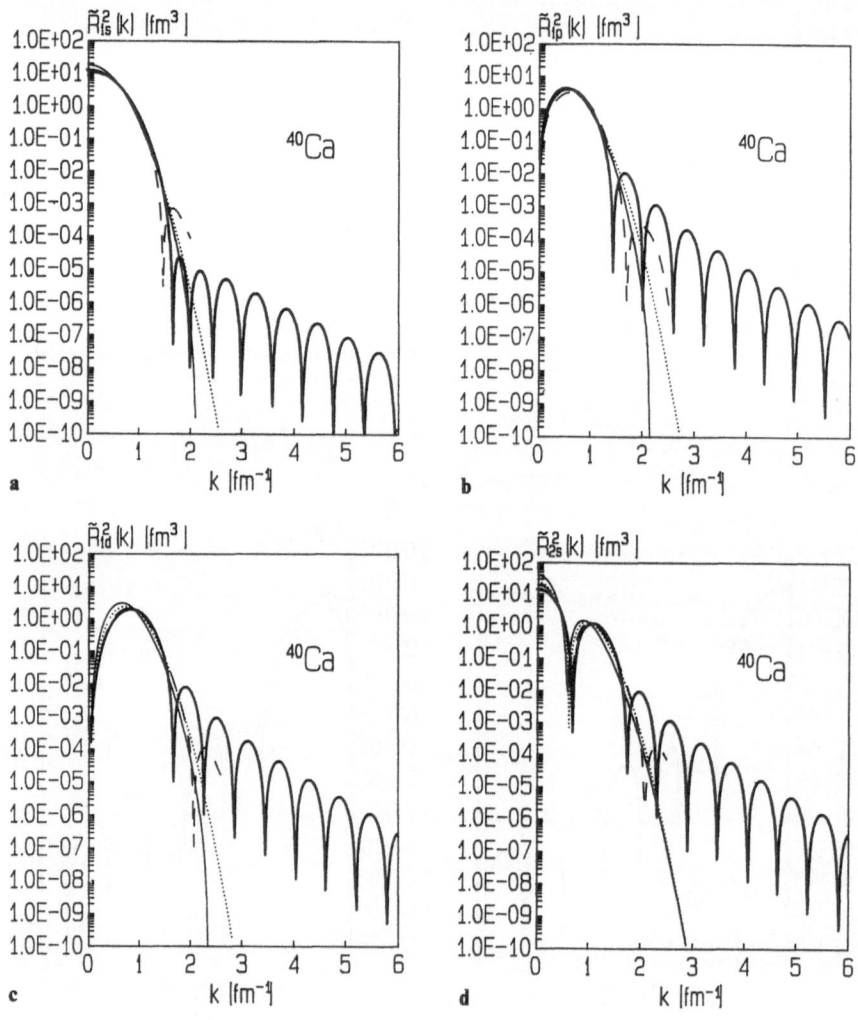

Fig. 6.17. Natural orbitals in momentum space for ^{40}Ca, 1s (**a**), 1p (**b**), 1d (**c**) and 2s (**d**) states calculated as in Fig. 6.13

account of N–N correlations and the available experimental results. This is not the case when a harmonic oscillator construction potential is used. It can be seen, indeed, that the corresponding natural orbitals in momentum space are very close to the Hartree–Fock and harmonic oscillator wave functions obtained in methods where uncorrelated nucleons are used (see Figs. 6.15–17). On the other hand, the natural orbitals in GCM with a square-well construction potential differ substantially from the uncorrelated wave functions both in magnitude and shape and are apparently strongly influenced by the short-range correlations. This can be expected, because when a square-well potential with infinite walls is used the intermediate state is characterized by all A nucleons confined in a sphere with finite radius x. This is not the case with the harmonic oscillator construction potential. The comparison with the natural orbitals in momentum space and with the nucleon momentum distributions obtained in the CDFM shows that in this model the short-range correlations affect both the occupation numbers and the wave functions while in the GCM with square-well construction potential the main effect is on the natural orbitals.

6.5 Comparison of CDFM and GCM Results with Experimental Data for Occupation Numbers

As was shown in Sects. 6.3, 6.4 the depletion of the Fermi sea obtained in the CDFM varies from 25.9% for ^{208}Pb up to 39.9% for ^{16}O, while in the GCM approach with square-well potential it varies between 1.6% (in ^{4}He) and 0.9% (in ^{40}Ca). For comparison the total depletion of the Fermi sea in the case of ^{208}Pb obtained by combining the prediction of RPA calculations [6.44b] with those of nuclear matter accounting for short-range and tensor correlations [6.11] is 20.6%. Jaminon et al. [6.56] choose a depletion of 11.6% determined in order to obtain better agreement with some empirical data [6.74b, 6.75] concerning the proton momentum distribution $(n(k)/Z)$ up to $k \simeq 2.2\,\mathrm{fm}^{-1}$ as well as with the experimentally deduced value of the occupation probability of the $3\,s_{1/2}$ state in [6.16]. The empirical values are extracted from analyses of the backward inclusive cross-sections due to 642 MeV electron and 200 MeV proton scattering on ^{12}C and ^{58}Ni in the framework of the "quasi two-body scaling" approach. As shown in Fig. 7.14 the CDFM results for the nucleon momentum distribution in ^{208}Pb are close to the results of Jaminon et al. [6.56] (with an 11.6% depletion) though the CDFM occupation numbers give a larger depletion: 25.9%. This is due to the fact that the CDFM wave functions differ from the uncorrelated Hartree–Fock ones in [6.45–47, 6.56].

The depletions of the Fermi sea which correspond to the sets of natural occupation numbers obtained in the CDFM are consistently larger than those extracted from one-nucleon transfer data by means of DWBA analysis (see, for instance, [6.53] and references therein). These data indicate a comparatively small total depletion of about 3–4% which is in a good agreement with RPA predictions. It has to be noted, however, that the RPA calculations do not include tensor and short-range N–N correlations and the inclusion of these correlations ([6.11]) leads to an increase of the depletion from 3.6% up to 20.6% in the case of ^{208}Pb. It should also be emphasized that in the DWBA calculations uncorrelated wave functions are used. Therefore, the extraction of the occupation numbers from the experimental data is strongly model-dependent since the wave functions used are evaluated within a particular uncorrelated model. Thus, the procedure is not consistent in the sense that the model wave functions do not contain the dynamical short-range, tensor, and other N–N correlations. This is the main deficiency of the procedure for the extraction of the occupation numbers from experiments on nuclear reactions. It is extremely important, therefore, for both the occupation numbers and the wave functions to be extracted from the experimental data on one and the same footing.

Experimental data on the occupation probabilities for single-particle states around the Fermi level in lead isotops have been obtained studying (e, e′p) and $\vec{d}$, ^{3}He) reactions. They were considered in Sect. 6.1. As was shown, an occupation probability of $n_{3s} = 0.82 \pm 0.09$ for the $3s_{1/2}$-state in ^{208}Pb has been extracted [6.16] by means of a sum-rule analysis of (e, e′p) reaction data. In this analysis the occupation probability is proportionally scaled by an effective number Z reflecting the contribution of $3s_{1/2}$ protons to the charge density distribution. If the effects of the depletion of other orbits as well as the core polarization effects are neglected the effective number Z can be expressed (6.5) simply in terms of the experimentally-deduced charge density difference $\Delta \rho$ between the density of the neighbouring nuclei with the same neutron numbers and the single-particle charge density ρ_{3s} of $3s_{1/2}$ protons evaluated within a particular model. For the case of ^{206}Pb and ^{205}Tl the mean-field analysis shows [6.6] that Z varies in the range $0.6 \leqslant Z \leqslant 0.8$. In [6.16] the value $Z = 0.7$ is used and the uncertainty just mentioned for Z was not taken into account. The latter would lead to a larger uncertainty of the $3s_{1/2}$ occupation probability $n_{3s_{1/2}} = 0.82 \pm 0.21$ in ^{208}Pb. Apparently, the determination of Z is the crucial point for the model independence of the sum-rule analysis [6.37b] so far as the particular value of Z is extracted from calculations using the mean-field approximation in which dynamical correlations are not included. In order to check the consistency of the CDFM natural orbitals and occupation numbers, the experimental data [6.6] concerning the charge density difference of ^{206}Pb and ^{205}Tl nuclei has been analyzed using the CDFM $3s_{1/2}$ wave functions. The fit of the experimental charge density difference with the CDFM wave functions leads to the new value of $Z = 0.48 \pm 0.10$. The corresponding occupation probability of the ^{208}Pb $3s_{1/2}$-state is $n_{3s_{1/2}} = 0.56 \pm 0.18$. This number is in agreement with the CDFM occupation number for this state: $n_{3s_{1/2}} = 0.46$.

7 Momentum Distributions in Nuclei

This chapter is devoted to nucleon–nucleon correlation effects on the two main ground state characteristics, nucleon momentum ($n(k)$) and density distributions ($\rho(r)$). The functional relation between them founded on the *Hohenberg–Kohn* theorem [7.19, 7.20], the formulation of a variational principle using both equivalent physical quantities (ρ and n), and the results of wide range of correlation methods and models of the momentum distribution in various nuclei and nuclear matter are presented in Sect. 7.1. The theoretical results for $n(k)$ are compared with the experimental data from different reactions on ^{2}H, ^{3,4}He, ^{12}C, ^{56}Fe nuclei and with the single-particle momentum distributions in ^{16}O and ^{208}Pb. The problems of the nucleon momentum distribution in nuclei at finite temperatures and its dependence on nuclear deformation are also considered. Special attention is paid on the extraction of the nucleon momentum distribution from the experimental data obtained in particle–nucleus reactions.

The correlation effects on the two-body- and alpha-cluster momentum distributions in nuclei are discussed in Sect. 6.2 using various correlation models and methods. Theoretical and experimental results for $d + p$-momentum distributions in ^{3}He, as well as for $t + p$- and $d + d$-momentum distributions in ^{4}He are also considered.

7.1 Nucleon Momentum and Density Distributions

The two main bulk characteristics of the ground state of nuclei, namely the nucleon momentum $n(k)$ and density $\rho(r)$ distributions, will be considered in this section. Until the mid-seventies more attention in nuclear theory has been paid to the study of the nuclear density distribution. This is related to the ability of the widely used Hartree–Fock theory to describe successfully this quantity, which is not very sensitive to the dynamical short-range correlations not included in this theory. It was shown, however, by *Jaminon* et al. [7.1] that it is in principle impossible to describe correctly both density and momentum distributions simultaneously in the framework of the Hartree–Fock approximation (see Sect. 3.1). The reason is that the nucleon momentum distribution (NMD) is sensitive to short-range and tensor nucleon–nucleon correlations reflecting the peculiarities of the nucleon–nucleon forces at small distances. It follows that the correct unified description of these related distributions has to be looked for in the framework of nuclear correlation methods.

We start with the definitions of the nucleon density and momentum distributions [7.2]. The local density operator $\hat{\rho}$ and the NMD operator $\hat{n}$ can be written in the spatial (r) and momentum (q) representation respectively in the forms:

$$\hat{\rho}_r = \sum_{j=1}^{A} \delta(r - r_j), \tag{7.1}$$

$$\hat{\rho}_q = \sum_{j=1}^{A} \exp(iq \cdot r_j), \tag{7.2}$$

$$\hat{n}_r = \frac{1}{(2\pi)^3} \sum_{j=1}^{A} \exp(ir \cdot q_j), \tag{7.3}$$

$$\hat{n}_q = \sum_{j=1}^{A} \delta(q - q_j). \tag{7.4}$$

The nucleon density $\rho(r)$ and NMD $n(q)$ are the expectation values of the operators (7.1) and (7.4) in the ground state specified by the many-particle wave function Ψ in the space- and momentum representations:

$$\rho(r) = \langle \Psi_r | \hat{\rho}_r | \Psi_r \rangle, \tag{7.5}$$

$$n(q) = \langle \Psi_q | \hat{n}_q | \Psi_q \rangle. \tag{7.6}$$

Using Eqs. (7.2) and (7.3) the form factor $F(q)$ and the inverse form factor $\mathscr{F}(r)$ can be determined:

$$F(q) = \langle \Psi_r | \hat{\rho}_q | \Psi_r \rangle = \int \rho(r) \exp(iq \cdot r) \, dr, \tag{7.7}$$

$$\mathscr{F}(r) = \langle \Psi_q | \hat{n}_r | \Psi_q \rangle = \int n(q) \exp(iq \cdot r) \frac{dq}{(2\pi)^3}. \tag{7.8}$$

The nucleon density $\rho(r)$ and momentum $n(q)$ distributions can be expressed by the form factor and the inverse form factor in the following form:

$$\rho(r) = \int F(q) \exp(- iq \cdot r) \frac{dq}{(2\pi)^3}, \tag{7.9}$$

$$n(q) = \int \mathscr{F}(r) \exp(- iq \cdot r) \, dr. \tag{7.10}$$

It is useful to define these distributions by the one-body density matrix $\rho(r, r')$, the Wigner distribution function and the one-particle Green function. Using the definition of the one-body density matrix

$$\rho(r, r') = A \int \Psi^*(r, r_2, \ldots) \Psi(r', r_2, \ldots) \, dr_2 \ldots dr_A, \tag{7.11}$$

where Ψ is the many-particle wave function, the local density $\rho(r)$ is given by the diagonal elements

$$\rho(r) \equiv \rho(r, r' = r),\tag{7.12}$$

while the NMD is:

$$n(q) = \rho(q, q')|_{q' = q} = \lim_{q' \to q} \int \Psi^*(q, q_2, \ldots) \Psi(q', q_2, \ldots)$$

$$\times \frac{dq_2}{(2\pi)^3} \cdots \frac{dq_A}{(2\pi)^3},\tag{7.13}$$

or

$$n(q) = \int \exp[iq(r - r')]\rho(r, r') \, dr \, dr'.\tag{7.14}$$

Defining the Wigner distribution function

$$W(r, q) = \int \rho(r + \tfrac{1}{2}\eta, r - \tfrac{1}{2}\eta) \exp(iq \cdot \eta) \, d\eta,\tag{7.15}$$

the density and momentum distributions can be expressed in the form:

$$\rho(r) = \int W(r, q) \frac{dq}{(2\pi)^3},\tag{7.16}$$

$$n(q) = \int W(r, q) \, dr.\tag{7.17}$$

These distributions can also be related to the one-particle Green function $G(r, t; r', t')$ by means of the relationship:

$$\rho(r, r') = -i \lim_{t' \to t} G(r, t; r', t').\tag{7.18}$$

Many nuclear theoretical methods use the local density distribution $\rho(r)$ as a fundamental variable in the theory. One of them is the Thomas–Fermi model [7.3, 7.4], in which the average kinetic energy per particle in small volume of the fermion system (with a radius-vector r) can be expressed by the local density or by the Fermi momentum $k_F(r)$ [7.5]:

$$\bar{E}_{kin} = \frac{3\hbar^2}{10m} \left(\frac{3\pi^2}{s}\right)^{2/3} \rho^{2/3}(r) = \frac{3}{5} \frac{\hbar^2 k_F^2(r)}{2m},\tag{7.19}$$

s being equal to 1 or 2 for electron or nucleon (proton and neutron) systems, respectively.

The total energy of the system is given by

$$E = \int dr \{\tau_{TF}(r) + V(r)\rho(r)\},\tag{7.20}$$

where

$$\int dr \tau_{TF}(r) \equiv \int \bar{E}_{kin}(r)\rho(r) \, dr = \frac{3\hbar^2}{10m} \left(\frac{3\pi^2}{s}\right)^{2/3} \int \rho^{5/3}(r) \, dr\tag{7.21}$$

is the total kinetic energy and

$$V(r) = V_{ext}(r) + V_{int}(r) \qquad (7.22)$$

is the potential energy of the interaction of a particle with the other particles (V_{int}) and eventually with an external field (V_{ext}). Equation (7.20) is valid if the potential energy is a slowly varying function of r.

The condition for the minimum of the total energy (7.20) leads to the Euler–Lagrange equation:

$$\frac{\delta E}{\delta \rho} - \lambda = 0, \qquad (7.23)$$

which together with the normalization condition

$$\int dr\, \rho(r) = A \qquad (7.24)$$

determines the equilibrium density of the system ρ and the particle separation energy λ.

The application of the Thomas–Fermi method to nuclear systems faces various difficulties and is not so successful as in the case of many-electron atoms. In the approach of *Bethe* [7.6] the total energy of the nucleus has the form:

$$E = \int W(\rho)\rho(r)\,dr + \tfrac{1}{2}\int u(r, r')[\rho(r) - \rho(r')]^2\,dr\,dr', \qquad (7.25)$$

where $W(\rho)$ is the energy per nucleon in nuclear matter, which includes the Thomas–Fermi kinetic energy $\tau_{TF}(r)$. The function $u(r, r')$ is related to the effective N–N interaction at the points r and r'. It turns out [7.7] that there is no continuous solution of the variational equation and the reason for this is the replacement of the attractive exchange interaction by the effective local interaction.

Various phenomenological methods have been suggested for the description of nuclear characteristics. In the energy density method of *Brueckner* et al. [7.8–7.12] the total energy has the form:

$$E = \int \varepsilon[\rho(r)]\,dr, \qquad (7.26)$$

where

$$\varepsilon[\rho] = \frac{0.3\hbar^2}{M}\left(\frac{3\pi^2}{2}\right)^{2/3}\frac{1}{2}[(1 + \alpha)^{5/3} + (1 - \alpha)^{5/3}]\rho^{5/3} + \rho V(\rho, \alpha)$$
$$+ \frac{1}{2}e\rho_p\Phi_c(r) - 0.7386e^2\rho_p^{4/3} + \frac{\hbar^2}{8m}\eta(\nabla\rho)^2 + c\frac{(\nabla\rho)^2}{2}, \qquad (7.27)$$

$$\alpha = (\rho_n - \rho_p)/(\rho_n + \rho_p), \quad \rho = \rho_n + \rho_p,$$

$$V(\rho, \alpha) = b_1\rho + b_2\rho^{4/3} + b_3\rho^{5/3} + \alpha^2(b_4\rho + b_5\rho^{4/3} + b_6\rho^{5/3}). \qquad (7.28)$$

The right-hand side terms in (7.27) are respectively the Thomas–Fermi kinetic energy, the potential energy (with $V(\rho, \alpha)$ from (7.28), the b_i being parameters) from nuclear matter theory, the Coulomb direct and exchange terms ($\Phi_c(r)$ is the

Coulomb potential related to the charge distribution ρ_p) and the gradient corrections to the potential and kinetic energy.

In the extended Thomas–Fermi method [7.13–7.18] the energy functional is given by the expression:

$$E_{ETF} = \int dr \left[V(r)\rho(r) + \frac{\hbar^2}{2m} \tau_{ETF}[\rho] \right],$$ (7.29)

where the local potential $V(r)$ is related to effective Skyrme forces and the semiclassical expression for the kinetic energy

$$\tau_{ETF}[\rho] = \tau_{TF}[\rho] + \tau_2[\rho] + \tau_4[\rho]$$ (7.30)

contains the Thomas–Fermi term (τ_{TF}) and two additional terms τ_2 and τ_4 which depend on the density ρ and its gradient and are proportional to $\hbar^2$ and $\hbar^4$. It has been shown by *Kirzhnitz* [7.16] that the term proportional to $\hbar^6$ leads to divergences in the theory.

The density functional approaches are founded on the theorem of *Hohenberg–Kohn* [7.19, 7.20] which establishes the existence of a universal (perhaps unique) energy functional $E[\rho]$ of the local density of the fermion system. It has been shown in the theorem that the many-fermion wave function and hence all properties of the ground state are unique functionals of the density. The energy functional $E[\rho]$ has a minimum value for the exact density of the system. By minimizing the energy functional with respect to the density

$$\delta\{E[\rho] - \mu \int \rho(r)\,dr\} = 0$$ (7.31)

taking account of the condition

$$\int \rho(r)\,dr = A$$ (7.32)

by means of the Lagrange multiplier μ, one can obtain the density ρ and the binding energy E of the system.

Consequences of the *Hohenberg–Kohn* theorem concern the relation between both fundamental quantities, namely nucleon momentum and density distributions [7.21, 7.2]. As proved in the theorem the many-fermion wave function Ψ is a functional of the density ρ

$$\Psi \rightarrow \Psi(r_1, r_2, \ldots, r_A; [\rho]) \equiv \Psi[\rho].$$ (7.33)

It follows from (7.33) that the one-body density matrix

$$\rho(r, r') = \int \Psi^*(r, r_2, \ldots, [\rho])\Psi(r', r_2, \ldots, [\rho])\,dr_2 \ldots dr_A$$
$$\equiv \rho(r, r'; [\rho])$$ (7.34)

is also a functional of the density ρ (i.e. there is a functional relation between the nondiagonal and diagonal elements of the one-body density matrix $\rho(r, r')$). This result leads to the conclusion that the NMD

$$n(k) = \rho(k, k) = \int \exp[ik \cdot (r - r')]\rho(r, r'; [\rho])\,dr\,dr' \equiv n(k; [\rho])$$ (7.35)

is also a functional of the density ρ. It has been shown in [7.21, 7.2] that the functional is an unique one. As in the Hohenberg–Kohn-theorem there is no constructive way to build the functional $n(k; [\rho])$. Additional physical assumptions (such as those in the CDFM) are needed for this aim.

This consideration makes it possible to formulate the variational principle using both equivalent physical quantities (ρ and n) taking account of the functional relation between them. The following energy functional of ρ and n has been suggested in [7.21, 7.2]:

$$\tilde{E}[\rho, n] = E[\rho, n] - E_F \int \rho(r)\,dr - \int g(k)[n(k) - n(k; [\rho])]\frac{dk}{(2\pi)^3}, \quad (7.36)$$

where the second and the third terms in the right-hand side are related to the conditions:

$$\int dr\,\rho(r) = A \quad (7.37)$$

and

$$n(k) = n(k; [\rho]), \quad (7.38)$$

respectively.

The variational principle leads to the system of equations:

$$\frac{\delta \tilde{E}}{\delta \rho} = 0 \rightarrow \frac{\delta E}{\delta \rho} + \int \frac{dk}{(2\pi)^3} g(k) \frac{\delta n(k; [\rho])}{\delta \rho} = E_F, \quad (7.39)$$

$$\frac{\delta \tilde{E}}{\delta n} = 0 \rightarrow \frac{\delta E}{\delta n} = \frac{g(k)}{(2\pi)^3}, \quad (7.40)$$

$$\int dr\,\rho(r) = A, \quad (7.41)$$

$$n(k) = n(k; [\rho]). \quad (7.42)$$

Three functions ($\rho(r), n(k)$ and $g(k)$) and the Lagrange multiplier E_F which determines the particle separation energy are solutions of the system (7.39)–(7.42).

The proposed theoretical scheme is important both in the pure theoretical aspects and in physical applications. Since the NMD is a quantity sensitive to the nucleon–nucleon correlations at small distances, the energy functional $E[\rho, n]$ has to be determined in accord with the characteristic features of the nucleon–nucleon forces. This formulation of the theory in terms both of ρ and n is in principle more general than the approaches using only ρ (e.g. the Thomas–Fermi method) or only n (Landau–Fermi liquid theory) as a basic variable.

An example for a functional relation between n and ρ is given by *Hüfner* and *Nemes* [7.22] in the local Fermi-gas approximation:

$$n(k; [\rho]) = \int dr\,n(k, \rho), \quad (7.43)$$

where

$$n(k, \rho) = 4\theta[k_F(\rho(r)) - |k|].$$ (7.44)

The expression for the NMD obtained in the coherent density fluctuation model (CDFM) in finite nuclear systems with monotonic decreasing density distributions $\rho(r)$ (see Sect. 4.2) [7.23 and 7.24, 7.2]

$$n(k) = \left(\frac{4\pi}{3}\right)^2 \frac{4}{A} \left[6 \int_0^{\alpha/k} \rho(x)x^5 \, dx - (\alpha/k)^6 \rho(\alpha/k) \right] \equiv n(k; [\rho]),$$ (7.45)

$$\int n(k) \frac{dk}{(2\pi)^3} = A; \quad \alpha = (9\pi A/8)^{1/3}$$

is a more complicated example of a functional relation between n and ρ. The linear dependence on ρ in (7.45) leads to the uniqueness of the functional relation $n(k; [\rho])$. A very important property of the NMD (7.45) is the power-law asymptotic $n(k) \xrightarrow[k \to \infty]{} k^{-8}$ if the conditions $d\rho/dr|_{r=0} = 0$ and $d^2\rho/dr^2|_{r=0} \neq 0$ are fulfilled. This is in accord with the result of *Amado* and *Woloshyn* [7.26–7.28]. At $k \to 0$:

$$n(k \to 0)/A \simeq (128\pi^2/3A^2) \int_0^\infty \rho(x)x^5 \, dx.$$ (7.46)

This result is almost independent of the mass number A and its value is twice that in nuclear matter $(n(k \to 0)/A = 4/\rho_0)$.

We shall discuss in this section the results for NMD in various nuclei calculated both by independent-particle models and correlation methods and compared with the available experimental data. The aim is to study the influence of the N–N correlations at short-distances on the behaviour of the NMD. Qualitative arguments for the N–N correlation effects on the NMD are given by *Gottfried* [7.25]. It has been pointed out that the attractive long-range part of the N–N potential is responsible for the detailed form of $n(k)$ at momenta $k \leqslant k_F$ (k_F being the Fermi momentum), while the behaviour of $n(k)$ at large momenta $(k > k_F)$ depends on the effects of the core and the attractive part of the N–N force just outside it. This is the reason why the nuclear shell-model gives reasonable results for the low momentum $(k \leqslant k_F)$ part of the NMD. The average shell-model potential is rather smooth and the single-particle wave functions do not contain high-momentum components. In the shell model two nucleons, however, cannot acquire high momenta by interacting because it is impossible for them to come close enough to each other. This results in the nonrealistic predictions of the shell model for the high-momentum part $(k > k_F)$ of the NMD. This behaviour can be described only by including the N–N correlation effects at short distances.

It is of interest to illustrate the influence of the short-range part of the N–N forces on the nuclear ground state characteristics in a solvable, though rather schematic model of N bosons with a one-dimensional Hamiltonian and delta-function forces ($\hbar = 2m = 1$) [7.26–7.28]:

$$\hat{H} = -\sum_{i=1}^{N} \frac{\partial^2}{\partial x_i^2} - g \sum_{i,j=1}^{N} \delta(x_i - x_j).$$ (7.47)

The binding energy of the system and the corresponding wave function [7.29] are

$$E_N = -\frac{1}{48} g^2 N(N^2 - 1),$$ (7.48)

$$\Psi(x_1, x_2, \ldots, x_N) = M \exp\left(-\frac{1}{4} g \sum_{i<j=1} |x_i - x_j|\right).$$ (7.49)

The wave function (7.49) leads to the following explicit expressions for the density [7.30] and the form factor:

$$\rho(x) = \sum_{n=1}^{N-1} a_n \exp(-gnN|x|/2),$$ (7.50)

$$F(q) = gN \sum_{n=1}^{N-1} \frac{a_n n}{q^2 + (gnN/2)^2},$$ (7.51)

with

$$a_n = \frac{1}{2} \frac{g(-1)^{n+1} n(N!)^2}{(N+n-1)!(N-n-1)!}.$$ (7.52)

The exact expression for the NMD can be approximated successfully at $N \gg 1$ and $q^2 \leqslant \lambda^2/4$ ($\lambda = qN/2$) by the solution of the model in the Hartree-approximation [7.28]:

$$n(q) = [\cosh(\pi q/\lambda)]^{-2}.$$ (7.53)

We note the important result of *Amado* [7.26] concerning the power-law behaviour of $n(q)$ at high momenta:

$$n(q) \sim [\tilde{v}(q)]^2 q^{-4},$$ (7.54)

where $\tilde{v}(q)$ is the Fourier transform of the two-particle potential. For delta-function forces (7.54) becomes

$$n(q) \sim q^{-4}.$$ (7.55)

For large number of particles N and $q^2 \gg \lambda^2 \gg (q/N)^2$ the form factor has the form [7.26]:

$$F(q) = \frac{\pi q}{\lambda} [\sinh(\pi q/\lambda)]^{-1}$$ (7.56)

and coincides with its form in the Hartree-approximation [7.30]. The asymptotic value of the form factor at $q/N \gg 1$ [7.31]

$$F(q) \sim (\tilde{v}(q)/q^2)^{N-1} \tag{7.57}$$

differs significantly from the asymptotic behaviour of the NMD (7.54)–(7.55).

The present interest in the study of the NMD in nuclei has been stimulated by the experiments of *Frankel* et al. [7.32] on inclusive proton production at large angles in high energy (600 and 800 MeV) proton scattering on Be, C, Cu, Ta, Ag and Pt nuclei. *Amado* and *Woloshyn* [7.33] interpreted the particle production in the kinematically forbidden region in the elementary act of interaction between free particles in the framework of single-particle mechanism. According to this mechanism the incident proton scatters on a target nucleon with momentum k and the cross-section depends essentially on the NMD of the target nucleons $n(k)$. It has been shown that the experimental data can be described successfully using the phenomenological NMD suggested in [7.33]:

$$n_s(k) = N_c k \gamma_s / \sinh(\gamma_s k), \tag{7.58}$$

$$n_c(k) = N_c / \cosh^2(\gamma_c k), \tag{7.59}$$

where γ_s and γ_c are parameters and N_c the normalization constant. The distributions (7.58)–(7.59) decrease exponentially at large momenta ($n(k) \sim \exp(-\gamma k)$) and this behaviour at $k = 2$–$5 \, \text{fm}^{-1}$ turns out to be decisive for the description of the data. The high-momentum tails of the realistic NMD at $k > 2 \, \text{fm}^{-1}$ is in drastic contrast with the predictions of the independent-particle models. As was shown by *Zabolitzky* and *Ey* [7.34] this high-momentum behaviour of the NMD is strongly determined by the short-range nucleon–nucleon correlations which are not included, in principle, in the many-nucleon wave function in the form of a single Slater determinant. The results for the NMD in various nuclei obtained by different correlation methods are compared with the predictions of the independent-particle models and with available experimental data in the figures presented in this section.

In Figs. 7.1–7.3 the NMD in ^{4}He, ^{16}O and ^{40}Ca calculated in the generator coordinate method (GCM) [7.35] with square-well construction potential and harmonic oscillator construction potential (Sect. 4.1), in the CDFM (Sect. 4.2 and Eq. (7.45)) [7.23, 7.24, 7.36] for ^{16}O and ^{40}Ca, in the exp(S)-method [7.34] (Sect. 3.2) using the Reid soft-core and de Tourreil–Sprung supersoft core B nucleon–nucleon potentials are compared with the results of the shell model with pure harmonic oscillator potential for ^{4}He and with the Hartree–Fock calculations using wave functions from [7.37] for ^{16}O and ^{40}Ca. The experimental data for $n(k)$ in ^{4}He extracted from the inclusive cross-section for ^{4}He(e, e') reaction up to momentum $k \simeq 3 \, \text{fm}^{-1}$ in [7.38] are also presented in Fig. 7.1. In the case of ^{40}Ca the results of *Dellagiacoma, Traini* and *Orlandini* [7.39, 7.40] from the two-body correlation operator method (see Sect. 3.3), as well as those of *Benhar* et al. [7.41] from the Jastrow-type correlation method are given in Fig. 7.3. The main characteristic feature of the NMD

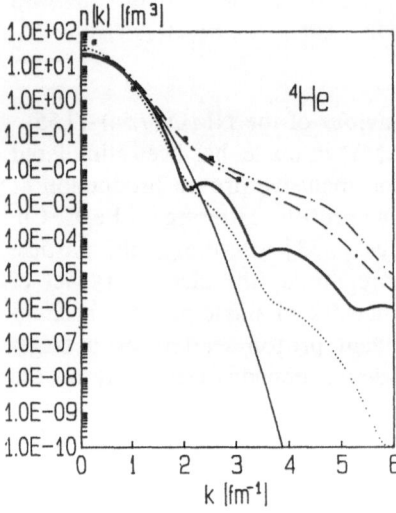

Fig. 7.1. Nucleon momentum distribution for ^{4}He calculated using the GCM with square-well construction potential (thick solid line), the GCM with a harmonic oscillator construction potential (dotted line) [7.35], using a pure harmonic oscillator wave functions (thin solid line), the exp (S)-method with the De Tourreil–Sprung supersoft core B (dash-one-dot) and the Reid soft core (dash-two-dots) N–N potentials [7.34]. Experimental data (black squares) from [7.38]. Normalization: $\int n(k)k^2 \, dk = A$

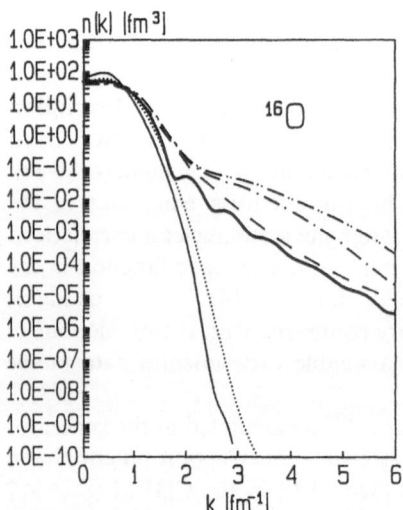

Fig. 7.2. Nucleon momentum distribution for ^{16}O calculated using the GCM with a square-well construction potential (thick solid line), the GCM with harmonic oscillator construction potential (dotted line) [7.35], the CDFM (dashed line) [7.23, 7.24], Hartree–Fock calculations using wave functions from [7.37] (thin solid line), the exp (S)-method with the De Tourreil-Sprung supersoft core B (dash-one-dot) and the Reid soft core (dash-two-dots) potentials [7.34]. Normalization as in Fig. 7.1

obtained by the correlation methods (namely the exp(S)-method, CDFM, GCM with square-well construction potential, methods from [7.39, 7.40] and [7.41]) is the existence of high-momentum components at momenta $k > 2 \, \text{fm}^{-1}$ which shows evidence that effective account has been taken of nucleon–nucleon correlations in these methods. The high-momentum components in the GCM with a square-well construction potential with infinite walls are due to the existence of intermediate states (i.e. states with a given value of the generator coordinate which is the radius of the well in this case) with high densities. This means that in

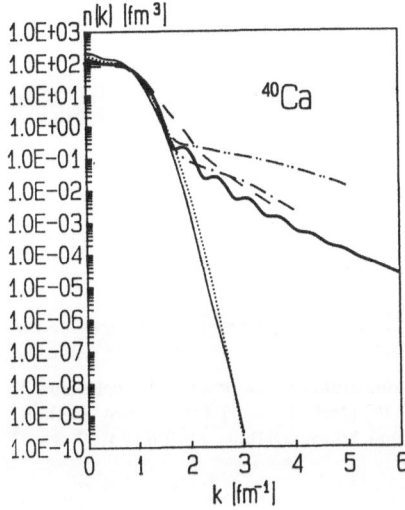

Fig. 7.3. Nucleon momentum distribution for ^{40}Ca calculated using the GCM with a square-well construction potential (thick solid line), the GCM with a harmonic oscillator construction potential (dotted line) [7.35], the CDFM (dashed line) [7.23, 7.24], Hartree–Fock calculations using wave functions from [7.37] (thin solid line), the two-body correlating operator method (dash-one-dot) [7.39, 7.40] and the correlation method [7.41] (dash-two-dots). Normalization as in Fig. 7.1

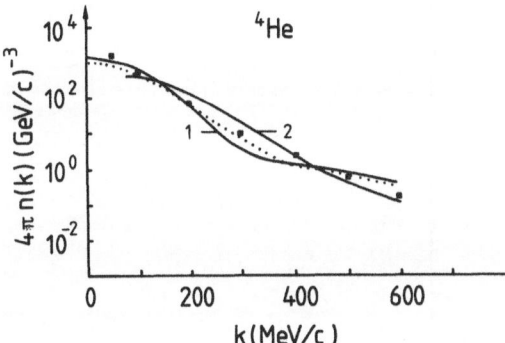

Fig. 7.4. Nucleon momentum distribution for ^{4}He: the black squares are the experimental data [7.38], the exp (S)-method [7.34] (dotted line), the correlation method from [7.42] (curve 1) and the CDFM [7.36] (curve 2). Normalization: $\int n(k)\,dk = 1$

these states the distances between the nucleons are small and short-range forces are operative. As the CDFM is related to this approach in the delta-function limit, the results in it are similar to the GCM ones. In the case of GCM with harmonic oscillator construction potential the high-density intermediate states are not possible and the NMD do not show high-momentum behaviour.

It can be seen from Fig. 7.4 that the results of the CDFM [7.36], the correlation method of *Akaishi* [7.42] and the exp(S)-method [7.34] are in good agreement with the experimental data for ^{4}He [7.38]. The result of the CDFM [7.36] is compared in Fig. 7.5 with the Jastrow method calculations of *Bohigas* and *Stringari* [7.43] for the NMD in ^{4}He.

The results of the variational Jastrow method (Sect. 3.3) using a pair-correlation operator containing central, tensor and spin correlations [7.47a] for the NMD in nuclei with mass number $A = 2, 3$ and 4 for nuclear matter are presented in Fig. 7.6. It is noted that these NMD $(n(k)/A)$ increase with A at

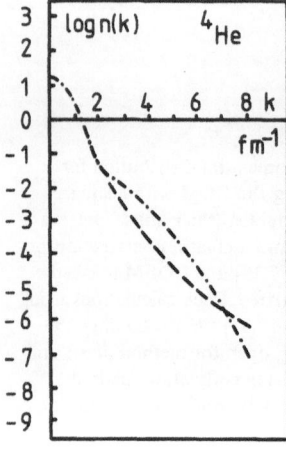

Fig. 7.5. Nucleon momentum distribution for ^{4}He calculated using the CDFM [7.36] (dashed line) and the Jastrow method [7.43] (dot-dashed line). Normalization as in Fig. 7.1

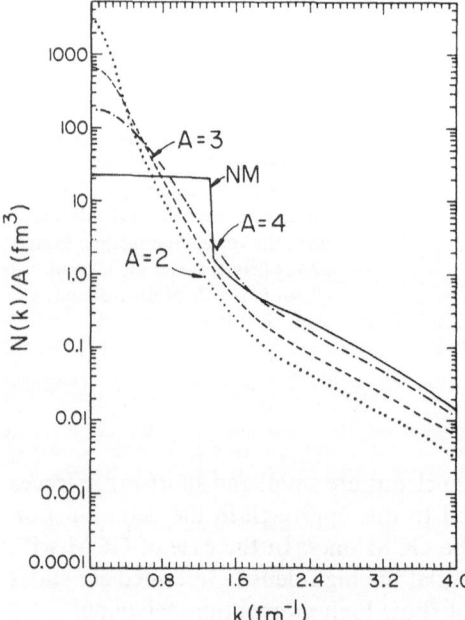

Fig. 7.6. The nucleon momentum distributions $N(k)/A$ for $A = 2, 3$ and 4 nuclei and nuclear matter, calculated by *Schiavilla* et al. [7.47a]

$k > 2.5\ \mathrm{fm}^{-1}$, which is in contrast with the calculations of *Traini* and *Orlandini* [7.40]. The charge form factors of the three- and four-body nuclei have been calculated by the variational Jastrow mehod in [7.47b]. Improved variational calculations using central, spin, isospin, tensor, spin–orbit and triplet-correlation operators have been carried out by *Wiringa* [7.47c] for the binding energies, density distributions and elastic form factors in ^{3}He and ^{4}He nuclei. The ground state characteristics of ^{16}O are calculated in [7.47d–e].

It can be seen from Fig. 7.7 that the results of the correlation methods, such as CDFM [7.23, 7.24] and the method of *Benhar* et al. [7.41] are also in accord with the experimental data for NMD in ^{12}C including the momenta $k > 2$ fm^{-1} [7.44–7.46], which is not the case for the results of Hartree–Fock calculations.

The phenomenological forms of the NMD (7.58) and (7.59) suggested by *Amado* and *Woloshyn* [7.33] which are close to the CDFM results for $n(k)$ in ^{12}C for large momentum interval, have been applied successfully by *Hatch* and *Koonin* [7.48] to the description of proton and pion spectra from high-energy ion–ion scattering (800 MeV per nucleon) in the reactions C + C, C + Pb, Ne + NaF assuming the mechanism in which a nucleon from the incident nucleus strikes a target nucleon.

In the work of *Di Toro* [7.49] the CDFM result for NMD in ^{12}C [7.23, 7.24, 7.36] describes satisfactorily the sub-barrier neutral pion production in the ^{12}C + ^{12}C reaction. As can be seen from Fig. 7.8 the cross-sections obtained by means of the shell-model NMD (using the harmonic oscillator potential), as well as from the $n(k)$ of the work of *Benhar* et al. [7.41] turn out to be unrealistic.

The NMD in ^{12}C obtained in the CDFM [7.23, 7.24, 7.36] and in [7.33] have been used by *Zulkarneev* et al. [7.50] to calculate the angular dependence of the polarization of protons emitted in the process p + ^{12}C → p + X at

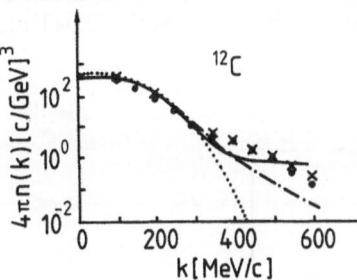

Fig. 7.7. Nucleon momentum distributions in ^{12}C. The points and crosses are the experimental data from [7.44–46], and the curves show the result of calculations using the correlation method [7.41] (solid line); the CDFM [7.23, 7.24] (dash-dotted line) and the Hartree–Fock method (dotted line). Normalization is as in Fig. 7.4

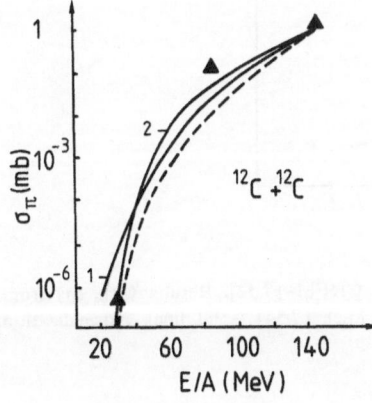

Fig. 7.8. Sub-barrier π^0-production in ^{12}C + ^{12}C reaction [7.49]. Experimental data (line 1); calculations with $n(k)$: from CDFM [7.23, 7.24, 7.36] (line 2), from the shell model (dashed line), from [7.41] (▲)

640 MeV. The results are in a qualitative agreement with the experimental data for the angular interval 60°–140°.

It is shown in [7.51] that a realistic NMD of the type suggested in [7.33] (7.59) can be used for the successful description of processes in which protons, neutrons, deuterons and alpha-particles are emitted when μ^- are captured by various nuclei ($23 \leqslant A \leqslant 209$).

The CDFM has been used [7.52] to calculate the proton and neutron momentum distributions:

$$n_a(k) = \left(\frac{4\pi}{3}\right)^2 \frac{1}{a}\left[6 \int\limits_0^{\alpha_a/k} \rho_a(x)x^5 \, dx - (\alpha_a/k)^6 \rho_a(\alpha_a/k)\right],$$

(7.60)

$$\alpha_a = \left(\frac{9\pi a}{4}\right)^{1/3}, \quad 2\int n_a(k)\frac{dk}{(2\pi)^3} = a, \quad (a = Z, N),$$

in the case of nuclei with monotonically-decreasing proton and neutron density distributions. The nuclei ^{39}K, ^{40}Ca and ^{48}Ca have been considered in [7.52]. The results for the proton momentum distribution in ^{40}Ca using the charge densities (types I and II) of *Sick* [7.53] obtained by a model-independent analysis of electron–nucleus scattering and muonic atoms, as well as densities obtained from the method of the single-particle potentials (SPP) [7.54] are presented in Fig. 7.9.

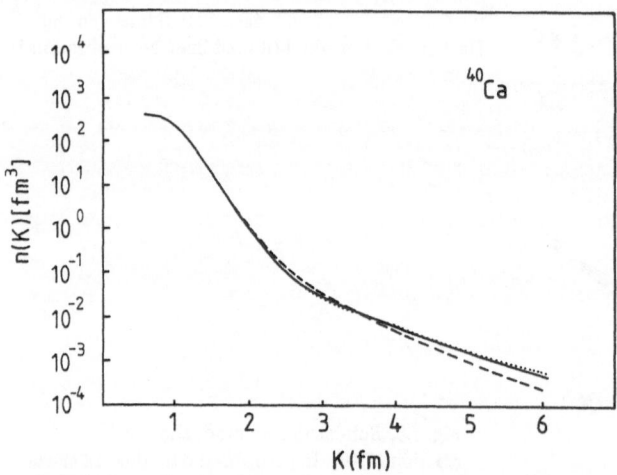

Fig. 7.9. Proton momentum distribution in ^{40}Ca in CDFM [7.52]. Results with $\rho(r)$ from: [7.53]-type I (dashed line), [7.53] - type II (dotted line), [7.54] (solid line). Normalization:
$$2\int n_Z(k)\frac{dk}{(2\pi)^3} = Z$$

The method of natural orbitals developed within the SPP method (Sect. 6.2) has been used in calculations of the proton momentum distribution in ^{40}Ca [7.55]. The natural orbitals (single-particle functions) (6.18) are included in the calculations of the proton momentum distribution using three different sets of occupation numbers: (i) $\tilde{n}_i = 1$ below the Fermi level and $\tilde{n}_i = 0$ above it (i.e. the total wave function is a single Slater determinant constructed by natural orbitals); (ii) a case with 3.55% of protons above the Fermi level and (iii) a case with 20.5% of protons above the Fermi level.

The results of the calculations of $n_p(k)$ in ^{40}Ca are presented in Fig. 7.10 and compared with: (i) $n_p(k)$ from [7.40] including short-range and tensor correlations; (ii) $n_p(k)$ from CDFM [7.52]; (iii) $n_p(k)$ from the variational approach in [7.41]. It can be noted that the distributions $n_p(k)$ from the natural orbital method contain larger high-momentum components than those from the harmonic oscillator shell model and from [7.40] with inclusion of tensor correlations only. This is due partially to the depletion of the Fermi sea as can be seen from the difference between curves 3), 4) and 5) in Fig. 7.10. The differences in the forms of the single-particle wave functions have, however, a much larger effect. The curve 3) corresponding to a Slater determinant wave function built up with natural orbitals is quite different from the result in the shell model and in [7.40]

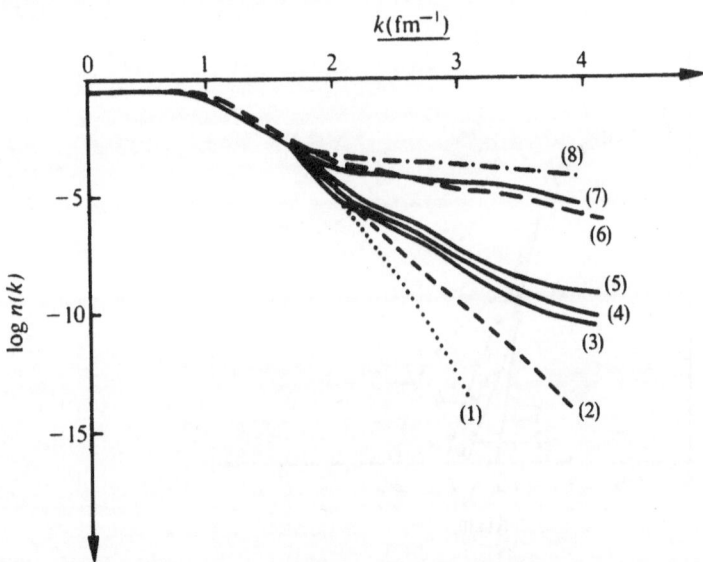

Fig. 7.10. Proton momentum distribution in ^{40}Ca. Results of: i) the shell model with harmonic oscillator potential [7.40] (curve 1); ii) the model from [7.40] including tensor (curve 2) and tensor-plus short-range correlations (curve 7); iii) the method of natural orbitals [7.55] realized in the single-particle potential model [7.54]: Slater determinant (curve 3), with 3.55% of protons above the Fermi sea (curve 4), with 20.5% of protons above the Fermi sea (curve 5); iv) the CDFM [7.52] (curve 6); v) the variational approach [7.41] (curve 8). Normalization $\int n_p(k)\,dk = 1$

with tensor correlations included. This shows that the natural orbitals contain the essential part of the short-range nucleon–nucleon correlations.

The results of calculations of $n_p(k)/Z$ in ^{40}Ca [7.55] are compared in Fig. 7.11 with those for $n_p(k)/Z$ in ^{208}Pb within the natural orbital method using different sets of occupation numbers [7.56]: (i) from the Hartree–Fock method with $\tilde{n} = 1$ up to $3s_{1/2}$ and $\tilde{n} = 0$ above this state; (ii) from RPA calculations where 3.5% of the protons are above the Fermi level [7.57]; (iii) from RPA [7.57] plus the calculated occupation numbers in nuclear matter with an account for short-range and tensor correlations [7.58] (20.5% of the protons are above the Fermi level) and (iv) with 11.5% of protons above the Fermi level. Comparison of the results leads to the conclusion that the occupation numbers have little effect on the SPP proton momentum distribution, which in the case of ^{40}Ca is between the Hartree–Fock result and the model with 11.6% of protons above the Fermi sea.

It is important to show the effects of the nucleon–nucleon correlations not only on the NMD but also on the density distribution. In the case of ^{40}Ca this is shown in Fig. 7.12. It turns out that the main effect of the short-range cor-relaions is a decrease of the central part of the density. These changes corres-

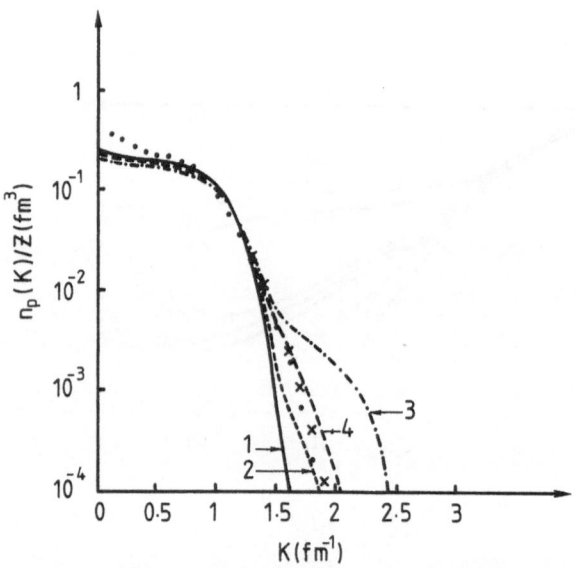

Fig. 7.11. Proton momentum distributions $n_p(k)/Z$ in ^{40}Ca from [7.55] with 3.55% of protons above the Fermi sea (••••) and with 20.6% of protons above the Fermi sea (x x x) compared with $n_p(k)/Z$ in ^{208}Pb with different sets of occupation numbers [7.56]: i) the Hartree–Fock method ($\tilde{n} = 1$ up to $3s_{1/2}$ state and $\tilde{n} = 0$ above it) (curve 1); ii) from RPA calculations [7.57] (3.5% of protons above the Fermi level) (curve 2); iii) from RPA + nuclear matter calculations including short-range and tensor correlations [7.58] (20.5% of protons above the Fermi level) (curve 3); iv) 11.6% of protons above the Fermi level (curve 4). Normalization: $\int n_p(k)\,dk = Z$

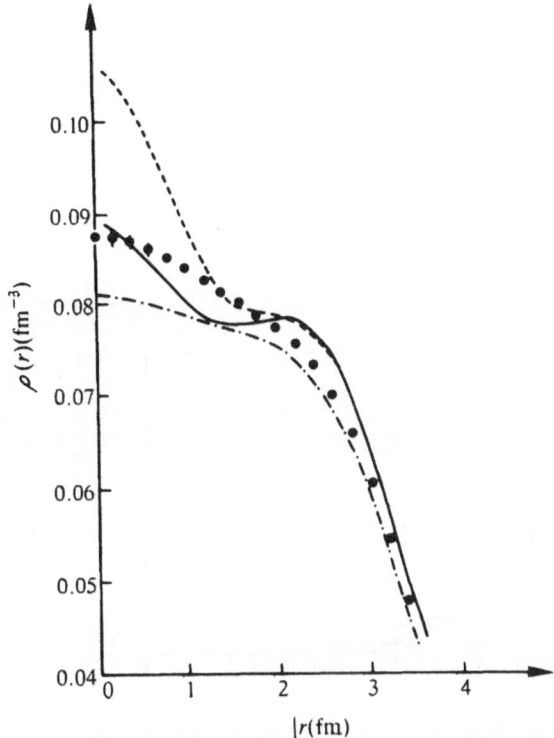

Fig. 7.12. Charge density distribution in ^{40}Ca. The experimental data (points) are from [7.53, 7.54] (Sick, type I). Results of calculations in [7.55]: i) Slater determinant constructed with natural orbitals (dashed line); ii) with 3.55% of protons above the Fermi level (solid line); iii) with 20.5% of protons above the Fermi level (dot-dashed line). Normalization: $\int \rho(r)\,dr = Z$

pond to the high-momentum behaviour of the NMD. The comparison between the correlation effects on the density and on the NMD shows that the latter are much more sensitive to the nucleon–nucleon correlations.

The effects on the proton density distributions in ^{4}He, ^{16}O and ^{40}Ca of the nucleon–nucleon correlations included within the generator coordinate method with different construction potentials are shown in Fig. 7.13 [7.35]. It should be noted that the resulting proton density distributions have the correct asymptotic behaviour.

The CDFM has been extended to the study of the NMD in nuclei whose density distributions $\rho(r)$ are not monotonically decreasing functions of r [7.36, 7.59]. This is the case with the experimental density distribution of the ^{208}Pb nucleus [7.60]. The CDFM calculations of the proton momentum distribution $n_p(k)/Z$ in ^{208}Pb [7.59] using the experimental charge distribution [7.60] are compared in Fig. 7.14 with the proton momentum distribution from [7.56, 7.61] and with the experimental data extracted from the inclusive electron (640 MeV)

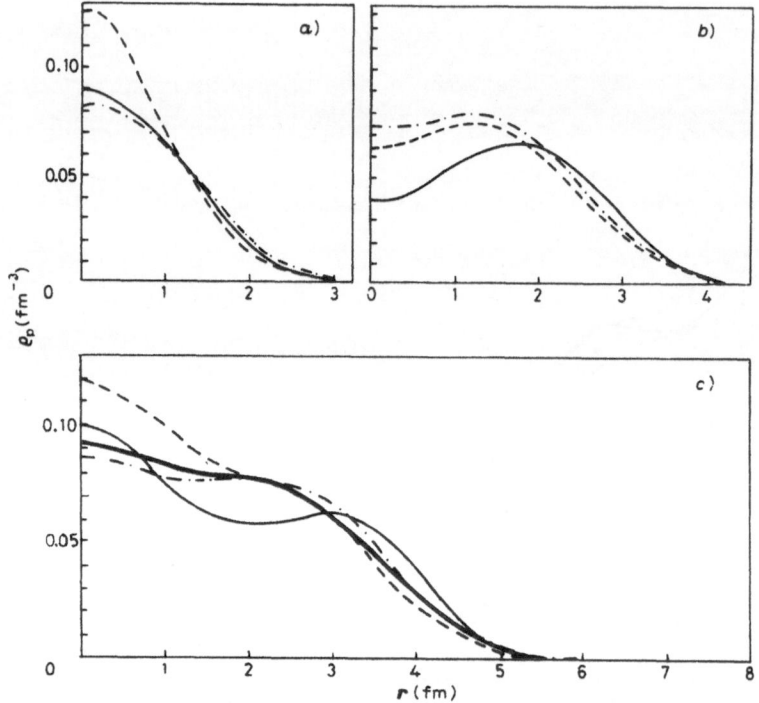

Fig. 7.13. Proton density distributions in ^{4}He(a), ^{16}O(b) and ^{40}Ca(c). Results of: GCM with square-well construction potential (thin solid line) and with harmonic oscillator construction potential (dashed line) [7.35]; Hartree–Fock calculations (dot-dashed line). In the case of ^{40}Ca the experimental density of *Sick* [7.53, 7.54] is shown (thick solid line)

and proton (200 MeV) scattering on ^{12}C and ^{58}Ni [7.62, 7.63]. It can be noted that the CDFM result describes satisfactorily the empirical data in the region $1.5 < k < 2.2 \, \text{fm}^{-1}$. This result is in accord with the calculations in [7.56, 7.61] in the case where 11.5% of protons are above the Fermi level and differs from the case with 20.5% of protons above the Fermi level as well as from the Hartree–Fock result.

The proton momentum distributions in ^{12}C and ^{90}Zr have been calculated within the CDFM using the experimental non-monotonic charge density distributions in [7.36].

The study of the relation between the NMD and the density distribution enables one to obtain additional information on the role of the short-range nucleon–nucleon correlations in nuclei. This relation is studied in [7.64] on the basis of the integral representation of both quantities:

$$\rho(r) = \int\limits_0^\infty dx \, w_\rho(x) (\tfrac{4}{3}\pi x^3)^{-1} \theta(x - |r|) \tag{7.61}$$

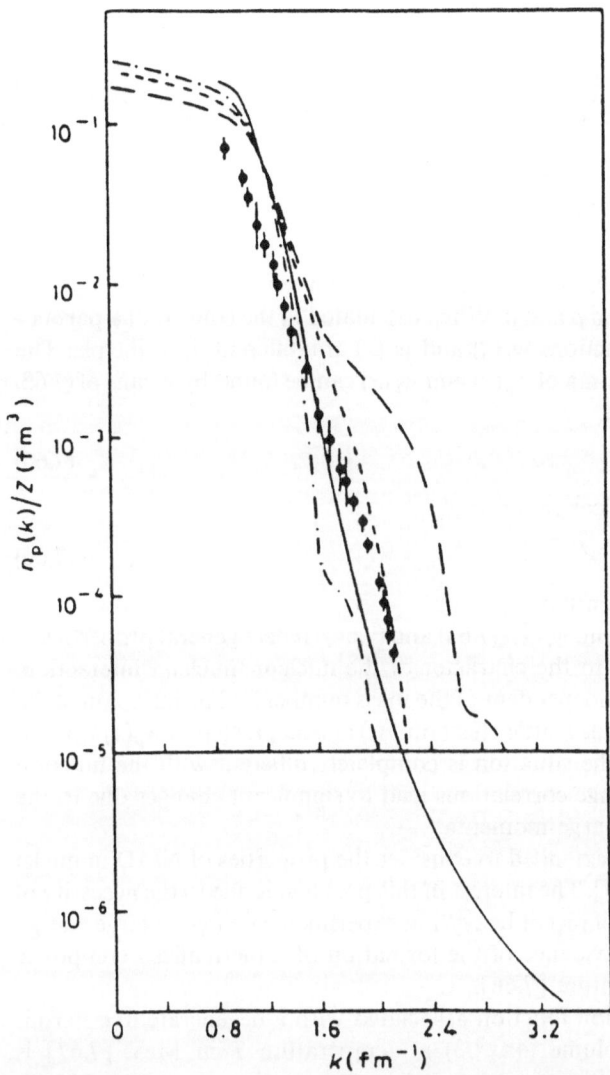

Fig. 7.14. Proton momentum distribution $n_p(k)/Z$ in ^{208}Pb. Results from: i) CDFM [7.59] using experimental charge distribution from [7.60] (solid line); ii) [7.56, 7.61] : Hartree–Fock (dot-dashed line); 20.5% of protons above the Fermi level (long-dashed line); 11.6% of protons above the Fermi level (short-dashed line). Experimental data (points) are taken from [7.61]. Normalization: $\int n_p(k)\,dk = Z$

and

$$n(k) = \frac{1}{(2\pi)^3} \int\limits_0^\infty dx\, w_n(x) \frac{4}{3}\pi x^3 \theta\left(\frac{\alpha}{x} - |k|\right) \qquad (7.62)$$

with weight functions

$$w_\rho(x) = -\frac{4}{3}\pi x^3 \frac{d\rho}{\rho r}\Big|_{r=x} \tag{7.63}$$

and

$$w_n(x) = -(2\pi)^3 \frac{3\alpha}{4\pi x^5} \frac{dn}{dk}\Big|_{k=\alpha/x} \tag{7.64}$$

$(\alpha = (9\pi/2)^{1/3})$.

The comparison between $w_\rho(x)$ and $w_n(x)$ can give information on the functional relation between ρ and n. When calculated in the context of a particular nuclear model the functions $w_\rho(x)$ and $w_n(x)$ are different, in principle. The positions of the main maxima of $w_\rho(x)$ and $w_n(x)$ can be found by means of (7.63, 7.64) to be respectively at:

$$R_A \simeq (A/4)^{1/3} r_0, \tag{7.65}$$

which is close to the nuclear radius and

$$r_0 \simeq (9\pi/2)^{1/3}/k_F, \tag{7.66}$$

k_F being the Fermi momentum.

The maxima of $w_\rho(x)$ and $w_n(x)$ ((7.65) and (7.66)) reflect general properties of nuclear structure related to the character of the nucleon–nucleon interaction. The value of r_0 is almost independent of the mass number A. The inclusion of the short-range correlations has a little effect on $\rho(r)$ at small r, so that $w_\rho(x)$ will not be changed essentially. The situation is completely different with the function $w_n(x)$, where the short-range correlations lead to significant changes due to the changes of the NMD at large momenta.

The CDFM has been extended to consider the properties of NMD in nuclei at finite temperature [7.65]. The interest in this problem is due to the necessity of a theoretical description of recent heavy-ion experiments involving large energy transfers which provide evidence of the formation of a thermalized composite system with high temperature [7.66].

The Wigner distribution function associated with a degenerate free Fermi-gas of nucleons in a volume $(4\pi x^3/3)$ at temperature T in MeV [7.67] is considered:

$$W_x(r, k, T) = \frac{4\theta(x-r)}{1 + \exp[(p^2/2M - \mu(x, T))T^{-1}]}, \tag{7.67}$$

where the chemical potential $\mu(x, T)$ has to be determined at any x and T by the normalization condition:

$$\rho_0(x, T) = \frac{2}{\pi^2} \int_0^\infty \frac{k^2\, dk}{1 + \exp[(\hbar^2 k^2/2M - \mu(x, T))T^{-1}]} = \frac{3A}{4\pi x^3}. \tag{7.68}$$

If the Wigner function (7.67) is introduced instead of the CDFM zero temperature function:

$$W_x(r, k) = 4\theta(k_F(x) - k)\theta(x - r),$$ (7.69)

then the generalized CDFM ansatz for $T \neq 0$ is:

$$W_{\text{CDFM}}(r, k, T) = \int_0^\infty |f(x, T)|^2 W_x(r, k, T)\, dx$$ (7.70)

and the expressions for the density $\rho(r, T)$ and the NMD follow from (7.16) and (7.17):

$$\rho(r, T) = \int_0^\infty |f(x, T)|^2 \rho(x)\theta(x - r)\, dx$$ (7.71)

and

$$n(k, T) = 4 \int_0^\infty |f(x, T)|^2 \frac{(4\pi x^3/3)\, dx}{1 + \exp[(\hbar^2 k^2/2M - \mu(x, T))T^{-1}]}.$$ (7.72)

As in the case $T = 0$, the weight function $f(x, T)$ can be expressed by the local density $\rho(r, T)$ as:

$$|f(x, T)|^2 = -\frac{1}{\rho_0(x)} \frac{d\rho(r, T)}{dr}\bigg|_{r=x}.$$ (7.73)

Substituting (7.73) into (7.72) one can obtain the relation between $n(k, T)$ and $\rho(r, T)$:

$$n(k, T) = \int_0^\infty \frac{d\rho(x, T)}{dx} \frac{4A}{\rho_0^2(x)} \frac{dx}{1 + \exp[(\hbar^2 k^2/2M - \mu(x, T))T^{-1}]},$$ (7.74)

which holds for a given temperature T and monotonically decreasing local densities $\rho(r, T)$.

This extension of the CDFM reproduces in the limit $T \to 0$ all the ground-state CDFM results at zero temperature.

The NMD $n(k, T)$ calculated for ^{208}Pb [7.65] using the symmetrized Fermi density distribution obtained with SkM forces [7.68]:

$$\rho_{\text{SF}}(r, T) = \rho_0(T)[\cosh[R(T)/b(T)] + \cosh[r/b(T)]]^{-1}$$ (7.75)

is presented in Fig. 7.15 for temperatures from 0 to 10 MeV. It is seen that the temperature dependence does not destroy the two different slopes of the NMD curves which are known from the $T = 0$ case. The flat interior of $n(k, T)$ $(0 < k < 1-2 \text{ fm}^{-1})$, where the Pauli correlations are mainly important is almost temperature-independent. On the other hand, the outer region $(k > 1.35 \text{ fm}^{-1})$, where the short-range and tensor correlations are dominant, is strongly affected by the temperature and the high-momentum distribution tail

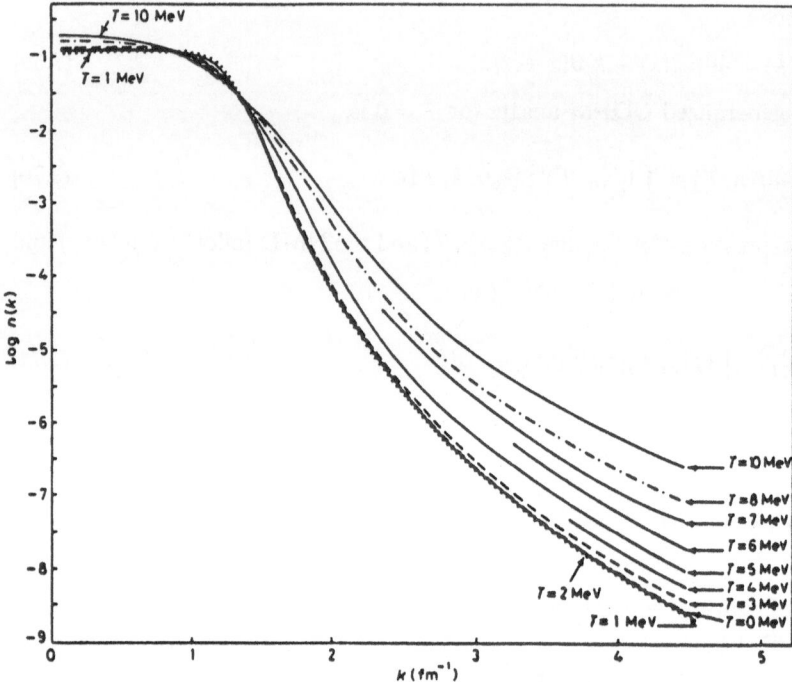

Fig. 7.15. The nucleon momentum distribution $n(k, T)$ calculated for ^{208}Pb (SkM forces) using the local density $\rho_{SF}(r,T)$ (7.75) at various temperatures [7.65]

increases with temperature above $T = 3$ MeV. It has been shown in [7.65] that the temperature dependence of $n(k, T)$ coming from the temperature dependence of the local density $\rho(r, T)$ dominates in the high-momentum region of $n(k, T)$.

Investigations of the temperature dependence of the nuclear hole-state spectral functions within the CDFM are also made in [7.65].

It is known that the finite size effects are important for the NMD in the interior region ($k < 1$ fm^{-1}). This is confirmed by the mean-field calculations using the density-dependent Hartree–Fock method (DDHF) of *Casas* et al. [7.69] which show quite different behaviour of the NMD for various heavy nuclei (^{90}Zr, ^{120}Sn and ^{208}Pb) for $k < 1$ fm^{-1} and a striking difference from the result for the infinite system. Just opposite is the behaviour of the local density distribution which is almost constant in the interior of the nucleus and is close to that of infinite nuclear matter. It is shown in [7.69] that the DDHF results for $n(k)$ can be reproduced well in the domain $0.5 < k < 1$ fm^{-1} if calculations with different forms of $\rho(r)$ (trapezoidal, cubic trapezoidal and diffuse Fermi density) are used in the local Fermi-gas approximation for finite nuclei ((7.43) and (7.44)). In this momentum region the NMD differ substantially.

The calculations of the exact momentum distributions in the spherical nuclei performed with different mean-field models by *Casas* et al. [7.70] show an almost universal shape of the NMD with little dependence on the effective interactions used.

The finiteness of the potential and its effect on the NMD have been studied by *Krivine* [7.71], who showed that the effects due to the finite size of the system are comparable with those due to the finiteness of the nucleons.

It has been shown by *Zverev* and *Saperstein* [7.72] that the NMD which contain a quasiparticle part (calculated within the Lagrangian quasiparticle method [7.73] for finite nuclei) and a regular tail insensitive to the finite-size effects are close to the predictions of the DDHF method for small momenta $(k < 1 \text{ fm}^{-1})$.

Another important aspect of the problem of NMD is their behaviour in deformed nuclei. This question is analyzed in [7.74a, b] using a phenomenological one-body Hamiltonian (the Nilsson Hamiltonian) as well as the density-dependent Hartree–Fock method [7.74c] for the Ne and Nd isotopes. It was found that the overall momentum distributions are not very sensitive to the value of the deformation parameter δ. The increase of δ leads to an increase of the NMD in the internal region $(0 \leqslant k \leqslant 1)$. The differences of the NMD are of the same order of magnitude as those found in [7.70] for various spherical nuclei. The effects of the deformation in the coordinate space are larger on the momentum distributions in different single-particle states. They are particularly strong for $k \approx 0$ and significant for momenta k below 1 fm^{-1}. It is concluded in [7.74] that the deformation effects can be observed in exclusive $(e, e'p)$ experiments which single out particular single-particle states, but not in inclusive (e, e') experiments in which these effects become small after averaging over all occupied orbitals. The theoretical predictions from [7.74] are confirmed in the $(e, e'p)$ experiments [7.75] on the spherical ^{142}Nd and non-spherical ^{146}Nd nuclei.

The results for the density and momentum distributions in a one-dimensional nuclear model with a two-nucleon potential with a repulsive core and a long-range attraction are obtained by Monte Carlo calculations for systems of 3, 4 and 5 particles and a non-interacting Fermi gas [7.76] and [7.77], respectively.

Correlated charge form factors of light nuclei $(4 \leqslant A \leqslant 16)$ are calculated in the cluster expansion within the Jastrow correlation method by *Massen* and *Panos* [7.78]. The results reproduce well the experimental charge form factors over quite a wide range of momentum transfer values.

A theoretical study of the ^{4}He NMD has been performed by *Grypeos* and *Ypsilantis* [7.79] in the framework of a single-particle model with a phenomenological mean-field potential containing both an attractive $(\sim r^2)$ and a repulsive $(\sim r^{-2})$ part. The analytical form of $n(k)$ they obtained shows high-momentum components, due to the presence of the repulsive mean-field part, which are close to those obtained by the correlation methods going beyond the independent-particle approximation. The NMD obtained has a power-law fall-off $(\sim k^{-7})$ at large momenta.

The calculations using the correlations ATMS (Amalgamation of Two-body correlation into the Multiple Scattering process) method from [7.42] have been continued to obtain a realistic momentum distribution in the case of ^{4}He by *Morita* et al. [7.80] and realistic spectral functions by *Tadokoro* et al. [7.81] using the Reid soft-core V_8 potential and the supersoft core nucleon–nucleon potential. Single-nucleon, two-nucleon-cluster relative motion, two-nucleon-cluster centre-of-mass, triton–proton and deuteron–deuteron relative motion momentum distributions have been studied [7.80]. The realistic NMD has a shoulder-like shape at $k = 2.5 \, \text{fm}^{-1}$ which comes from the short-range correlations.

A detailed review of the results obtained for the characteristics of ^{4}He within the ATMS method [7.42, 7.82], the exp(S)-method [7.34], the variational Monte Carlo method [7.83] as well as the hyperspherical harmonic expansion method [7.84, 7.85] is given by *Akaishi* [7.86, 7.87]. Evidence is found for marked high-momentum components around $2 \, \text{fm}^{-1}$ in the NMD, due to the short-range and D-state correlations. It is pointed out that realistic four-nucleon systems could soon be treated as accurately as three-nucleon systems.

Realistic calculations of ^{4}He density functions using the correlation ATMS method with the Reid V_8 potential have been made by *Morita* et al. [7.88]. They show that the one-body density distribution $\rho(r)$ has very high density in the central region in spite of the short-range repulsion of the realistic nucleon–nucleon force and that the D-state distribution is more compact than the S-state one.

The local density approximation to the NMD ((7.43) and (7.44)) [7.22] has been developed to study $n(k)$ for light and heavy nuclei by *Stringari* et al. [7.89]. The NMD is written in the form

$$n^A(k) = n_{\text{SD}}^A(k) + \delta n^A(k) \equiv n_{\text{SD}}^A(k) + \frac{1}{(2\pi)^3} \int \delta v(k_{\text{F}}^A(r), k) \, \mathrm{d}r, \qquad (7.76)$$

where the mean-field part is included in $n_{\text{SD}}^A(k)$ and $\delta n^A(k)$ includes the quantity $\delta v(k_{\text{F}}^A(r), k)$ which is calculated for nuclear matter and can be obtained from calculations with realistic nucleon–nucleon forces. More simple estimates of this quantity are made on the basis of the Lowest Order Cluster (LOC) approximation [7.90a] which enables effects of dynamical correlations to be included. The calculations of the NMD for ^{4}He, ^{16}O, ^{40}Ca and nuclear matter including short-range correlation effects are in agreement with the results of microscopic calculations for nuclear matter [7.90b], for ^{16}O and ^{40}Ca from [7.41] and for ^{4}He from [7.47]. They also provide a systematic description of the NMD for heavier nuclei such as ^{90}Zr and ^{208}Pb.

The properties of the nuclear matter using a separable representation of the Paris potential have been studied by *Baldo* et al. [7.91]. It is shown that the nucleon momentum distribution $n(k)$ for $0 < k < 2 \, \text{fm}^{-1}$ depends mainly on the ratio k/k_{F}, while at $2 < k < 4.5 \, \text{fm}^{-1}$ it can be reproduced by $1/7k_{\text{F}}^5 \exp(-1.6k)$ (k and k_{F} in fm^{-1}). The contributions to the mass operator up to $k = 4.5 \, \text{fm}^{-1}$ and for $k_{\text{F}} = 1.10, 1.36, 1.55$ and $1.75 \, \text{fm}^{-1}$ are determined by evaluating the

BHF-, the second order- and one of the third order terms in the reaction matrix. The average kinetic energy per nucleon for the correlated ground state (T) is larger than for the non-correlated system (T_0) $(T/T_0 \approx 2$ for all k_F considered).

A semiclassical method has been used by *Viñas* and *Guirao* [7.92] for calculations of density and momentum distributions, as well as the binding energy of nuclei at zero and finite temperatures. The nucleons are assumed to be an ensemble of fermions moving in a harmonic oscillator potential. The results are obtained using the Bloch density and its Wigner–Kirkwood expansion in powers of $\hbar^2$. It is shown that at finite temperature, when shell-effects are washed out, the semiclassical results can be used instead of the quantum mechanical ones.

Now we discuss briefly the problems of extracting the NMD from the experimental data obtained in particle–nucleus reactions. It has to be emphasized that the information of NMD strongly depends on the suggested reaction mechanism. For instance, the use of the mechanism proposed by *Weber* and *Miller* [7.93] in which the incident particle scatters on a group of $A - 1$ nucleons, and of realistic form of $n(k)$ with high-momentum components, leads also (as the single-particle mechanism from [7.33]) to a satisfactory description of the inclusive proton production in the experiments of *Frankel* et al. [7.32] on proton–nuclei scattering.

Other methods describing the processes of inclusive particle production in high-energy proton–nucleus scattering are those of [7.94, 7.95] that assume that the incident particle strikes clusters of several nucleons $(k = 2, 3, 4, \ldots)$ [7.94] or "correlated clusters" [7.95].

The extraction of the information on the NMD from nuclear reaction experimental data is complicated essentially by the effects of the final state interactions (FSI) as well as of the nucleon resonance excitations. It has been shown by *Amado* and *Woloshyn* [7.96] that the data from the experiments with high momentum transfers cannot be interpreted unambiguously by means of the NMD of the target nucleons. These problems are studied by [7.97–100]. It is noted by *Boffi* et al. [7.100] that the lack of orthogonality between the final states in the continuum and the initial bound states is decisive when the transferred momentum is small in comparison with the outgoing proton momentum. This effect is practically absent for $(e, e'p)$ processes but is essential for (γ, p) reactions at photon energies about 100 MeV.

The problem of experimental determination of the NMD has been considered by *Frankel* in [7.101] where it has been shown that the relation of the cross-section to the NMD can be expressed in the form:

$$\frac{d\sigma}{d^3q} = \frac{G(k_{min})}{|p - q|} C(p, k_{min}). \tag{7.77}$$

In (7.77) p and q are the momenta of the initial and outgoing particles,

$$|k_{min}| = |p - q| - |p'|, \tag{7.78}$$

where p' is the momentum of the initial particle after the scattering and

$$G(k_{min}) = \int_{k_{min}}^{\infty} n(k)k \, dk. \tag{7.79}$$

The quantity $G(k_{min})/|p - q|$ is related to the probability of obtaining a residual nucleus with $A - 1$ nucleons with a momentum k_{min} and $C(p, k_{min})$ describes the dependence of the cross-section for the inclusive process $p + A \rightarrow p + p' + (A - 1)$ on p and k_{min}. The account of the role of the FSI leads to the conclusion that $n(k)$ in (7.79) has to be considered as an effective one $(n_{eff}(k))$, which is proportional to the actual momentum distribution $n(k)$.

The NMD in few-body nucleon systems (^{3}He and ^{4}He) can be extracted from the inclusive (e, e') cross-sections analysed by the method of y-scaling developed by *Ciofi degli Atti* et al. [7.38, 7.45, 7.46, 7.102–105]. It has been shown that within the Plane Wave Impulse Approximation (PWIA) the inelastic cross-section depends on nuclear structure peculiarities through the spectral function $P(k, E)$ (see Chap. 5). The NMD can be expressed using $P(k, E)$ by the relation:

$$n(k) = \int_{E_{min}}^{\infty} P(k, E) \, dE = n_{gr}(k) + n_{ex}(k), \tag{7.80}$$

where $k = k_N - q$, k_N being the momentum of the struck nucleon in the laboratory system and q the momentum transfer,

$$E_{min} = |E_A| - |E_{A-1}| \tag{7.81}$$

with E_A and E_{A-1} being the ground state energies of the initial and final nuclei. The spectral function can be written in the form:

$$P(k, E) = P_{gr}(k, E) + P_{ex}(k, E). \tag{7.82}$$

Then

$$n_{gr}(k) = \int_{E_{min}}^{\infty} P_{gr}(k, E) \, dE \tag{7.83}$$

is the NMD corresponding to the $A -$ body configuration with the spectator $(A - 1)$ system in the ground state, and

$$n_{ex}(k) = \int_{E_{min}}^{\infty} P_{ex}(k, E) \, dE \tag{7.84}$$

is the NMD corresponding to the A-body configuration with the spectator $(A - 1)$ system in all possible virtual excited states. The distribution n_{ex} can be considered as the correlation contribution to the NMD $n(k)$. In PWIA the cross-section for the exclusive two-body channel is related to n_{gr}:

$$\sigma_3^{gr} = \frac{d^3\sigma}{d\Omega_2 \, dE_N \, d\Omega_N} = C\sigma_{eN} \, n_{gr}(k), \tag{7.85}$$

where σ_{eN} is the free e–N cross-section and C is a kinematical factor. It turns out that the inclusive cross-section is not directly related to the NMD but depends on the knowledge of the full spectral function:

$$\sigma_2 \equiv \frac{d^2\sigma}{d\varepsilon_2\,d\Omega_2} = \{Z\bar{\sigma}_{ep} + N\bar{\sigma}_{en}\} \left|\frac{\partial\omega}{k\,\partial\cos\alpha}\right|^{-1}$$

$$\times 2\pi \int\limits_{E_{min}}^{E_{max}(q,\,\omega)} dE \int\limits_{k_{min}(q,\,\omega,\,E)}^{k_{max}(q,\,\omega,\,E)} P(k,E)k\,dk. \tag{7.86}$$

In (7.86) the kinematical factor is related to the integration over the direction of the unobserved nucleon, $\omega(q)$ is the energy transfer,

$$\frac{\partial\omega}{k\,\partial\cos\alpha} = \frac{q}{E_2}, \tag{7.87}$$

$$E_2(k_N) = [M^2 + k^2 + q^2 + 2kq\cos\alpha]^{1/2} \tag{7.88}$$

and the limits of integration are determined by energy conservation.

It has been shown by *West* [7.106] that the nuclear structure function of inclusive cross-section by a system of non-interacting particles becomes, at large momentum transfer, the probability distribution of finding in a nucleus a nucleon with momentum $y = y_0 = (k \cdot q)/q = M(\omega - q^2/2M)/q$. Considering realistic systems it has been pointed out [7.105] that y is the lowest longitudinal momentum of a nucleon bound with minimal removal energy ($E_{A-1}^{*f} = 0$). In Eq. (7.86) the quantities with an overbar are evaluated at $E_{A-1}^{*f} = 0$ (i.e. $E = E_{min}$) and $k = k_{min}(q, \omega, E_{min})$.

Introducing the scaling variable y as a new kinematical variable, any dependence upon ω in the nuclear structure function $F(q, \omega)$:

$$F(q,\omega) = 2\pi \int\limits_{E_{min}}^{E_{max}(q,\,\omega)} dE \int\limits_{k_{min}(q,\,\omega,\,E)}^{k_{max}(q,\,\omega,\,E)} P(k,E)k\,dk \tag{7.89}$$

can be expressed as a dependence on q and y:

$$F(q,y) = 2\pi \int\limits_{E_{min}}^{E_{max}(q,\,y)} dE \int\limits_{k_{min}(q,\,y,\,E)}^{k_{max}(q,\,y,\,E)} P(k,E)k\,dk. \tag{7.90}$$

At finite but large enough values of q the structure function (7.90) can be rewritten for complex nuclei in the form:

$$F(q,y) = 2\pi \int\limits_{E_{min}}^{\infty} dE \int\limits_{k_{min}(q,\,y,\,E)}^{\infty} P(k,E)k\,dk. \tag{7.91}$$

Using (7.82) the scaling function (7.91) becomes [7.46]:

$$F(q,y) = 2\pi \int\limits_{|y|}^{\infty} n_{gr}(k)k\,dk + 2\pi \int\limits_{E_{min}}^{\infty} dE \int\limits_{k_{min}(q,\,y,\,E)}^{\infty} P_{ex}(k,E)k\,dk, \tag{7.92}$$

where the first term scales in y, but the second term represent a "scaling violation" (k_{min} depends on q) due to the nucleon binding. In the asymptotic limit, $q \to \infty$:

$$\lim_{q \to \infty} k_{\min}(q, y, E) \equiv k_{\min}^{\infty}(y, E) \cong |y - (E - E_{\min})| \qquad (7.93)$$

one has:

$$F(y) = 2\pi \int_{|y|}^{\infty} n_{gr}(k)k \, dk + 2\pi \int_{E_{\min}}^{\infty} dE \int_{k_{\min}^{\infty}(y, E)}^{\infty} P_{ex}(k, E)k \, dk, \qquad (7.94)$$

or

$$F(y) = f(y) - B(y), \qquad (7.95)$$

where $f(y)$ is the longitudinal momentum distribution

$$f(y) = 2\pi \int_{|y|}^{\infty} n(k)k \, dk \qquad (7.96)$$

and

$$B(y) = 2\pi \int_{E_{\min}}^{\infty} dE \int_{|y|}^{k_{\min}^{\infty}(y, E)} P_{ex}(k, E)k \, dk. \qquad (7.97)$$

After taking the derivative of both sides of (7.95) one gets:

$$n(k) = -\frac{1}{2\pi y}\left[\frac{dF(y)}{dy} + \frac{dB}{dy}\right], \quad k = |y|. \qquad (7.98)$$

Hence, the extraction of the NMD in this y-scaling approach needs the asymptotic scaling function $F(y)$ to be obtained from the experimental data and the binding corrections term dB/dy to be estimated in a realistic way.

The experimental inclusive cross-sections σ_2 for ^{3}He [7.107, 7.108], ^{4}He, ^{12}C and ^{56}Fe [7.44] and for nuclear matter [7.109] gives the possibility of determining the experimental scaling function [7.45–46]:

$$F_1^{exp}(q, y) = \frac{\sigma_2^{exp}(q, \omega)}{(Z\bar{\sigma}_{ep} + N\bar{\sigma}_{en})}\left|\frac{\partial \omega}{k\partial \cos \alpha}\right|. \qquad (7.99)$$

The experimental knowledge of the asymptotic scaling function $F(y)$ and the estimate of the binding corrections for ^{3}He and for complex nuclei by *Ciofi degli Atti* et al. [7.110, 7.111] allow the NMD for ^{2}H, ^{3}He, ^{4}He, ^{12}C, ^{56}Fe and nuclear matter to be obtained [7.46] using Eq. (7.98). The experimental results for ^{4}He and ^{12}C are presented in Fig. 7.4 and Fig. 7.7 and compared with the theoretical results from different many-body and mean-field theories. The results for various nuclei [7.46] are shown in Fig. 7.16 and compared with theoretical calculations. They confirm the conclusion that $n(k)$ at $k \lesssim 1$ fm^{-1} can be predicted by the mean-field methods, but for $k \gtrsim 2$ fm^{-1} the NMD behaviour depends on correlation effects in nuclei and is almost independent of the mass number A.

It has to be noted that the method suggested [7.46] for the determination of the asymptotic scaling function can be used even if the experimental data are

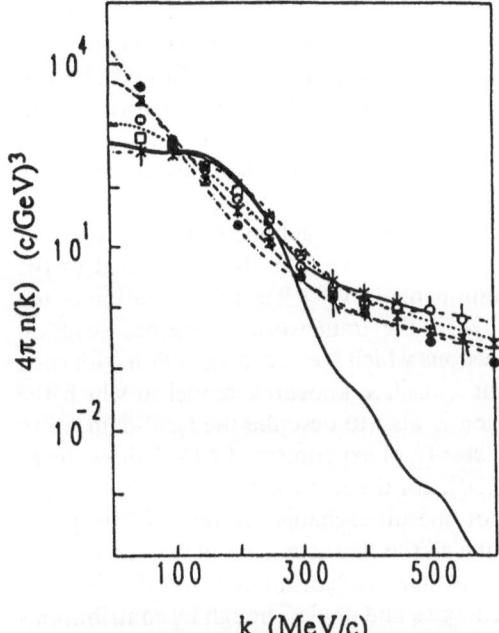

Fig. 7.16. The nucleon momentum distribution $n(k)$ for ^{2}H ($\bullet$), ^{3}He (x), ^{4}He (o), ^{12}C ($\square$) and ^{56}Fe(x) obtained from the asymptotic scaling function [7.46]. Dot-dashed line: theoretical $n(k)$ for ^{2}H obtained from the RSC [7.112] interaction; dashed line: $n(k)$ for ^{3}He [7.113]; dotted line: $n(k)$ for ^{4}He [7.34]; double-dot-dashed line: $n(k)$ for ^{12}C [7.41]; thin solid line: $n(k)$ for ^{56}Fe [7.111]; and thick solid line: $n(k)$ for ^{56}Fe calculated within the Hartree–Fock method [7.46]. The normalization is $\int n(k)\,dk = 1$

affected by FSI effects and that the binding correlations can be explicitly evaluated using the spectral functions. The available experimental data allow one to determine the NMD only up to $k \simeq 500$ MeV/c (~ 2.50 fm^{-1}) for light nuclei and $k \simeq 450$ MeV/c (~ 2.25 fm^{-1}) for heavy nuclei. These limits, however, are already sufficient to show the failure of the mean-field methods. It is also shown in [7.46] that the experimental results for $n(k)$ obtained from inclusive (e, e′) reactions are in qualitative overall agreement with the data obtained from the exclusive electron–nuclei processes (^{2}H (e, e′p) n [7.114], ^{3}He (e, e′p) X [7.115–117], ^{4}He (e, e′p) ^{3}H [7.118–120] and ^{4}He (e, e′p) X [7.119, 7.120], ^{12}C (e, e′p) ^{11}B [7.121]).

A new method for extracting the NMD in ^{56}Fe from the experimental scaling function using a modified nuclear matter spectral function has been introduced by *Ji* and *McKeown* [7.122] for the analysis of quasielastic electron scattering data.

In addition to the experimental data related to the extraction of $n(k)$ mentioned above, we should note other studies of particle–nuclei scattering using various incident particles: i) leptons [7.123–127], ii) protons [7.62, 7.63,

7.128–135], iii) alpha-particles [7.136], iv) deuterons [7.137]; various detected particles [7.138, 7.139], as well as various reactions: $(\pi^+, \pi^+ p)$ and $(\pi^+, 2p)$ [7.140], $(\alpha, {}^3He)$ and $({}^{16}O, {}^{15}O)$ [7.22, 7.141–144], $(p, 2p)$ [7.145] and charged particle emission in a muon capture by nuclei [7.146]. The kinematical conditions for which the single-particle knockout reaction mechanism and the form assumed for the high-momentum components of the NMD are reliable, are considered by *Fujita* [7.147].

Another class of experiments, namely photoreactions (γ, p) and (γ, n) at photon energies above that of the giant resonance can also be applied to the study of high-momentum NMD components [7.148]. The interpretation of the (γ, p)-reaction data (e.g. [7.149–161]) in the framework of various reaction mechanisms leads, however, to conclusions which are in contradiction with each other. It is shown in [7.150] that the quasifree knockout model in which the photon interacts with a target nucleon is able to describe the (γ, p)-data up to $E_\gamma \simeq 120$ MeV. The analysis of the later (γ, p) experiments [7.157] show, however, that neither the model from [7.150], nor the Jastrow-type model [7.162], as well as the model that accounts for meson-exchange currents [7.163] and self-consistent RPA are able to describe all the characteristics of the process. It is concluded in [7.157] that the single-particle mechanism does not dominate in photoreactions at large intervals of energies and angles though its contributions are not negligible. The advantages of the quasideuteron model, in which a correlated proton–neutron pair absorbs the incident photon [7.164, 7.165] are emphasized in [7.158]. Systematic studies of (γ, n) processes (e.g. [7.158, 7.166–172]) show that the (γ, p) and (γ, n) cross-sections are comparable at similar energies and angles and this is in contrary to the predictions of the single-nucleon absorption mechanism and supports the interpretation by models of quasideuteron type or of the two-step mechanism [7.173].

It can be concluded that the present theoretical and experimental studies of photoreactions cannot give reliable information on the nucleon momentum distributions in nuclei. A more extended review of photoreaction studies related to the problem of the nucleon–nucleon correlations is given in Chap. 10.

Particular attention has been paid to the study of the NMD in a given single-particle state [7.121]. This can be achieved by studying the hole-state spectral functions (Chap. 5) of the target nucleus in $(e, e'p)$ reactions. The $p_{1/2}$ and $p_{3/2}$-hole states in ${}^{16}O$ have been analysed by *Boffi* et al. [7.174] using the data from the ${}^{16}O$ $(e, e'p)$ ${}^{15}N$-reaction in the framework of the generalized DWIA. The NMD in the $1p_{3/2}$-state in ${}^{12}C$ is obtained from the analyses of $(e, e'p)$-reaction [7.175] and (γ, p)-reaction [7.154]. The NMD in $p_{1/2}$- and $p_{3/2}$-states in ${}^{16}O$ are determined from $(e, e'p)$ [7.176] and (γ, p) [7.153, 7.160, 7.155, 7.156] processes. The distributions in ${}^{12}C$ and ${}^{16}O$ are presented and discussed in the review of *Frullani* and *Mougey* [7.121] (see also Chap. 10).

The measurements of the spectral functions from the $(e, e'p)$ reaction on ${}^{208}Pb$ [7.177] at missing momentum values between 50 and 300 MeV/c give the possibility of determining the experimental momentum distributions of protons in $3s_{1/2}$, $2d_{3/2}$, $2d_{5/2}$, $1h_{11/2}$ and $1g_{7/2}$-states. These distributions are shown in

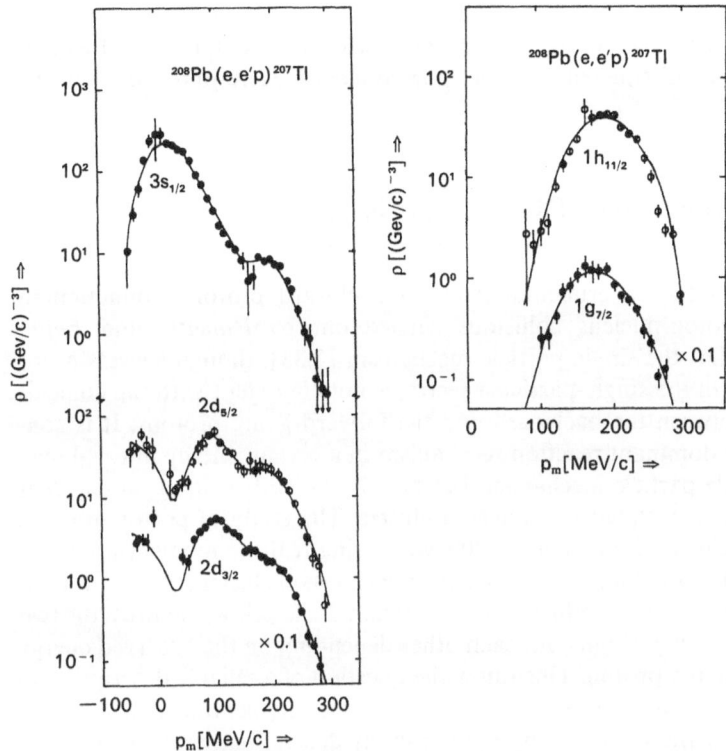

Fig. 7.17. Momentum distributions for $3s_{1/2}$, $2d_{3/2}$, $2d_{5/2}$, $1h_{11/2}$, and $1g_{7/2}$ proton knockout in the reaction ^{208}Pb (e, e′ p)^{207}Tl (taken from [7.178]). The solid lines show the fits to the data using Woods–Saxon bound-state wave functions

Fig. 7.17 taken from [7.178]. The results for the single-particle momentum distributions in ^{12}C, ^{51}V, ^{90}Zr and ^{208}Pb obtained from high-resolution (e, e′p) experiments are reviewed by *de Witt Huberts* [7.127b] and *Dieperink* and *de Witt Huberts* [7.127c].

The special interest of measurements of the reaction $A(e, e′2N)$ $(A − 2)$ is discussed by *Lightbody Jr.* [7.179], *Ciofi degli Atti* [7.180] and *Mougey* [7.181]. Because of the combination of low (<1 GeV) energies and ~1% duty factors of the existing electron accelerators, the (e, e′2N) reactions are difficult to study [7.181]. Future studies of these processes should provide the most direct way to determine the two-nucleon correlation function and the high-momentum components of the nuclear wave function. As pointed out in [7.181], effects of the exchange of heavier mesons ($\rho, \omega, ...$) and of nucleon structure should be important at high momenta. These experiments should clarify the yet unsolved problems of the processes taking place in (e, e′) and (e, e′p) reactions that are related to the quenching of the longitudinal response function, the filling in of the "dip-region" between the quasi-free and the Δ-peaks in inclusive scattering, and also to the behaviour of the deep inelastic cross-section in the $x > 1$ as well as the anomalous missing energy spectra from inclusive (e, e′p) reactions in the dip-region [7.179, 7.180].

7.2 Cluster Momentum Distributions in Nuclei

The analyses of the experimental data on backward proton production in high-energy proton-nucleus collisions carried out by *Haneishi* and *Fujita* [7.182] show that the single-particle mechanism [7.33], though successful for the description of the single-particle spectra, is not consistent with the coincidence spectra between the backward and the forward-going protons. It is concluded that the dominant reaction mechanism in a certain energy interval may not be the single-particle mechanism but mainly the scattering of the incident proton with the correlated two-nucleon cluster. The study of proton–nucleus collisions at incident energies of 35–300 MeV using realistic momentum distributions [7.183] shows that at 300 MeV the two-nucleon cluster reaction dominates the inclusive proton production cross-section while below 200 MeV the two processes are in competition with each other depending on the observed energy of the backscattered proton. This raises the question of a reliable description of the two-nucleon centre-of-mass and relative state momentum distributions. These quantities are defined using the two-body density matrix:

$$\rho^{(2)}(\xi_1,\xi_2;\xi_1',\xi_2') = \tfrac{1}{2}A(A-1)\sum_{\eta_3\ldots\eta_A}\int d\boldsymbol{r}_3\ldots d\boldsymbol{r}_A\,\Psi^+(\xi_1,\xi_2,\xi_3,\ldots,\xi_A)$$
$$\times\,\Psi(\xi_1',\xi_2',\xi_3,\ldots,\xi_A), \tag{7.100}$$

where $\xi \equiv (\boldsymbol{r};\sigma,\tau) \equiv (\boldsymbol{r};\eta)$, and Ψ is the total many-body wave function.

The two-nucleon momentum distribution (TNMD) $n^{(2)}(\xi_1,\xi_2)$ is defined by the diagonal terms of the two-body density matrix in momentum space:

$$n^{(2)}(\zeta_1,\zeta_2) = \rho^{(2)}(\zeta_1,\zeta_2;\zeta_1,\zeta_2). \tag{7.101}$$

Introducing the relative $\boldsymbol{q}$ and the centre of mass $\boldsymbol{p}$ momenta of the nucleon–nucleon pair by

$$\boldsymbol{q} = (\boldsymbol{k}_1 - \boldsymbol{k}_2)/2, \qquad \boldsymbol{p} = \boldsymbol{k}_1 + \boldsymbol{k}_2 \tag{7.102}$$

the following expressions for the centre of mass and the relative state TNMD for N–N pair inside the nucleus can be obtained:

$$n_{\mathrm{NN}}^{\mathrm{c.m.}}(\boldsymbol{p}) = \int\frac{d\boldsymbol{q}}{(2\pi)^3}\,n^{(2)}\left(\frac{\boldsymbol{p}}{2}+\boldsymbol{q},\frac{\boldsymbol{p}}{2}-\boldsymbol{q}\right), \tag{7.103}$$

$$n_{\mathrm{NN}}^{\mathrm{rel.}}(\boldsymbol{q}) = \int\frac{d\boldsymbol{p}}{(2\pi)^3}\,n^{(2)}\left(\frac{\boldsymbol{p}}{2}+\boldsymbol{q},\frac{\boldsymbol{p}}{2}-\boldsymbol{q}\right). \tag{7.104}$$

The phenomenological model developed by *Haneishi* and *Fujita* [7.182] gives the following expressions for the TNMD:

$$n^{\text{c.m.}}(p) = N_2 \left\{ e^{-(p^2/2p_0^2)} + 2\varepsilon_0 \left(\frac{2q_0^2}{p_0^2 + q_0^2} \right)^{3/2} e^{-(p^2/(p_0^2 + q_0^2))} \right.$$

$$\left. + \varepsilon_0^2 (q_0/p_0)^3 e^{-(p^2/2q_0^2)} \right\}, \tag{7.105}$$

$$n^{\text{rel.}}(q) = N_{\text{rel}} \left\{ e^{-(2q^2/p_0^2)} + 2\varepsilon_0 \left(\frac{2q_0^2}{p_0^2 + q_0^2} \right)^{3/2} e^{-(4q^2/(p_0^2 + q_0^2))} \right.$$

$$+ \varepsilon_0^2 (q_0/p_0)^3 e^{-(2q^2/q_0^2)} + C_0 \left(\frac{p_0^2}{p_0^2 + 2\gamma^2} \right)^{3/2} \cdot e^{-[q^2/(\gamma^2 + p_0^2/2)]}$$

$$\left. + O(\varepsilon_0 C_0, \varepsilon_0^2 C_0) \right\}, \tag{7.106}$$

where ε_0, q_0, p_0, C_0, γ are parameters and N_2 and N_{rel} are normalization constants. It is shown in [7.182] that the centre-of-mass TNMD (7.105) contains high-momentum components which play a dominant role in reproducing the high-energy part of the backward proton spectra. The calculations reproduce well the results for the contour plot of the coincidence spectra between the forward- and backward-going protons in the experiments of *Miake* et al. [7.184]. Evidence is obtained for the existence of high-momentum components in the nuclear Fermi motion which seems to be rather universal from light to heavy nuclei.

Akaishi [7.42] calculated the TMND in the ^{4}He nucleus within the ATMS correlation variational method mentioned in Sect. 7.1. The momentum distributions $n_{\text{NN}}^{\text{c.m.}}$ and $n_{\text{NN}}^{\text{rel.}}$ obtained are parametrized by

$$n_{\text{NN}}^{(\text{c.m.})(\text{rel.})} = N \{ \exp(-p^2/(2a)) + s \exp(-p^2/(2at)) \} \tag{7.107}$$

with $\{a(\text{fm}^{-2}), s, t\} = \{0.42 \times 3, 0.01, 8\}$ for $n_{\text{NN}}^{\text{c.m.}}$ and $\{0.42/4, 0.015, 6\}$ for $n_{\text{NN}}^{\text{rel.}}$. It is pointed out that the TNMD have prominent high-momentum components which reflect the role of the nucleon–nucleon correlations.

The coherent density fluctuation model (CDFM) (see Sect. 4.2) has been extended in [7.185] to calculations of both the two-nucleon centre-of-mass and the two-nucleon relative momentum distributions. In the CDFM the wave function Ψ is expressed in the form:

$$\Psi(\xi_1, \ldots, \xi_A) = \int_0^\infty f(x) \varphi(x; \xi_1, \ldots, \xi_A) \, dx, \tag{7.108}$$

where the generating function $\varphi(x; \xi_1, \ldots, \xi_A)$ describes the state of A-nucleons uniformly distributed in a sphere with radius x (the so-called "flucton"). It is assumed that for many-fermion systems the following relation for the wave functions φ approximately holds:

$$\frac{1}{2}A(A-1) \sum_{\eta_3,\ldots,\eta_A} \int d\mathbf{r}_3 \ldots d\mathbf{r}_A \, \varphi^+(x'; \xi_1, \xi_2, \xi_3, \ldots, \xi_A)$$

$$\times \varphi(x; \xi_1', \xi_2', \xi_3, \ldots, \xi_A)$$

$$= \delta(x - x') \rho^{(2)}(\xi_1, \xi_2; \xi_1', \xi_2'; x), \tag{7.109}$$

where $\rho^{(2)}(\xi_1, \xi_2; \xi_1', \xi_2'; x)$ is the two-body density matrix for a system of A-free nucleons confined in a sphere of radius x. The integration over r_1 and r_2 and summation over η_2 and η_1 at $\xi_1 = \xi_1'$ and $\xi_2 = \xi_2'$ leads to the well-known delta-function limit for the overlap kernel in the generator coordinate method (GCM):

$$\langle \varphi(x; \{\xi_i\}) | \varphi(x'; \{\xi_i\}) \rangle = \delta(x - x'), \quad i = 1, \ldots, A. \tag{7.110}$$

Using (7.100), (7.101) and (7.109) one gets for the TNMD:

$$n^{(2)}(\zeta_1, \zeta_2) = \int_0^\infty dx |f(x)|^2 n_x^{(2)}(\zeta_1, \zeta_2), \tag{7.111}$$

where $\zeta \equiv (k; \sigma, \tau)$ and $n_x^{(2)}(\zeta_1, \zeta_2)$ is the TNMD for a system of non-interacting nucleons confined in a volume $\Omega(x) = 4\pi x^3/3$:

$$n_x^{(2)}(\zeta_1, \zeta_2) = \frac{1}{2}[n_x^{(1)}(\zeta_1) n_x^{(1)}(\zeta_2) - |\rho^{(1)}(\zeta_1, \zeta_2; x)|^2]. \tag{7.112}$$

In (7.112)

$$\rho^{(1)}(\zeta_1, \zeta_2; x) = (2\pi)^3 \delta_{\eta_1 \eta_2} \delta(\mathbf{k}_1 - \mathbf{k}_2) \theta(k_F(x) - |\mathbf{k}_1|) \theta(k_F(x) - |\mathbf{k}_2|) \tag{7.113}$$

and

$$n_x^{(1)}(\zeta) \equiv \rho^{(1)}(\zeta, \zeta; x) = \Omega(x) \theta(k_F(x) - |\mathbf{k}|). \tag{7.114}$$

Substituting (7.113) and (7.114) in (7.112) and the latter in (7.111) one gets for the case of neutron–proton pairs:

$$n_{np}^{(2)}(\mathbf{k}_1, \mathbf{k}_2) = 4 \int_0^\infty dx |f(x)|^2 \Omega^2(x) \theta(k_F(x) - |\mathbf{k}_1|) \theta(k_F(x) - |\mathbf{k}_2|). \tag{7.115}$$

In the case of nuclei with $Z = N = A/2$ the p–n centre-of-mass and relative motion TNMD normalized to $A^2/4$ have the form:

$$n_{np}^{c.m.}(\mathbf{p}) = A \int_0^\infty dx |f(x)|^2 \Omega(x) \left[1 - \frac{3}{4} \frac{|\mathbf{p}|}{k_F(x)} + \frac{1}{16} \frac{|\mathbf{p}|^3}{k_F^3(x)} \right]$$

$$\times \theta(k_F(x) - |\mathbf{p}|/2), \tag{7.116}$$

$$n_{np}^{rel.}(\mathbf{q}) = 8A \int_0^\infty dx |f(x)|^2 \Omega(x) \left[1 - \frac{3}{2} \frac{|\mathbf{q}|}{k_F(x)} + \frac{1}{2} \frac{|\mathbf{q}|^3}{k_F^3(x)} \right]$$

$$\times \theta(k_F(x) - |\mathbf{q}|). \tag{7.117}$$

The calculations of $n_{np}^{c.m.}$ and $n_{np}^{rel.}$ for ^{4}He, ^{16}O and ^{40}Ca nuclei are made using the symmetrized Fermi-type density distribution, with parameters determined by analyses of high-energy electron–nucleus scattering, in the relation:

$$|f(x)|^2 = -\frac{1}{\rho_0(x)} \frac{d\rho(r)}{dr}\bigg|_{r=x} , \quad \rho_0(x) = \frac{3A}{4\pi x^3} . \tag{7.118}$$

The CDFM results for the $n_{np}^{c.m.}$ and $n_{np}^{rel.}$ are shown in Figs. 7.18–20 and compared with the results of *Akaishi* [7.42] for ^{4}He, with those of *Haneishi* and *Fujita* [7.182], with the uncorrelated case of harmonic oscillator shell model as well as with calculations in the GCM [7.186].

In the GCM the two-body density matrix (7.100) has the form:

$$\rho^{(2)}(\xi_1,\xi_2;\xi_1',\xi_2') = \int dx\, f^*(x) \int dx'\, f(x')\rho^{(2)}(x,x';\xi_1,\xi_2;\xi_1',\xi_2') \tag{7.119}$$

with

$$\rho^{(2)}(x,x';\xi_1,\xi_2;\xi_1',\xi_2') = \tfrac{1}{2}A(A-1) \sum_{\eta_3,\dots,\eta_A} \int d\mathbf{r}_3\dots d\mathbf{r}_A\, \varphi^+(x;\xi_1,\xi_2,\xi_3,\dots,\xi_A)$$

$$\times \varphi(x';\xi_1',\xi_2',\xi_3,\dots,\xi_A). \tag{7.120}$$

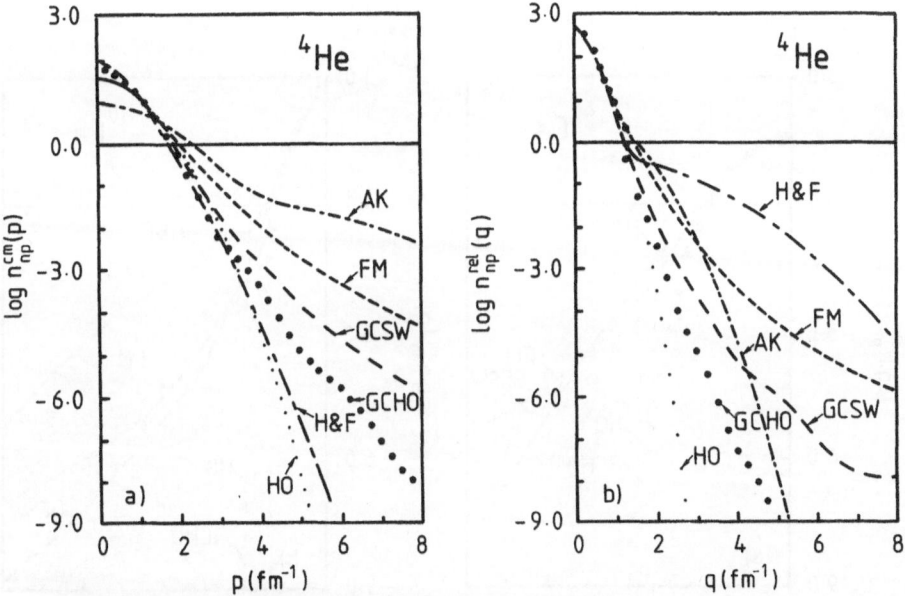

Fig. 7.18. Two-nucleon centre-of-mass (**a**) and relative motion (**b**) momentum distributions of n–p pairs in ^{4}He. HO – harmonic-oscillator shell-model result; GCHO – GCM with harmonic oscillator construction potential; GCSW – GCM with square-well construction potential [7.35]; FM – the CDFM result [7.185]; H&F – phenomenological distributions from [7.182]; AK – the result from [7.42]. The normalization is $\int n_{np}(k)\,dk/(2\pi)^3 = 1$

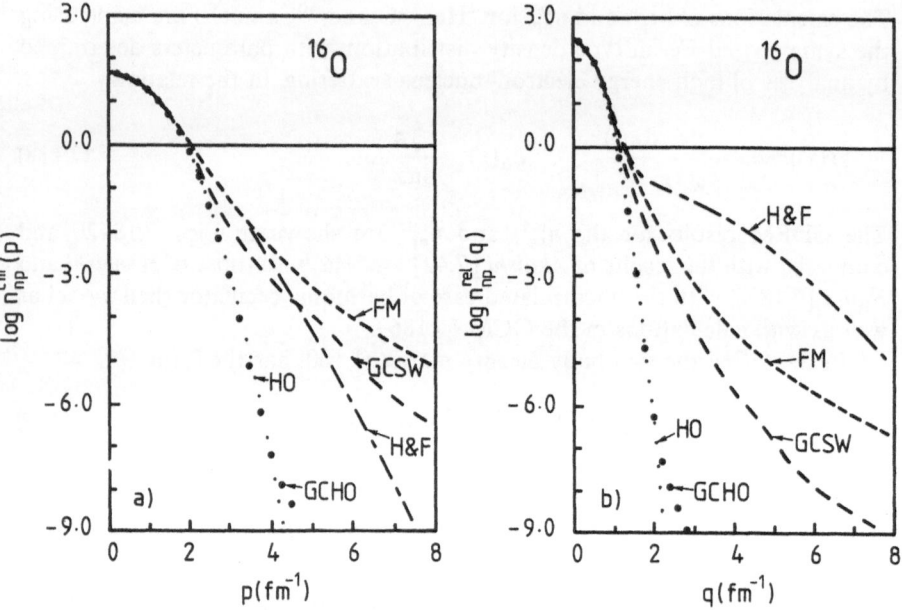

Fig. 7.19. The same as in Fig. 7.18. for ^{16}O

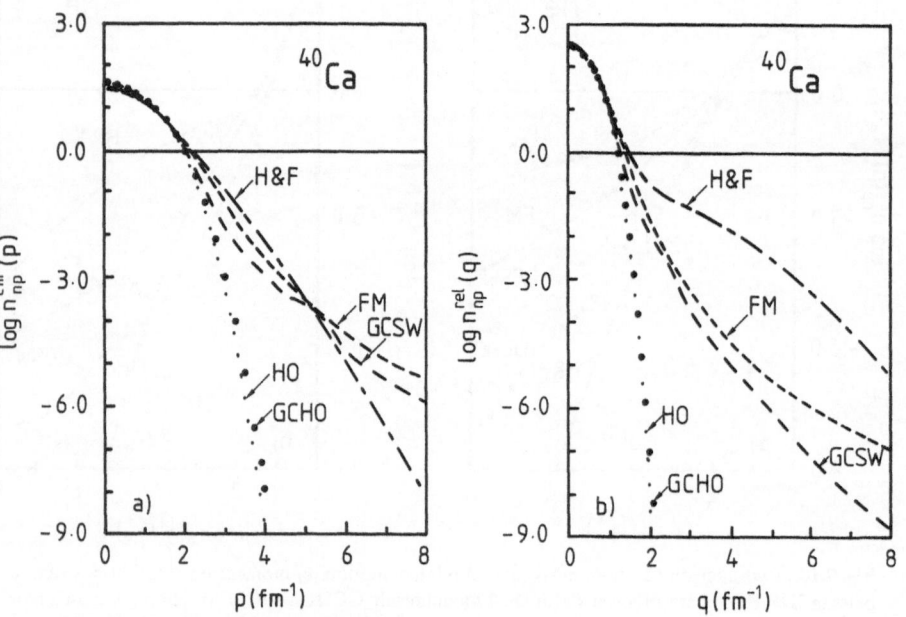

Fig. 7.20. The same as in Fig. 7.18. for ^{40}Ca

The Eqs. (7.119) and (7.120) have been obtained from (7.100) using the total wave function Ψ in the form (7.108). If the generating wave functions φ are Slater determinants constructed from a complete set of one-particle functions $\chi_i(x, \xi)$ the matrix (7.120) can be expressed as [7.187a–b]:

$$\rho^{(2)}(x, x'; \xi_1, \xi_2; \xi_1', \xi_2') = \frac{I(x, x')}{2} \det \begin{pmatrix} \rho(x, x'; \xi_1, \xi_1') & \rho(x, x'; \xi_1, \xi_2') \\ \rho(x, x'; \xi_2, \xi_1') & \rho(x, x'; \xi_2, \xi_2') \end{pmatrix}, \quad (7.121)$$

where

$$\rho(x, x'; \xi, \xi') = \sum_{k,l=1}^{A} (N^{-1})_{lk} \, \chi_k^*(x, \xi) \chi_l(x', \xi'), \quad (7.122)$$

$(N^{-1})_{lk}$ is the inverse matrix of the matrix:

$$N_{kl}(x, x') = \sum_{\eta} \int d\mathbf{r} \, \chi_k^*(x, \xi) \chi_l(x', \xi), \quad (7.123)$$

and

$$I(x, x') = \det(N_{kl}). \quad (7.124)$$

Thus the following expression for the two-body density matrix in the GCM is obtained:

$$\rho^{(2)}(\xi_1, \xi_2; \xi_1', \xi_2') = \frac{1}{2} \int dx \, f^*(x) \int dx' \, f(x') I(x, x')$$
$$\times [\rho(x, x'; \xi_1, \xi_1') \rho(x, x'; \xi_2, \xi_2')$$
$$- \rho(x, x'; \xi_1, \xi_2') \rho(x, x'; \xi_2, \xi_1')]. \quad (7.125)$$

The TNMD $n^{(2)}(\zeta_1, \zeta_2)$ (7.101) has the following form for the case of n–p pairs and $Z = N$ nuclei:

$$n_{np}^{(2)}(k_1, k_2) = \frac{1}{4} \int dx \, f^*(x) \int dx' \, f(x') I(x, x') \tilde{\rho}(x, x'; k_1) \tilde{\rho}(x, x'; k_2), \quad (7.126)$$

where

$$\tilde{\rho}(x, x'; k) = 4 \sum_{\lambda, \mu=1}^{A/4} (N^{-1})_{\mu\lambda} \, \tilde{\chi}_\lambda^*(x, k) \tilde{\chi}_\mu(x', k), \quad (7.127)$$

$\tilde{\chi}_\lambda(x, k)$ being the Fourier transform of the orbital $\chi_\lambda(x, r)$ and each orbital state is occupied by 4 nucleons. The centre-of-mass and relative motion TNMD can be obtained using Eqs. (7.102)–(7.104). The function $f(x)$ in Eqs. (7.125)–(7.126) is determined by solving the equation [7.187c–e]:

$$\int [\mathcal{H}(x, x') - E I(x, x')] f(x') dx' = 0, \quad (7.128)$$

where the overlap kernel $I(x, x')$ is given by (7.124) and the energy kernel $\mathcal{H}(x, x')$ in the case of Skyrme effective forces is determined in Sect. 4.1. Two different construction potentials (with corresponding single-particle functions

$\chi_\lambda(x, r))$ have been used in [7.186], namely the square-well potential with infinite walls and the harmonic oscillator potential. In these cases the generator coordinate is the radius of the well and the oscillator parameter, respectively.

The TNMD $n_{np}^{c.m.}$ and $n_{np}^{rel.}$ are calculated for ^{4}He, ^{16}O and ^{40}Ca nuclei within the proposed GCM-scheme (Figs. 7.18–20). The values of the Skyrme parameters for the case of the square-well construction potential are the same as those used for calculations of NMD in [7.35], where they are determined to fit the binding energies of ^{4}He, ^{16}O and ^{40}Ca: $t_0 = -2765.0$, $t_1 = 383.94$, $t_2 = -38.04$, $t_3 = 15865$ and $\sigma = 1/6$. These values lead to the following infinite nuclear matter characteristics: $E/A = -16.78$ MeV/N, $K = 235.1$ MeV, $m^*/m = 0.7$ and $\rho_0 = 0.148$ fm^{-3}. In the case of the harmonic oscillator construction potential the SkM* parameter set is used [7.188].

It can be noted that the existence of high-momentum components in the TNMD in the case of the square-well construction potential with infinite walls within the GCM (results noted by GCSW in Figs. 7.18–20), as well as in the CDFM (FM in the Figures), is evidence for the effective inclusion of the short-range nucleon–nucleon correlations in these methods. This is obviously due to the presence of intermediate states with high nuclear density in the case of the GCSW and CDFM (at small values of the generator coordinate x), where the nucleons are close to each other and the short-range correlations are operative. As can be seen this is not the case when the harmonic oscillator construction potential is used within the GCM (results noted by GCHO in Figs. 7.18–20). The large difference between the GCSW and GCHO calculations at momenta $k > 2.5$ fm^{-1} leads to the expected result that the use of the effective Skyrme forces cannot give a correct account of the nucleon–nucleon correlations at high momenta. The latter can be achieved mainly by the relevant choice of the GCM construction potential (i.e. of the intermediate states) within the different theories.

It has to be emphasized that the similarity of the GCSW and the CDFM results shows that the role of the delta-function limit in the CDFM ((7.109) and (7.110)) is not very important for the TNMD especially in the heavier nuclei.

It should also be noted that, as expected, the curves of $n_{np}^{rel.}(q)$ in ^{4}He, ^{16}O and ^{40}Ca calculated in GCSW are close to each other up to momenta $q \simeq 4$ fm^{-1}. They represent the behaviour of the proton–neutron relative motion TNMD in the deuteron when the latter is situated in nuclear medium. In the region $q > 4$–5 fm^{-1} other degrees of freedom, different from the nucleon ones, will also be of importance.

Many aspects of nuclear structure and reactions suggest that nucleons can combine to form transient sub-structures or clusters, and among these the alpha-particle is the most likely because of its high binding energy and symmetry (see the review of *Hodgson* [7.189a]). It is important to determine the degree of alpha-clustering not only to facilitate an economical description of nuclear structure and reactions, but also to learn more about the nucleon–nucleon correlations in the nuclear interior.

The character of alpha-clustering depends on the nuclear size. Light nuclei can be considered to consist of linked clusters of alpha-particles, deuterons and nucleons. In heavier nuclei alpha-clusters can be expected in the region of the nuclear surface since condensation into alpha-clusters is energetically favourable at densities around one-third of that in the nuclear interior [7.189b]. Many nuclei decay spontaneously by emitting alpha-clusters and heavy fragments. Fusion reactions are affected by clustering in intermediate states and breakup reactions provide evidence of cluster structure in the projectile. At higher energies some nuclear reactions preferentially proceed by cluster transfer or by knockout and pickup processes. At very high energies nuclei can be fragmented into a wide range of clusters of nucleons.

Considering nuclear reactions that involve the transfer of more than one nucleon, it might be expected that reaction channels involving the removal or replacement of a cluster will have an enhanced probability for nuclei with a particular cluster structure. Cluster momentum distributions are of importance in the studies of these reactions. The use of the PWIA in the study of the cross-sections of alpha-particle knockout processes, such as $(p, p\alpha)$ and $(\alpha, 2\alpha)$ [7.189c–192] or the DWIA [7.191, 7.193–195] needs a knowledge of the momentum distribution of the alpha-clusters in the target nucleus (in PWIA) or of their "distorted momentum distributions" (in DWIA). The inclusive backward production of protons and clusters in medium energy proton and alpha scattering on nuclei can be considered in the framework of the "quasi-two-body scaling" hypothesis [7.138, 7.196] in which the internal momentum distributions of the target particle is introduced.

The DWIA calculations of low energy (p, α) reactions show that the analysing power in the continuum can be reproduced by an alpha-particle knockout mechanism [7.197]. The angular distributions of the (α, α') reaction [7.198] can also be described by taking into account the interaction of the incoming alpha-particles with preformed alpha-particles in the target nucleus. Thus the alpha-particle momentum distribution is important for the understanding of these reactions.

Analyses of reactions using electron beams, such as $(e, e'd)$, $(e, e'\alpha)$ [7.199–203] and (e, α) [7.204, 7.205] are considerably simpler because the electromagnetic interaction does not strongly distort the structure of the target nucleus, the electron waves are not appreciably distorted and there are no three-body final state interactions. In the case of the Born approximation, where one photon is exchanged between the electron and the target nucleus, it has been shown [7.202, 7.206] that the cross-section for the $A(e, e'b) B$ reaction can be expressed in terms of the momentum distribution of the cluster b in the target nucleus and the cross-section of the elastic electron scattering on this cluster.

Since it is an essential component of calculations of the cross-sections of alpha-particle knockout reactions, the alpha-particle momentum distribution (AMD) also provides a sensitive probe of short-range and tensor N–N correlations in nuclei.

The AMD has been calculated in the framework of the CDFM [7.207]. The definition of the four-body density matrix [7.187a] has the form:

$$\rho^{(4)}(\xi_1,\xi_2,\xi_3,\xi_4;\xi_1',\xi_2',\xi_3',\xi_4') = \frac{A(A-1)(A-2)(A-3)}{4!}$$

$$\times \sum_{\eta_5,\ldots,\eta_A} \int d\mathbf{r}_5 \ldots d\mathbf{r}_A \, \Psi^+(\xi_1,\xi_2,\xi_3,\xi_4,\xi_5,\ldots,\xi_A)$$

$$\times \Psi(\xi_1',\xi_2',\xi_3',\xi_4',\xi_5,\ldots,\xi_A), \tag{7.129}$$

$$\xi = (r;\sigma,\tau) \equiv (r;\eta).$$

The substitution of the many-body wave function $\Psi(\xi_1,\ldots,\xi_A)$ (7.108) in (7.129) leads to the following expression for $\rho^{(4)}$:

$$\rho^{(4)}(\xi_1,\xi_2,\xi_3,\xi_4;\xi_1',\xi_2',\xi_3',\xi_4') = \frac{A(A-1)(A-2)(A-3)}{4!}$$

$$\times \int_0^\infty dx' f^*(x') \int_0^\infty dx \, f(x) \sum_{\eta_5,\ldots,\eta_A} \int d\mathbf{r}_5 \ldots d\mathbf{r}_A \, \varphi^+(x';\xi_1,\xi_2,\xi_3,\xi_4,\xi_5,\ldots,\xi_A)$$

$$\times \varphi(x;\xi_1',\xi_2',\xi_3',\xi_4',\xi_5,\ldots,\xi_A). \tag{7.130}$$

A generalization of the delta-function limit in the CDFM concerning the 4-body density matrix in the many-nucleon system has been made by assuming that the following relation for the generating functions φ holds:

$$\frac{A(A-1)(A-2)(A-3)}{4!} \sum_{\eta_5,\ldots,\eta_A} \int d\mathbf{r}_5 \ldots d\mathbf{r}_A \, \varphi^+(x';\xi_1,\xi_2,\xi_3,\xi_4,\xi_5,\ldots,\xi_A)$$

$$\times \varphi(x;\xi_1',\xi_2',\xi_3',\xi_4',\xi_5,\ldots,\xi_A) = \delta(x-x')$$

$$\times \rho^{(4)}(x;\xi_1,\xi_2,\xi_3,\xi_4;\xi_1',\xi_2',\xi_3',\xi_4'), \tag{7.131}$$

where $\rho^{(4)}(x;\xi_1,\ldots,\xi_4;\xi_1',\ldots,\xi_4')$ is the 4-body density matrix in the plane wave case for a system with density $\rho_0(x) = 3A/(4\pi x^3)$ described by the function $\varphi(x;\{\xi_i\})$.

By means of subsequent integrations over $\mathbf{r}_i$, summing over η_i at $\xi_i = \xi_i'$ of (7.131) ($i = 4,3,2,1$) and using the properties of the density matrices [7.187a]:

$$\rho^{(p-1)}(\xi_1,\ldots\xi_{p-1};\xi_1',\ldots,\xi_{p-1}') = \frac{p}{A+1-p} \sum_{\eta_p} \int \rho^{(p)}(\xi_1,\ldots,\xi_{p-1},\xi_p;$$

$$\xi_1',\ldots,\xi_{p-1}',\xi_p) \, d\mathbf{r}_p \tag{7.132}$$

one can obtain the delta-function limit for the overlap kernel in the GCM (Eq. (7.110)).

Using Eqs. (7.130) and (7.131) the following expression for the 4-body density matrix is obtained in the CDFM:

$$\rho^{(4)}(\xi_1, \xi_2, \xi_3, \xi_4; \xi_1', \xi_2', \xi_3', \xi_4')$$

$$= \int_0^\infty dx \, |f(x)|^2 \rho^{(4)}(x; \xi_1, \ldots \xi_4; \xi_1', \ldots, \xi_4'). \tag{7.133}$$

The four-body momentum distribution $n^{(4)}(\zeta_1, \zeta_2, \zeta_3, \zeta_4)$ is expressed by the diagonal elements of the four-body density matrix in momentum space

$$n^{(4)}(\zeta_1, \zeta_2, \zeta_3, \zeta_4) = \rho^{(4)}(\zeta_1, \zeta_2, \zeta_3, \zeta_4; \zeta_1, \zeta_2, \zeta_3, \zeta_4), \tag{7.134}$$

where $\zeta_i \equiv (k_i; \sigma_i, \tau_i) \equiv (k_i; \eta_i)$, k_i being the momentum of the ith nucleon. In the CDFM this quantity has the form:

$$n^{(4)}(\zeta_1, \zeta_2, \zeta_3, \zeta_4) = \int_0^\infty dx \, |f(x)|^2 n_x^{(4)}(\zeta_1, \zeta_2, \zeta_3, \zeta_4), \tag{7.135}$$

where

$$n_x^{(4)}(\zeta_1, \zeta_2, \zeta_3, \zeta_4) \equiv \rho^{(4)}(x; \zeta_1, \zeta_2, \zeta_3, \zeta_4; \zeta_1, \zeta_2, \zeta_3, \zeta_4) \tag{7.136}$$

is the four-nucleon momentum distribution of the flucton.

In the case of a single Slater determinant wave function (e.g. the flucton state wave function φ) the many-body density matrices are expressed by means of a determinant built up by one-body density matrices [7.187b]. In the four-body case:

$$\rho^{(4)}(x; \zeta_1, \ldots \zeta_4; \zeta_1', \ldots, \zeta_4')$$

$$= \frac{1}{4!} \begin{vmatrix} \rho^{(1)}(x; \zeta_1; \zeta_1') & \rho^{(1)}(x; \zeta_1; \zeta_2') & \rho^{(1)}(x; \zeta_1; \zeta_3') & \rho^{(1)}(x; \zeta_1; \zeta_4') \\ \rho^{(1)}(x; \zeta_2; \zeta_1') & \rho^{(1)}(x; \zeta_2; \zeta_2') & \rho^{(1)}(x; \zeta_2; \zeta_3') & \rho^{(1)}(x; \zeta_2; \zeta_4') \\ \rho^{(1)}(x; \zeta_3; \zeta_1') & \rho^{(1)}(x; \zeta_3; \zeta_2') & \rho^{(1)}(x; \zeta_3; \zeta_3') & \rho^{(1)}(x; \zeta_3; \zeta_4') \\ \rho^{(1)}(x; \zeta_4; \zeta_1') & \rho^{(1)}(x; \zeta_4; \zeta_2') & \rho^{(1)}(x; \zeta_4; \zeta_3') & \rho^{(1)}(x; \zeta_4; \zeta_4') \end{vmatrix}, \tag{7.137}$$

where

$$\rho^{(1)}(x; \zeta_i; \zeta_j) = (2\pi)^3 \delta_{\eta_i \eta_j} \delta(k_i - k_j) \Theta\left(k_F(x) - \frac{|k_i + k_j|}{2}\right) \tag{7.138}$$

is the one-body density matrix in the case of a single Slater determinant wave function φ built up with plane wave functions. In (7.138):

$$k_F(x) = (3\pi^2 \rho_0(x)/2)^{1/3} \equiv \alpha/x, \quad (\alpha \equiv (9\pi A/8)^{1/3}) \tag{7.139}$$

is the Fermi-momentum. It follows from (7.138) that the single-nucleon momentum distribution of a flucton with radius x has the form:

$$n_x^{(1)}(\zeta) = \rho^{(1)}(x; \zeta; \zeta) = V(x)\theta(k_F(x) - |k|)\delta_{\eta\eta} \tag{7.140}$$

with $V(x) = 4\pi x^3/3$.

In the alpha-particle case (two protons and two neutrons with antiparallel spins) the diagonal elements of the matrix (7.137) are

$$\rho^{(\alpha)}(x;\zeta_1,\ldots,\zeta_4;\zeta_1,\ldots,\zeta_4) = \frac{1}{4!}\rho^{(1)}(x;\zeta_1;\zeta_1)\rho^{(1)}(x;\zeta_2;\zeta_2)$$

$$\times \rho^{(1)}(x;\zeta_3;\zeta_3)\rho^{(1)}(x;\zeta_4;\zeta_4). \tag{7.141}$$

According to (7.134) and using (7.135), (7.136) and (7.141), the four-nucleon momentum distribution for alpha-clusters in CDFM can be obtained in the form:

$$n^{(\alpha)}(k_1,k_2,k_3,k_4) = \sum_{\eta_1,\eta_2,\eta_3,\eta_4} \rho^{(\alpha)}(\zeta_1,\zeta_2,\zeta_3,\zeta_4;\zeta_1,\zeta_2,\zeta_3,\zeta_4)$$

$$= \int_0^\infty dx\,|f(x)|^2 V^4(x) \prod_{i=1}^4 \Theta(k_F(x) - |k_i|) \tag{7.142}$$

with the normalization condition:

$$\frac{1}{(2\pi)^{12}} \int \prod_{i=1}^4 (dk_i)\, n^{(\alpha)}(k_1,k_2,k_3,k_4) = (A/4)^4. \tag{7.143}$$

The quantity $(A/4)^4$ is the total number of alpha-particles in the case when any four nucleons with different values of the quantum numbers $\{\sigma\tau\}$ form an alpha-cluster in nuclei with even number of protons and neutrons and $Z = N = A/2$ [7.208].

Introducing Jacobi momenta P, p_1, p_2, p_3 [7.209]:

$$P = k_1 + k_2 + k_3 + k_4, \quad p_1 + \tfrac{1}{2}p_2 + \tfrac{4}{3}p_3 = k_1 - k_4,$$

$$-p_1 + \tfrac{1}{2}p_2 + \tfrac{4}{3}p_3 = k_2 - k_4, \quad -p_2 + \tfrac{4}{3}p_3 = k_3 - k_4, \tag{7.144}$$

the centre-of-mass AMD $n_{\text{c.m.}}^{(\alpha)}(P)$ can be obtained:

$$n_{\text{c.m.}}^{(\alpha)}(P) = \int \frac{d\Omega_P}{(2\pi)^3} \int n^{(\alpha)}(P,p_1,p_2,p_3) \prod_{i=1}^3 (dp_i) \frac{1}{(2\pi)^9} \tag{7.145}$$

with the normalization condition:

$$\int_0^\infty n_{\text{c.m.}}^{(\alpha)}(P) P^2\,dP = 1 \tag{7.146}$$

and the notation

$$\Omega_P = \{\theta_P, \varphi_P\}. \tag{7.147}$$

Taking account of the explicit form of $n^{(\alpha)}(P,p_1,p_2,p_3)$, Eq. (7.145) can be written in the form:

$$n_{\text{c.m.}}^{(\alpha)}(P) = \left(\frac{4}{A}\right)^4 \frac{1}{(2\pi)^{12}} \int d\Omega_P \int dp_1 \int dp_2 \int dp_3 \int_0^a dx\,|f(x)|^2 (\tfrac{4}{3}\pi x^3)^4 \tag{7.148}$$

with $a = \alpha/\max\{S_1, S_2, S_3, S_4\}$, where $\max\{S_1, S_2, S_3, S_4\}$ is the largest of the quantities

$$S_1 \equiv |P/4 + p_1 + p_2/2 + p_3/3|,$$
$$S_2 \equiv |P/4 - p_1 + p_2/2 + p_3/3|,$$
$$S_3 \equiv |P/4 - p_2 + p_3/3|,$$
$$S_4 \equiv |P/4 - p_3|$$

(7.149)

and α is given by Eq. (7.139).

As in the case of TNMD, the calculations of $n_{c.m.}^{(\alpha)}$ in the CDFM have been made [7.207] using Eq. (7.118) for the weight function $|f(x)|^2$ and the symmetrized Fermi density distribution. The results for $n_{c.m.}^{(\alpha)}$ calculated for ^{9}Be, ^{12}C, ^{16}O, ^{20}Ne, ^{24}Mg, ^{28}Si, ^{32}S and ^{40}Ca using Monte Carlo method are given in Figs. 7.21 and 7.22. The values of the parameters R (the half-radius) and

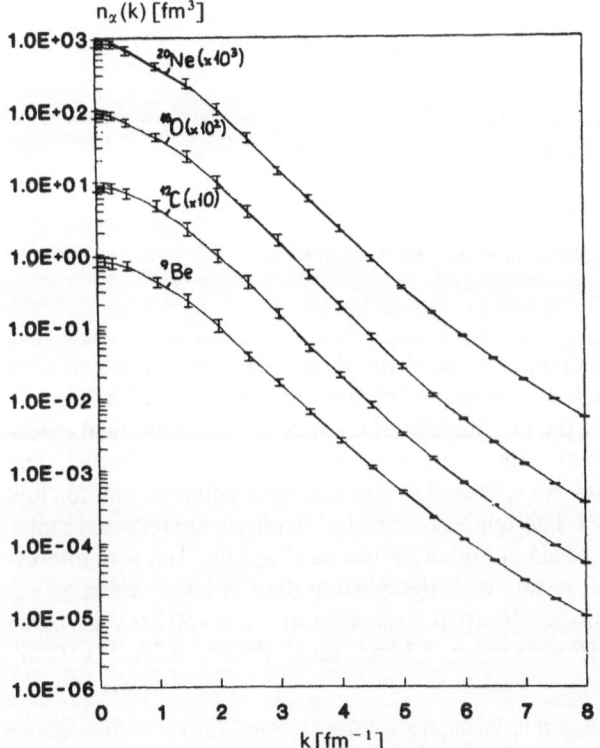

Fig. 7.21. The alpha-particle momentum distributions for ^{9}Be, ^{12}C, ^{16}O and ^{20}Ne. The error bars indicate the uncertainties in the Monte Carlo calculations. The normalization is: $\int_0^\infty n_{cm}^{(\alpha)}(k)k^2\,dk = 1$.
[7.207]

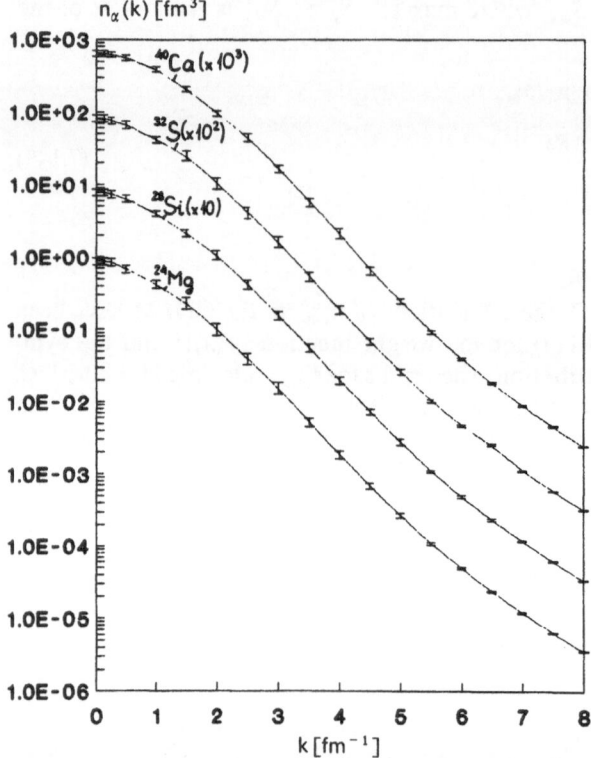

Fig. 7.22. The same as in Fig. 7.21 but for ^{24}Mg, ^{28}Si, ^{32}S and ^{40}Ca

b (surface diffuseness) of the symmetrized Fermi density distribution which have been obtained from elastic electron scattering data are given in Table 7.1.

It should be noted that the possibilities of comparing the theoretical calculations of $n_{\text{c.m.}}^{(\alpha)}(P)$ in CDFM with some experimental results are limited. The available experimental data are scarce, they are mainly qualitative and for low momenta $(0 < P < 1 \text{ fm}^{-1})$. Though N–N correlation effects are reflected mainly in the behaviour of the AMD at higher momenta $(P \gtrsim 2 \text{ fm}^{-1})$ it is of interest to compare the theoretical results with the existing data at low momenta.

The PWIA analysis of the ^{12}C (p, pα) reaction at $E_{\text{p}} = 150$ MeV [7.189c] leads to values of $n^{(\alpha)}(P = 0) = 1.9$–$5.3 \text{ fm}^3$ and $n^{(\alpha)}(P = 1 \text{ fm}^{-1}) = 0.56$–

Table 7.1. Values of the parameters R (half-radius) and b (surface diffuseness) of the symmetrized Fermi density distribution in the calculations of $n_{\text{c.m.}}^{(\alpha)}$ in CDFM [7.207] (in fm)

Nuclei	^{9}Be	^{12}C	^{16}O	^{20}Ne	^{24}Mg	^{28}Si	^{32}S	^{40}Ca
R	1.80	2.214	2.562	2.61	2.934	3.085	3.255	3.556
b	0.46	0.488	0.497	0.55	0.569	0.563	0.601	0.578

1.60 fm^3. These values depend on the uncertainties in the determination of the effective number $N_{\text{eff}}(\alpha)$ of alpha-clusters in ^{12}C (in [7.189c] which is estimated to be $N_{\text{eff}}(\alpha) = 0.30^{+0.23}_{-0.11}$). The .CDFM results are $n^{(\alpha)}(P = 0) = 0.9$ fm^3 and $n^{(\alpha)}_{\text{c.m.}}(P = 1\ \text{fm}^{-1}) = 0.45$ fm^3, which are of similar magnitude to those from [7.189c].

The values of $n^{(\alpha)}_{\text{c.m.}}$ at zero momenta extracted from the PWIA analysis of $(\alpha, 2\alpha)$ $(E_\alpha = 700\ \text{MeV})$ reaction [7.192] for ^{12}C and ^{16}O are $n^{(\alpha)}_{^{12}\text{C}}(P = 0) = 1.36$ fm^3 and $n^{(\alpha)}_{^{16}\text{O}}(P = 0) = 1.21$ fm^3. The CDFM result for both nuclei is $n^{(\alpha)}(P = 0) \simeq 0.9$ fm^3. Thus these theoretical predictions are not in contradiction with the experimental data considered. The theoretical results for $n^{(\alpha)}_{\text{c.m.}}(P = 0)$ are almost constant for $12 < A < 28$ and decrease monotonically for larger A in agreement with the result of *Dollhopf* et al. [7.192].

In the region $P \simeq 4$–7 fm^{-1} the CDFM AMD $n^{(\alpha)}_{\text{c.m.}}$ can be reproduced well by the function $\exp(-P/P_0)$ with P_0 independent of A within 10%. This agrees with the results from (p, α) inclusive reactions at intermediate energies $(E_p = 210, 300$ and $480\ \text{MeV})$ [7.138] and proton-$(E_p = 90\ \text{MeV})$ and alpha-$(E_\alpha = 140\ \text{MeV})$ induced inclusive reactions [7.196]. In the case of ^{9}Be $P_0 \simeq 130\ \text{MeV}/c$. The estimate from [7.138] gives $P_0 \simeq 75\ \text{MeV}/c$. For ^{12}C, P_0 is $122\ \text{MeV}/c$.

Some additional information on the low-momenta $n^{(\alpha)}_{\text{c.m.}}(P)$ behaviour $(P < 0.5\ \text{fm}^{-1})$ can also be obtained from ^{16}O and ^{28}Si $(\alpha, 2\alpha)$ reactions at $E_\alpha = 650\ \text{MeV}$ and $850\ \text{MeV}$ [7.191], from ^{12}C $(\alpha, 2\alpha)$ [7.208], as well as from $(p, p\alpha)$ reactions on light nuclei [7.210].

The alpha-particle centre-of-mass momentum distributions obtained in the CDFM may be used in calculations of cross-sections of alpha knockout reactions.

The two-body d + p amplitudes in ^{3}He, as well as the t + p and d + d amplitudes in ^{4}He and the corresponding momentum distributions are calculated by *Schiavilla* et al. [7.47a] within the improved variational Jastrow-type correlation method (see Chap. 3) using the realistic Urbana and Argonne two-nucleon interactions and their three-nucleon interaction model VII. The d + p momentum distribution in ^{3}He is also calculated using Faddeev wave function. As can be seen from Fig. 7.23, the calculated d + p momentum distribution in ^{3}He at low momenta is in good agreement with the results obtained by the Saclay analysis of the ^{3}He (e, e'p)d and ^{3}He (e, e'p) (n + p) reactions [7.115] in the PWIA. It is concluded that final-state interactions might be responsible for the disagreement between the calculated and the PWIA results [7.211].

The proton and triton momentum distributions in ^{4}He have been extracted in Dubna experiments [7.212] on ^{12}C(^{4}He, p) and ^{12}C(^{4}He, t) reactions at beam momenta of 4.52 and 2.69 GeV/N, respectively, with a fragment emission angle $\theta° < 0.4°$. A relativistic impulse approximation has been used in the analysis of the data. The triton momentum distribution $n_t(k)$ extracted from these experiments [7.212] exceeds for $k > 0.2\ \text{GeV}/c$ the theoretical calculations of *Schiavilla* et al. [7.47a] and the ATMS results of *Morita* et al. [7.80]. The results do not confirm the existence of a dip at $k \simeq 0.4\ \text{GeV}/c$ predicted theoretically in [7.47a, 7.80].

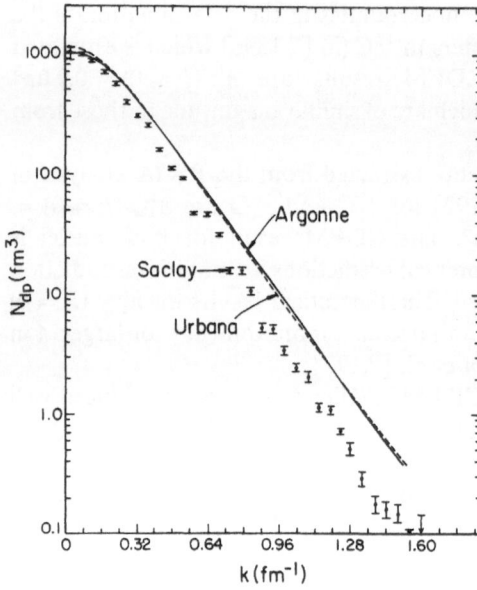

Fig. 7.23. The d + p momentum distribution in ³He calculated in [7.47a] compared with the PWIA analysis results of the Saclay electron scattering data [7.115]

The p + t-momentum distribution for recoil momenta up to 350 MeV/c has been extracted from (e, e′p) reaction for ⁴He by *Van den Brand* et al. [7.213].

The breakup of a nucleus and the momentum distribution of one of the pieces (a fragment of K nucleons) have been considered within a purely kinematical model in the spirit of the Thomas–Fermi approximation by *Gan* et al. [7.214]. Momentum distributions of fragments with nearly Gaussian-like form have been obtained. It is found that Gaussian widths agree well with the data on ¹⁶O fragmentation.

8 Ground and Excited Collective Nuclear States within the Coherent Density Fluctuation Model and the Generator Coordinate Method

In this chapter we extend the coherent density fluctuation model and the generator coordinate method and use them to make detailed studies of the energies and density distributions of the ground and collective states in nuclei.

8.1 Energies and Density Distributions in Ground and Excited Monopole States within the Coherent Density Fluctuation Model

The basic dynamical equation of the coherent density fluctuation model (CDFM) [8.1–3] (Sects. 4.1,2)

$$-\frac{\hbar^2}{2m_{\text{eff}}} \frac{d^2 f(x)}{dx^2} + V(x) f(x) = E f(x) \tag{8.1}$$

has been obtained from the generator coordinate method (GCM) integral equation (4.4) using the relations ((4.9)–(4.11)) or the delta-function limit ((4.15)–(4.16)). The Eq. (8.1) enables us to calculate the energies E_n and the corresponding functions $f_n(x)$ $(n = 0, 1, 2, \ldots)$ of the ground state $(n = 0)$ and the excited nuclear states $(n = 1, 2, \ldots)$. The density distributions $\rho_n(r)$ have the form:

$$\rho_n(r) = \int_0^\infty |f_n(x)|^2 \rho_0(x) \Theta(x - |r|) \, dx \tag{8.2}$$

with

$$\rho_0(x) = 3A/4\pi x^3, \tag{8.3}$$

A being the mass number.

The effective mass m_{eff} is considered to be a constant in the CDFM. Its value is determined by the fit of the theoretical calculations of the ground state energy of various nuclei to the experimental data.

The function $V(x)$ corresponds to the potential energy of the coherent motion of all A nucleons. It is determined by the matrix element of the nuclear

Hamiltonian $\hat{H}$ (Eq. (4.13)):

$$V(x) = \langle \Phi(\{r_i\}, x) | \hat{H} | \Phi(\{r_i\}, x) \rangle, \tag{8.4}$$

where the CDFM generating function $\Phi(\{r_i\}, x)$ describes the state of A nucleons uniformly distributed in a sphere with radius x and the density $\rho_0(x)$ (Eq. (8.3)) (the so-called "fluctons"). Considering the fluctons as pieces of nuclear matter with density $\rho_0(x)$ one can use for $V(x)$ the corresponding nuclear matter energy by the method of *Brueckner* et al. [8.4–8.5] (Eqs. (7.26)–(7.28)). For convenience we give the corresponding expressions in the present Section as well:

$$V(x) = A V_0(x) + V_C - V_{CO}, \tag{8.5}$$

where

$$V_0(x) = 37.53 \left[(1 + \alpha)^{5/3} + (1 - \alpha)^{5/3}\right] \rho_0^{2/3}(x) + b_1 \rho_0(x) + b_2 \rho_0^{4/3}(x)$$
$$+ b_3 \rho_0^{5/3}(x) + \alpha^2 \left[b_4 \rho_0(x) + b_5 \rho_0^{4/3}(x) + b_6 \rho_0^{5/3}(x)\right] \tag{8.6}$$

with

$$\alpha = (N - Z)/(N + Z) \tag{8.7}$$

and

$$b_1 = -741.28; \qquad b_2 = 1179.89; \qquad b_3 = -467.54;$$
$$b_4 = 148.26; \qquad b_5 = 372.84; \qquad b_6 = -769.57.$$

$V_0(x)$ in (8.5) corresponds to the energy per nucleon in nuclear matter (in MeV) taking account of the neutron–proton asymmetry. V_C is the Coulomb energy of protons in a flucton:

$$V_C = \frac{3}{5} \frac{Z^2 e^2}{x} \tag{8.8}$$

and the flucton Coulomb exchange energy is:

$$V_{CO} = 0.7386 \, Z e^2 (3Z/4\pi x^3)^{1/3}. \tag{8.9}$$

As an example, the function $V(x)$ in the case of the ^{40}Ca nucleus is shown in Fig. 8.1. The minimum of $V(x)$ at $x_0 \simeq 3.6$ fm corresponds to the density $\rho_0(x_0) = 0.204 \, \text{fm}^{-3}$. The value of the minimum $|V_{\min}(x)| = 576.7$ MeV corresponds to the binding energy per nucleon of 14.4 MeV/N. We note the anharmonic form of the function $V(x)$ in the vicinity of x_0 and the repulsive part at $x < 2.7$ fm. The unphysical behaviour of $V_0(x)$ (8.5) [8.4–8.5] at small x (or large densities): $V_0(x)_{x \to 0} \to -\infty$ should also be mentioned.

The analysis of the solutions of (8.1) can be carried out approximating the function $V(x)$ [8.3] by the Morse potential [8.6] in the vicinity of x_0:

$$V(x) \to U_M(x) = U_0 \{\exp[-2\alpha(x - x_0)] - 2 \exp[-\alpha(x - x_0)]\}. \tag{8.10}$$

The value of the parameter U_0 is chosen to be equal to $|V(x_0)|$. The value of

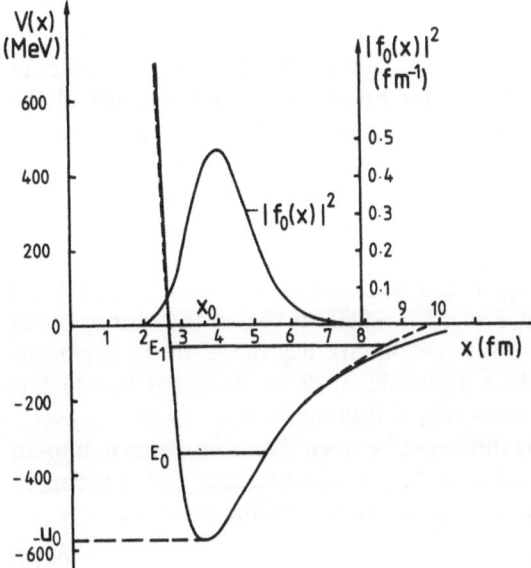

Fig. 8.1. The function $V(x)$ (8.5–8.9) (solid line), the Morse potential $U_M(x)$ (dashed line), the function $|f_0(x)|^2$ and the energies E_0 and E_1 in the case of ^{40}Ca nucleus

α can be found by the fit of $U_M(x)$ to $V(x)$ in the vicinity of x_0. The approximation of $V(x)$ by $U_M(x)$ is shown in Fig. 8.1 by the dashed line. The use of the Morse potential leads to the well-known analytical expressions for $f_n(x)$ and E_n. For instance, the normalized function $f_0(x)$ corresponding to the ground state has the form:

$$f_0(x) = \left(\frac{\alpha}{\Gamma(2s)}\right)^{1/2} \{k\exp[-\alpha(x-x_0)]\}^{s_0} \exp\left\{-\frac{k}{2}\exp[-\alpha(x-x_0)]\right\},$$

$$(8.11)$$

where

$$k = 2(2m_{\text{eff}}U_0)^{1/2}/\alpha\hbar, \tag{8.12}$$

$$s_0 = k/2 - 1/2 \tag{8.13}$$

and $\Gamma(2s)$ is the gamma function.

The function $f(x)$ (8.11) is a two-parameter function of α and U_0 ($f(x,\alpha,U_0)$). This allows one to determine the value of the parameter m_{eff} in the following variational way. Multiplying Eq. (8.1) by $f^*(x)$ and integrating over x one can obtain:

$$-\frac{\hbar^2}{2m_{\text{eff}}}\int f^*(x)\frac{d^2f(x)}{dx^2}\,dx + \int f^*(x)V(x)f(x)\,dx = E. \tag{8.14}$$

The left-hand side of (8.14) is minimized with respect to the parameters U_0 and

α of $f(x, \alpha, U_0)$ for each fixed value of m_{eff}, $V(x)$ having the form ((8.5)–(8.9)) and $\int |f(x)|^2\, dx = 1$. The particular value of E which coincides with the experimental binding energy determines the value of the parameter m_{eff}. The values of m_{eff} and the ground state energies E_0 for nuclei in different regions of the periodic table are given in Table 8.1 together with the approximation $m_{\text{eff}} = \text{const. } m_N/A$ (m_N being the nucleon mass). It can be seen that m_{eff} can be approximated by

$$m_{\text{eff}} \simeq 3\, m_N/A. \tag{8.15}$$

Concerning the small value of m_{eff} it can be noted that G. *Brown* et al. [8.8] obtained the nucleon ground state energy (~ 980 MeV) and the energy of the Roper-resonance (~ 1440 MeV) using the 3-quark bag model in the generator coordinate method and obtained a value of the nucleon effective mass which is much smaller than the nucleon mass: $m_{\text{eff}} = 0.48\, m_N$.

It is of interest to consider the difference between the value of the minimum of $V(x)$ ($|V(x_0)| = U_0$) and the value of the ground state energy E_0. It can be seen from Table 8.2 that for a wide range of nuclei this difference is:

$$U_0 - |E_0| \simeq (17.5\text{--}20)\, A^{2/3} \text{ MeV}. \tag{8.16}$$

This means that the existence of the collective vibrational motion (Eq. (8.1)) leads to a change of the average nucleon separation energy from 14–16 MeV/N (corresponding to the minimum value of $V(x)$ (8.5) for nuclear matter) to 8–9 MeV for finite nuclei, i.e. to the corrections of the type $\sim A^{2/3}$ in the Weizsäcker formula for the binding energy. Usually this correction is related to the existence of a surface in the finite nuclear system.

We shall call these vibrational motions "rigid vibrations". The problems related to the choice of the particular form of the function $V(x)$ in (8.1) can be studied by fixing the universal dependence of m_{eff} on A in the form (8.15) and then looking for an effective potential energy which is able to reproduce the experimental binding energy and the density distribution. The Morse potential $U_M(x)$ (8.10) gives certain possibilities for this study. The depth U_0 can be chosen to coincide with $|V(x_0)|$ and the value of α can be determined from the fit of the theoretical ground state energy E_0 to the experimental one. We shall consider

Table 8.1. The values of m_{eff} and the ground state energies E_0 (in MeV) in the CDFM

Nuclei	m_{eff}	$m_{\text{eff}} = \text{const. } m_N/A$	E_0	E_0 experiment [8.7]
^{16}O	186.0	3.17 m_N/A	-127.721	-127.624
^{40}Ca	70.5	3.00 m_N/A	-342.090	-342.063
^{90}Zr	31.0	2.97 m_N/A	-780.904	-783.916
^{116}Sn	24.6	3.04 m_N/A	-988.143	-988.714
^{140}Ce	22.0	3.28 m_N/A	-1177.446	-1172.696
^{208}Pb	16.0	3.54 m_N/A	-1635.800	-1636.492

the solution of Eq. (8.1) using the potential of Morse $(V(x) = U_M(x))$ (8.10). Introducing the notation

$$s = (- 2m_{\text{eff}} E)^{1/2}/(\alpha \hbar), \tag{8.17}$$

$$\xi = k \exp[- \alpha(x - x_0)], \tag{8.18}$$

$$\xi_0 = k \exp(\alpha x_0), \tag{8.19}$$

$$a = s + 1/2 - k/2 \tag{8.20}$$

(with k from (8.12)) the solution $f(x)$ of (8.1) which satisfies the conditions:

$$f(x) = 0 \text{ at } x \to \infty$$

and $\tag{8.21}$

$$f(x) = 0 \text{ at } x \to 0$$

has the form

$$f(x) = N\xi^s \exp(- \xi/2) \, {}_1F_1(a, 2s + 1, \xi). \tag{8.22}$$

In Eq. (8.22) N is a normalization constant and ${}_1F_1$ is the degenerate hypergeometrical function. The eigenvalues E_n are determined by the condition:

$${}_1F_1(a, 2s + 1, \xi)|_{\xi = \xi_0} = 0. \tag{8.23}$$

It turns out that the values of the parameters αx_0 and k for all the nuclei considered allow one to find good approximate solutions of the complicated Eq. (8.23). Following the method for solving (8.23) suggested in [8.9] it is shown in every particular case that the value of a practically coincides with one of the negative integers (or zero):

$$a = - n, \quad n = 0, 1, 2, \ldots . \tag{8.24}$$

In other words, the particular form of the potential $U_M(x)$ enables us to use as a good approximation the solutions of the problem with this potential when $- \infty < x < \infty$. In this case the energy levels are determined by [8.10]:

$$- E_n = U_0 \left[1 - \frac{\alpha \hbar}{(2m_{\text{eff}} U_0)^{1/2}} (n + 1/2) \right]^2, \tag{8.25}$$

when $n = 0, 1, 2, \ldots$ up to the maximum value for which the condition

$$\frac{(2m_{\text{eff}} U_0)^{1/2}}{\alpha \hbar} > n + 1/2 \tag{8.26}$$

is still satisfied (so that the parameter s_n according to (8.17), (8.20) and (8.24):

$$s_n = k/2 - (n + 1/2) \tag{8.27}$$

remains positive). The normalized functions $f_n(x)$ have the form:

$f_0(x)$ – see Eq. (8.11),

$$f_1(x) = \left(\frac{\alpha(2s_1 + 1)}{\Gamma(2s_1)}\right)^{1/2} [ke^{-\alpha(x-x_0)}]^{s_1} e^{-[\frac{k}{2}e^{-\alpha(x-x_0)}]}$$

$$\times \left[1 - \frac{ke^{-\alpha(x-x_0)}}{2s_1 + 1}\right], \tag{8.28}$$

$$f_2(x) = \frac{\alpha(2s_2 + 1)(2s_2 + 2)}{2\Gamma(2s_2)} [ke^{-\alpha(x-x_0)}]^{s_2} \cdot e^{-[\frac{k}{2}e^{-\alpha(x-x_0)}]}$$

$$\times \left\{1 - \frac{2ke^{-\alpha(x-x_0)}}{2s_2 + 1} + \frac{[ke^{-\alpha(x-x_0)}]^2}{(2s_2 + 1)(2s_2 + 2)}\right\}. \tag{8.29}$$

At the fixed value of E_0 (equal to the experimental binding energies) and using $U_0 = |V(x_0)|$ (from (8.5)) one can get from (8.25) (for $n = 0$) the following expression for the parameter α:

$$\alpha = \frac{2(2m_{\text{eff}} U_0)^{1/2}}{\hbar}\left[1 - \left(\frac{|E_0|}{U_0}\right)^{1/2}\right]. \tag{8.30}$$

From Eqs. (8.26) and (8.30) we obtain the relation

$$\frac{|E_0|}{U_0} > \left(\frac{2n}{2n + 1}\right)^2, \quad (n = 1, 2, 3, \ldots) \tag{8.31}$$

which gives the condition for the existence of monopole "rigid vibrations" within the CDFM [8.2]. It can be seen that (8.31) contains an A-dependence through the binding energy E_0 (which in CDFM is the energy of the zero-

Table 8.2. Energies of extreme breathing states within the CDFM (in MeV)

Nuclei	$\dfrac{U_0 - |E_0|}{A^{2/3}}$	$|E_0|/U_0$	$E_1^* = |E_0| - |E_1|$	$E_2^* = |E_0| - |E_2|$
^{4}He	14.6	0.434	—	—
^{12}C	18.4	0.488	90	—
^{16}O	18.9	0.514	122	—
^{24}Mg	19.7	0.546	181	—
^{28}Si	19.6	0.565	209	—
^{32}S	19.9	0.575	236	—
^{40}Ca	20.0	0.593	286	—
^{48}Ca	19.1	0.622	326	—
^{56}Fe	20.1	0.626	382	—
^{58}Ni	20.0	0.627	392	—
^{66}Zn	20.2	0.637	437	—
^{90}Zr	19.8	0.663	551	777
^{116}Sn	19.5	0.680	660	966
^{140}Ce	18.8	0.698	741	1119
^{208}Pb	17.3	0.730	928	1473
^{208}Pb[8.11]			(900)	(1840)
^{235}U	17.4	0.730	1011	1606

vibrational motion at the potential $V(x) \simeq U_M(x)$ and $m_{eff} = 3\, m_N/A$) and the depth of the potential U_0 ($U_0 \simeq$ const. A). The values of the ratio $|E_0|/U_0$ for various nuclei are presented in Table 8.2.

The first excited state appears for all nuclei with $A \geqslant 12$; from (8.31) the condition for this is ($|E_0|/U_0) > \frac{4}{9} = 0.444 \ldots$. The second excited state appears for nuclei with $A \gtrsim 70$; the condition is ($|E_0|/U_0) > \frac{16}{25} = 0.640$. The approximate dependence of the excited state energies $E_n^* = |E_0| - |E_n|$ ($n = 1, 2$) on A can be found in the form:

$$E_1^* \cong 9.6\, A - E_s A^{2/3} + 115, \quad A = 12\text{--}80,$$

$$E_1^* \cong 4.6\, A - E_s A^{2/3} + 575, \quad A > 80, \qquad\qquad (8.32)$$

$$E_2^* \cong 7.7\, A - E_s A^{2/3} + 484, \quad A \gtrsim 70,$$

$$U_0 - |E_0| = E_s A^{2/3},$$

where $E_s \cong 19$ MeV. The numerical results for E_1^* and E_2^* are given in Table 8.2. The binding energies per nucleon $|E_n|/A$ ($n = 1, 2, \ldots$) are also given in Fig. 8.2.

The solutions of Eq. (8.1) $f_n(x)$ ((8.11), (8.28, 8.29)) determine from (8.2) the distributions of the nuclear density $\rho_n(r)$. The functions $|f_n(x)|^2$ and the corresponding densities $\rho_n(r)$ (for $n = 0, 1$) in the case of ^{40}Ca nucleus are given in Fig. 8.3 and in Fig. 8.4, respectively. The ground state density $\rho_0(r)$ is in a qualitative agreement with the symmetrized Fermi density ρ_{SF} with some increase of the half-radius (for ^{40}Ca: from $R_{1/2} = 3.556$ fm for ρ_{SF} to 3.70 fm for $\rho_0(r)$) and a decrease of the diffuseness (from 0.578 fm for ρ_{SF} to 0.534 fm for

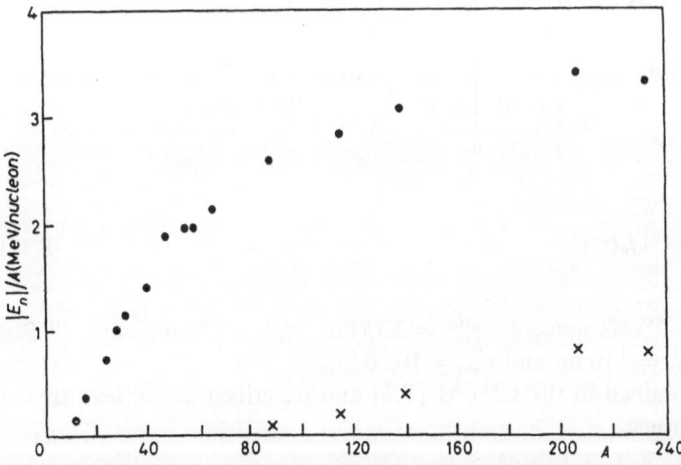

Fig. 8.2. Energies per nucleon in the first $|E_1|/A$ (points) and in the second $|E_2|/A$ (crosses) excited states

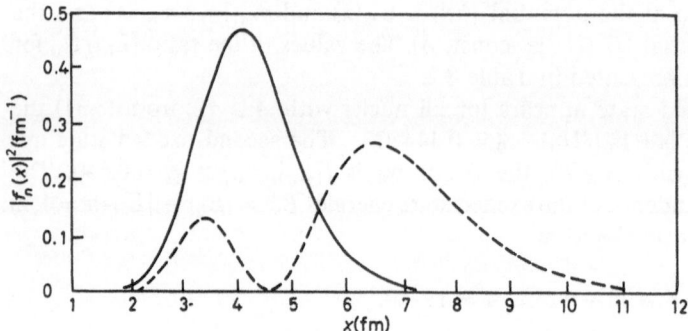

Fig. 8.3. The functions $|f_0(x)|^2$ (solid line) and $|f_1(x)|^2$ (dashed line) for ^{40}Ca

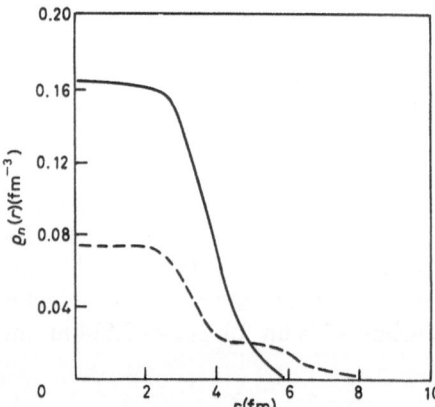

Fig. 8.4. The nuclear density distributions in the ground state ($\rho_0(r)$ – solid line) and in the excited state ($\rho_1(r)$ – dashed line) for ^{40}Ca

$\rho_0(r)$). The rms radius can be calculated using the expression:

$$r_{rms}^{(n)2} = \frac{\int r^2 \rho_n(r)\, dr}{\int \rho_n(r)\, dr} = \frac{4\pi}{A} \int_0^{\infty} r^4\, dr \int_0^{\infty} |f_n(x)|^2 \frac{3A}{4\pi x^3} \Theta(x - r)\, dx$$

$$= \frac{3}{5} \int_0^{\infty} dx \cdot x^2 |f_n(x)|^2 . \tag{8.33}$$

In the case of ^{40}Ca nucleus: $r_{rms}^{(0)} = 3.35$ fm; $r_{rms}^{(1)} = 5.35$ fm; for ^{208}Pb: $r_{rms}^{(0)} = 5.55$ fm; $r_{rms}^{(1)} = 7.10$ fm and $r_{rms}^{(2)} = 10.20$ fm.

The results obtained in the CDFM [8.2] and described above lead to the following conclusions:

i) The energies of the breathing vibrational states ("rigid vibrations") are comparable with the nuclear binding energies. This type of vibration corresponds to a high incompressibility of the medium.

ii) The CDFM predicts the existence of a threshold with respect to the mass number A for the appearance of the vibrational collective states. The first excited level appears in all nuclei with $A \geqslant 12$ and the second one appears only in the nuclei beyond $A \approx 70$.

iii) The nuclear size is essentially larger in the excited states than in the ground state. The nuclear density in the excited state decreases in the centre and increases in the nuclear surface region.

The CDFM results for ^{208}Pb nucleus are in good agreement with those obtained by *Balazs* and *Pauli* [8.11] by a method generalizing the Thomas–Fermi model; these are given in Table 8.2.

Now we shall discuss the experimental possibilities of observing the "rigid vibrational" states predicted by the CDFM. It can be expected that such states can be excited in reactions with large transferred energies, such as deep-inelastic particle-nucleus interactions, pion absorption by nuclei, and high energy interactions of complex nuclei. If the breathing states are excited in such processes one can expect some specific consequences of their decay. The nuclear density in the excited states is significantly lower than that in the ground state. This gives possibilities for clusterization in nuclei (mainly alpha-clusters) [8.12]. An isotropic emission of alpha-particles and fast neutrons can be expected [8.13]. The average energy of the charge fragment from ^{208}Pb is roughly estimated to be around 5 MeV. Such isotropically emitted low-energy alpha-particles have been observed by *Schlepütz* et al. [8.14].

Another possibility is related to the abnormal nuclear fission of heavy nuclei. In reactions of 1 GeV protons with ^{238}U, ^{232}Th, ^{197}Au and ^{184}W, events with a large nucleon and two-fragment emission with kinematics quite different from the fission kinematics have been observed [8.15–18]. It is found in [8.18] that such events differ from the binary fission by the enhanced total kinetic energy of the two fragments and by the significant value of the transferred transverse momentum. These results cannot be explained as ternary fission. It is concluded in [8.18] that the experimental data can be interpreted as the production of highly excited unstable "third fragment" and a successive fission of the residual nucleus. Data confirming the existence of such a mechanism for collective excitation accompanied by a binary fission have also been observed in 11.5 GeV proton scattering on ^{238}U [8.19] and in the experimental investigations of the deep spallation in heavy nuclei. This has been done by measuring the angular distributions of ^{128}Ba from ^{238}U [8.20] and isotopes of Xe and Kr from ^{232}Th [8.21].

The existence of a highly excited unstable "third fragment" in the reactions considered above can be interpreted by means of the excited collective states predicted in the CDFM as a result of high-energy vibrations of the nuclear density.

It is of interest to analyze the isoscalar giant monopole resonance or the so-called breathing vibrational states (with $I^{\pi} = 0^+$ at energies of approximately 13 to 20 MeV) in the framework of the CDFM. These states have been well established experimentally [8.22–29] and are considered to be compressional nuclear vibrations. Their study concerns the problem of the compressibility of finite nuclei and nuclear matter (e.g. in [8.30–34]). The description of such states is related mainly to general characteristics of atomic nuclei and weakly to the peculiarities of nuclear structure. This gives the possibility of using both microscopic and macroscopic methods to study these states [8.33].

The appearance of such collective states in the framework of the CDFM [8.35] is related to another type of nuclear density vibrations which will be called "soft vibrations" in contrast with the "rigid vibrations" considered above.

Following *Brueckner* et al. [8.30] the energy of these vibrations can be written

$$\hbar\omega_{\text{s.v.}} \simeq \frac{\hbar}{x_0 A^{1/3}} \left(\frac{K_A}{m_{\text{N}}}\right)^{1/2}, \tag{8.34}$$

where $x_0 A^{1/3}$ is the nuclear radius in the equilibrium state of the nucleus. The incompressibility of finite nuclei K_A can be expressed by analogy with the nuclear matter incompressibility (K_∞) in the form [8.31, 8.36]:

$$K_A = r_0^2 \frac{1}{A} \frac{\partial^2 E_b(r_{\text{rms}})}{\partial r_{\text{rms}}^2}\bigg|_{r_{\text{rms}}=r_0}. \tag{8.35}$$

In (8.35) $E_b(r_{\text{rms}})$ is the nuclear binding energy as a function of the rms radius r_{rms}. The value of r_0 is determined by the minimum of the function $E_b(r_{\text{rms}})$. The binding energy can be expanded in the vicinity of r_0. Keeping the first two terms only:

$$E_b(r_{\text{rms}}) = E_b(r_0) + \frac{\beta}{2}(r_{\text{rms}} - r_0)^2, \tag{8.36}$$

the following expression for the energy of the soft monopole vibrations can be obtained by means of (8.34–36):

$$\hbar\omega_{\text{s.v.}} = \left(\frac{\hbar^2}{Am_{\text{N}}} \frac{\partial^2 E_b(r_{\text{rms}})}{\partial r_{\text{rms}}^2}\bigg|_{r_{\text{rms}}=r_0}\right)^{1/2} = \left(\frac{\hbar^2\beta}{Am_{\text{N}}}\right)^{1/2}. \tag{8.37}$$

The values of the energies of the breathing monopole states for different nuclei can be calculated by using (8.14) and the function (8.11) in the following way. The energy $E_b(r_{\text{rms}})$ is evaluated from (8.14) at a fixed value of the parameter $m_{\text{eff}} = 3m_{\text{N}}/A$ for each pair of parameter values (α, U_0) in $f_0(x)$ (8.11). The rms radius can be determined for the pair (α, U_0) by (8.33) for $n = 0$. Let $(\bar{\alpha}, \bar{U}_0)$ be the values of α and U_0 at which the energy E_b has a minimum that coincides with the experimental binding energy (e.g., for ^{16}O: $\bar{\alpha} = 0.98$ fm^{-1} and $\bar{U}_0 = 216$ MeV). This minimum is the lowest value of the energy for a given set of values of α and U_0 at $m_{\text{eff}} = 3m_{\text{N}}/A$. The function $E_b(r_{\text{rms}})$ in the form (8.36) and the value $\beta_{(U)}$ are determined at a fixed value of $\bar{\alpha}$ and with variation of U_0. The same procedure is made at fixed $\bar{U}_0$ with variations of α in order to determine $\beta_{(\alpha)}$. It turns out that the first procedure (with variations of U_0 at fixed $\bar{\alpha}$) gives $\beta_{(U)}$ which leads to energy values $\hbar\omega_{\text{s.v.}}$ (8.37) being in accord with the energies of 0^+ monopole states calculated in other ways [8.30, 8.31, 8.37] and with some available experimental data [8.24, 8.28, 8.29, 8.33, 8.38]. The energies determined by $\beta_{(\alpha)}$ are sufficiently larger and correspond obviously to another type of nuclear motion. The energies of the 0^+ vibrational states (the "soft vibrations") are given and compared with other calculations and experimental data in Table 8.3. It is seen that the CDFM results are in good agreement with the result of *Brink* and *Nash* [8.37] for ^{16}O nucleus and exceed the values of the isoscalar giant monopole resonance obtained in different experiments for heavier nuclei. For nuclei with $A \gtrsim 40$ the CDFM breathing state energies can be described approximately by $\hbar\omega_{s.v.} \simeq 94\ A^{-1/3}$ MeV, whilst the energies predicted by *Brueckner* [8.30] are $\hbar\omega \simeq 97\ A^{-1/3}$ MeV. The experimental values can be approximated by $\hbar\omega \simeq 80\ A^{-1/3}$ MeV [8.33].

The incompressibility of finite nuclei K_A has been calculated in the CDFM [8.35] by means of (8.34), where the energies obtained are used instead of $\hbar\omega_{s.v.}$ and the $x_0 A^{1/3}$ are taken to be equal to the experimentally-known mean-square radii. The values of K_A for various nuclei obtained in this way are given in Table 8.4. Following the evaluation of the nuclear matter incompressibility K_∞ from [8.33]:

$$K_\infty = 15 K_A/\pi^2 \tag{8.38}$$

and substituting for K_A the approximate value $K_A \approx 200$ MeV (for nuclei with $A \gtrsim 40$) we obtain $K_\infty \simeq 300$ MeV. This value of the nuclear matter incompressibility is in accord with the result of *Sharma* et al. [8.28]: $K_\infty = 300 \pm 25$ MeV.

Table 8.3. Energies of 0^+ breathing excited states (in MeV)

Nuclei	$\hbar\omega_{s,v}$ CDFM [8.35]	$\hbar\omega$ [8.30]		$\hbar\omega$ [8.37]	$\hbar\omega$ [8.31]	$\hbar\omega_{exp}$ [8.24, 8.28, 8.29, 8.33]
		$\eta = 8$	$\eta = 12$			
^{16}O	23.2	30.5	30.0	24		22.9 $(T = 2)$ [8.38]
^{40}Ca	27.3	29.5	26.5		18.3	20.0
^{90}Zr	21.0	22.0			16.4	16.2
^{116}Sn	18.0	21.0				15.6
^{140}Ce	18.6	19.5				14.8
^{208}Pb	17.0	16.5	15.5		13.2	13.7

Table 8.4. The incompressibility of finite nuclei K_A in the CDFM [8.35] (in MeV)

Nuclei	^{16}O	^{40}Ca	^{90}Zr	^{116}Sn	^{140}Ce	^{208}Pb
K_A	96	216	194	167	196	206

To conclude this Section we emphasize that the two qualitatively different types of monopole vibrational states (related to "rigid" [8.2] and "soft" [8.35] vibrations) occur in the CDFM. Comparing these two types of collective motions we note that the "rigid vibration" energies depend on the nuclear matter incompressibility which is larger [8.31, 8.33] than the finite nuclei incompressibility which determine the "soft vibration" energies. In contrast with the "rigid" vibrational states which are related to volume density vibrations, the "soft" monopole states originate from the vibrations of the nuclear surface.

8.2 Generator Coordinate Method Calculations of Energies and the Density Distributions of the Ground and Excited Monopole States

In this Section the calculations of ground and excited state characteristics in nuclei are based on the generator coordinate method (GCM) equation for the weight function $f(x)$ [8.39] (see Sect. 4.1):

$$\int [\mathscr{H}(x, x') - E I(x, x')] f(x') \, dx' = 0. \tag{8.39}$$

In the case of Slater determinant generating functions and effective Skyrme nucleon-nucleon forces the expressions for the energy kernel $\mathscr{H}(x, x')$ and the overlap kernal $I(x, x')$ are given for $Z = N$ nuclei by (4.34)–(4.40). The nuclear density distribution can be calculated by (4.43).

Two construction potentials, namely the square-well potential with infinite walls (4.47) and the harmonic oscillator potential (4.46) have been used in the calculations of the energies, the density distributions and rms radii of the ground and first monopole states in ^{4}He, ^{16}O and ^{40}Ca nuclei [8.40]. The values of the Skyrme parameters in the case of the square-well construction potential are determined to give an optimal fit to the binding energies of ^{4}He, ^{16}O and ^{40}Ca. The following parameter values have been obtained: $t_0 = -2765.0$; $t_1 = 383.94$; $t_2 = -38.04$; $t_3 = 15865$ and $\sigma = 1/6$. With this parameter set the following infinite nuclear matter characteristics have been obtained using ((4.30)–(4.33)): $E/A = -16.78$ MeV/N, $K = 235.1$ MeV, $m^*/m = 0.7$ and the equilibrium density $\rho_0 = 0.148$ fm^{-3}. In the case of the harmonic oscillator construction potential, SkM* parameter set values giving realistic binding energies have been used [8.41].

In Table 8.5 the calculated values of the energies of the ground and the first excited monopole states (without the Coulomb energy) as well as the rms radii are presented.

The energies and radii obtained in this GCM approach are in general agreement with the results of the GCM calculations of *Flocard* and *Vautherin* [8.42]. We emphasize that the difference in the energies $\Delta E = E_1 - E_0$ turns out to be not strongly dependent on the type of construction potential used. In [8.42] ΔE is 27.58 MeV for ^{4}He, 31.67 MeV for ^{16}O and 28.4 Mev for ^{40}Ca in

Table 8.5. Energies (MeV) and rms radii (fm) of the ground and first excited monopole states in GCM with the square-well construction potential with infinite walls and Skyrme effective forces

Nuclei	E_0	r_0	E_1	r_1	$\Delta E = E_1 - E_0$
^{4}He	-37.06	1.78	-9.83	2.75	27.23
^{16}O	-144.58	2.63	-111.37	2.90	33.21
^{40}Ca	-402.92	3.40	-369.21	3.52	33.71

the case of the harmonic oscillator potential. As in [8.42] the inclusion of monopole vibrations changes significantly the value of the rms radius from the ground to the first excited monopole state in ^{4}He nucleus. For ^{16}O and ^{40}Ca nuclei this change is smaller.

The calculated GCM proton density distributions in ^{4}He, ^{16}O and ^{40}Ca are shown in Fig. 7.13 for the cases of square-well and harmonic oscillator construction potentials. They are compared with the calculations within the Hartree–Fock method using the SkM* forces. It can be noted that in spite of the fact that the generating functions for the square-well construction potential with infinite walls are nonzero only for a finite space volume, the resulting density distributions have the correct asymptotic form. This is apparently due to the inclusion of the collective zero-motion dynamics in the GCM procedure. The evaluated densities are close to the Hartree–Fock ones in the important surface region for the nuclei considered. The differences are larger in the central region of the nuclei. The proton charge distribution of *Sick* [8.43] extracted from the experimental data by means of model-independent analysis is also shown in Fig. 7.13 for the ^{40}Ca nucleus.

The GCM calculations of nucleon and two-nucleon momentum distributions presented in Chap. 7, as well as of natural orbitals and occupation numbers (given in Sect. 6.4) show that the chosen square-well construction potential with infinite walls is responsible for the effective account of the short-range nucleon–nucleon correlations within the proposed GCM approach. The results of this Section show that it is also possible to describe satisfactorily the density distributions in the surface region as well as the energies and the rms radii in the ground and first excited monopole states in accord with the results from other GCM calculations [8.42].

Part 3
Nucleon Correlations and Nuclear Reactions

9 Electron–nucleus Scattering

The coherent density fluctuation model is applied in this chapter using the high-energy approximation to describe electron elastic scattering by nuclei (Sect. 9.1). The problems of explaining the cross-sections and the longitudinal and transverse response functions obtained in quasielastic electron scattering by various nuclei are discussed with different theoretical methods in Sect. 9.2. The results for the Coulomb sum rule are also presented in the same Section. The theoretical results obtained by methods in which nucleon binding and Fermi motion are taken into account are compared with the experimental results on deep-inelastic lepton-nucleus scattering in Sect. 9.3. The role of the short-range correlations for a reasonable explanation of this process in the region $0.2 \lesssim x \lesssim 0.7$ is particularly emphasized. The results of some quark models are also discussed.

9.1 Electron Elastic Scattering

The electromagnetic interaction of electrons with nuclei leads to great simplification of the analysis of the elastic and inelastic scattering of electrons compared with hadron–nucleus scattering. The electron–nucleus scattering gives important information on the static (e.g. the charge distribution, rms radii etc.) as well as the various dynamic properties of atomic nuclei. It is also a useful tool to study the nucleon–nucleon dynamic correlations at small distances (e.g. [9.1]).

One of the methods of analyzing the electron–nucleus scattering is to solve the Dirac equation and to determine the scattering phases followed by summing the partial amplitudes (the phase shift analysis method). The calculations with this method face difficulties related to the necessity of summing series of different partial amplitudes with opposite signs. The direct relation between the charge distribution and the scattering cross-section has often been lost in such calculations [9.2].

The study of nuclear structure imposes the following condition on the wave number k:

$$kR \gg 1, \tag{9.1}$$

where R is the nuclear size. The elastic scattering cross-section for electrons with wave number k satisfying (9.1) has characteristic diffractional maxima and minima at scattering angles $\theta \gg (kR)^{-1}$ which depend strongly on the size and surface diffuseness of the charge distribution.

The Born approximation can be used to study the electron scattering on light nuclei with charge number Z which satisfies the condition

$$Z\frac{e^2}{\hbar c} \ll 1, \quad \text{i.e. } Z \ll 137. \tag{9.2}$$

The distortion of the electron wave and the change of the wave length in the nuclear Coulomb field are not taken into account in this approximation. The Born approximation successfully describes electron scattering on light nuclei except around the diffractional cross-section minima.

For medium and heavy nuclei the distortions of the incident and the scatterred waves have to be taken into account. In these cases theoretical methods based on the high-energy approximation (HEA) have been used. The basic condition for the HEA is

$$E \gg V, \tag{9.3}$$

where V is the potential energy of the interaction between the electron and the nuclear charge, E being the incident electron energy. The calculation of the scattering amplitude by these methods are based on relationships which are similar to those used in the Born approximation but electron wave functions with distorted amplitude and phase have to be used instead of plane waves [9.1–7].

Now we shall discuss the application of the coherent density fluctuation model (CDFM) [9.8, 9.9] (see Sect. 4.2) to the problems of the electron elastic scattering by nuclei.

It has been shown in [9.8a,b] that in the Born approximation for the electron cross-section, the CDFM charge form factor is ((4.74) and (4.75)):

$$F(q) = \int_0^\infty dx |f(x)|^2 F_\Theta(qx), \tag{9.4}$$

where

$$F_\Theta(qx) = j_1(qx)/(qx) \tag{9.5}$$

is the flucton form factor.

It is essential to use the more realistic HEA for the analysis of the electron–nucleus elastic scattering in the framework of the CDFM. In this case the distortion of the electron wave in the HEA [9.2, 9.6] as it scatters on a flucton is taken into account by the averaging by the CDFM weight function $|f(x)|^2$. Following (4.73) the form factor can be written approximately in the form:

$$F(q) \cong \int_0^\infty dx |f(x)|^2 F_0(q, x), \tag{9.6}$$

where $F_0(q, x)$ is the form factor of a flucton with radius x in the HEA [9.2, 9.6]:

$$F_0(q, x) = \frac{3}{2x^3} \sum_{\varepsilon = \pm 1} \frac{G(x, \varepsilon)}{q_{\text{eff}}^2(x, \varepsilon)} (x + i\varepsilon/q) \exp[i(\varepsilon q x + \Phi(x, \varepsilon))]. \tag{9.7}$$

In Eq. (9.7):

$$G(x, \varepsilon) = \frac{q(1 - V(0)/k) + 3\varepsilon b q^2 x + aq(q^2 - 2k^2)x^2 - \frac{5}{2}\varepsilon c q^2 (4k^2 - q^2)x^3}{q(1 - V(0)/k) - aq\left(\frac{3}{2}k^2 - \frac{1}{2}q^2\right)x^2 + \varepsilon b\left(\frac{3}{2}q^2 - 2k^2\right)x + \varepsilon c\left(k^2 - \frac{5}{4}q^2\right)(4k^2 - q^2)x^3}, \tag{9.8}$$

$$\Phi(x, \varepsilon) = -\varepsilon \frac{V(0)}{k} qx - b(4k^2 - q^2)x^2/2 - \varepsilon a(qk^2/2 - q^3/12)x^3$$
$$+ c(4k^2 - q^2)^2 x^4/8, \tag{9.9}$$

$$q_{\text{eff}}(x, \varepsilon) = \varepsilon q(1 - V(0)/k) - \varepsilon aq(3k^2/2 - q^2/4)x^2 - b(4k^2 - q^2)x$$
$$+ c(4k^2 - q^2)^2 x^3/2, \tag{9.10}$$

where

$$V(0) = -\frac{3}{2}\frac{\gamma}{x}, \qquad \gamma = Ze^2/(\hbar c), \tag{9.11}$$

$$a = \gamma/(xk)^3, \tag{9.12}$$

$$b = \frac{3}{4}\gamma/(xk)^2, \tag{9.13}$$

$$c = \frac{3}{32}\gamma/(xk)^4. \tag{9.14}$$

The HEA cross-section for elastic electron–nucleus scattering is expressed by the point charge cross-section $(d\sigma/d\Omega)_{\text{Mott}}$ and the HEA form factor of the nucleus $F(q)$ [9.2]:

$$\frac{d\sigma}{d\Omega} = \left(\frac{d\sigma}{d\Omega}\right)_{\text{Mott}} |F(q)|^2, \tag{9.15}$$

where

$$\left(\frac{d\sigma}{d\Omega}\right)_{\text{Mott}} = \left(\frac{Ze^2}{2E}\right)^2 \frac{\cos^2 \theta/2}{\sin^4 \theta/2}. \tag{9.16}$$

The substitution of (9.6) and (9.7) into (9.15) gives the final expression for the CDFM electron elastic scattering cross-section in the HEA [9.8, 9.10, 9.11]:

$$\frac{d\sigma}{d\Omega} = \left(\frac{d\sigma}{d\Omega}\right)_{\text{Mott}} \left| \int_0^\infty dx |f(x)|^2 F_0(q, x) \right|^2. \tag{9.17}$$

The weight function $|f(x)|^2$ is expressed by the nuclear density distribution (Eq. (4.66)) for $d\rho/dr \leqslant 0$:

$$|f(x)|^2 = -\frac{1}{\rho_0(x)} \frac{d\rho(r)}{dr}\bigg|_{r=x}, \tag{9.18}$$

where $\rho_0(x) = 3A/4\pi x^3$ is the density of a flucton with radius x.

Two different charge densities have been used to calculate the differential cross-sections and the form factors using (9.15)–(9.18):

i) The Fermi-type density distribution:

$$\rho_F(r) = \rho_0 \frac{1}{1 + \exp[(r-R)/b]}; \qquad \rho_0 \simeq \frac{3}{4\pi R^3 [1 + (\pi b/R)^2]}, \tag{9.19}$$

ii) The symmetrized density distribution suggested in [9.12]:

$$\rho(r) = \rho_0 \int_{\frac{r-R}{s\sqrt{2}}}^{\frac{r+R}{s\sqrt{2}}} \exp(-y^2)\,dy, \qquad \rho_0 = \frac{3}{4\pi^{3/2}(R^3 + 3Rs^2)}. \tag{9.20}$$

By means of Eq. (9.17) the differential cross-sections for electron elastic scattering on ^{40}Ca ($E_e = 250$ MeV [9.8, 9.11], 500 MeV [9.11], 750 MeV [9.10, 9.11]) and ^{208}Pb ($E_e = 502$ MeV [9.11]) as well as the form factors of ^{12}C and ^{16}O [9.10, 9.11] have been calculated. The CDFM results are given in Figs. 9.1–9.4 and are compared with the experimental data for ^{40}Ca from [9.13, 9.14]

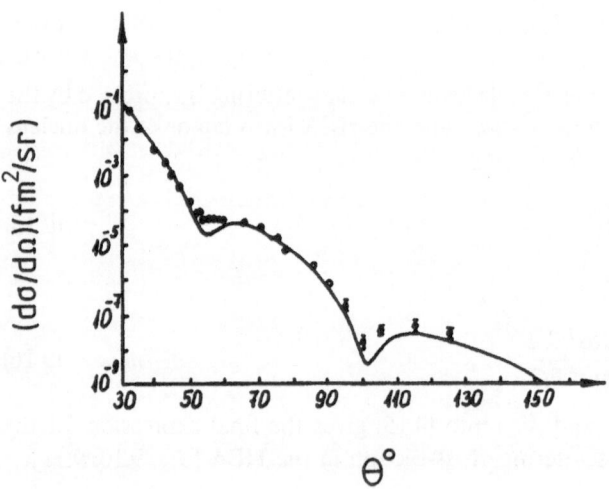

Fig. 9.1. Differential cross section for electron elastic scattering by ^{40}Ca at $E_e = 250$ MeV. The solid curve shows the HEA result in CDFM calculations [9.11] with the density (9.20). The experimental data are taken from [9.13]

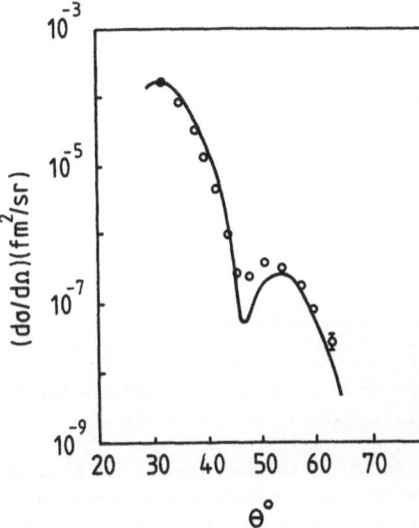

Fig. 9.2. The same as in Fig. 9.1. for $E_e = 500$ MeV

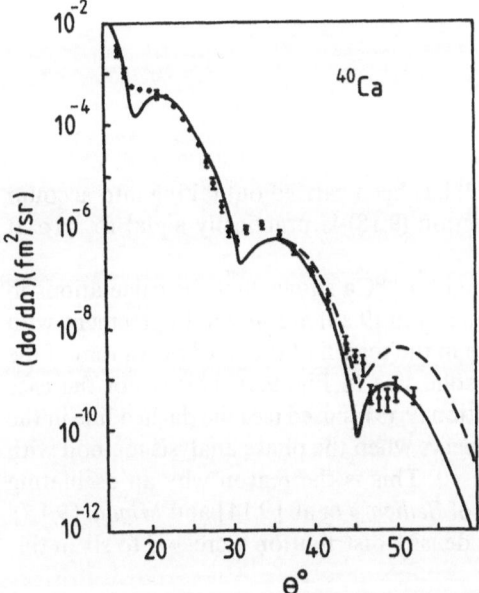

Fig. 9.3. Differential cross section of electron elastic scattering at $E_e = 750$ MeV on ^{40}Ca. The solid line shows the CDFM calculations [9.10, 9.11] using the density (9.20) and the dashed line the CDFM calculations [9.10, 9.11] using the density (9.19). The experimental data are taken from [9.14]

and for ^{12}C, ^{16}O and ^{208}Pb from [9.1]. The following values of the parameters for both densities (9.19) and (9.20) have been used:

   i) for (9.19): ^{40}Ca ($R = 3.60$ fm, $b = 0.576$ fm);
   ii) for (9.20): ^{12}C ($R = 2.29$ fm, $s = 0.90$ fm); ^{16}O ($R = 2.50$ fm, $s = 0.86$ fm);
^{40}Ca ($R = 3.55$ fm, $s = 0.94$ fm); ^{208}Pb ($R = 6.557$ fm, $s = 0.88$ fm).

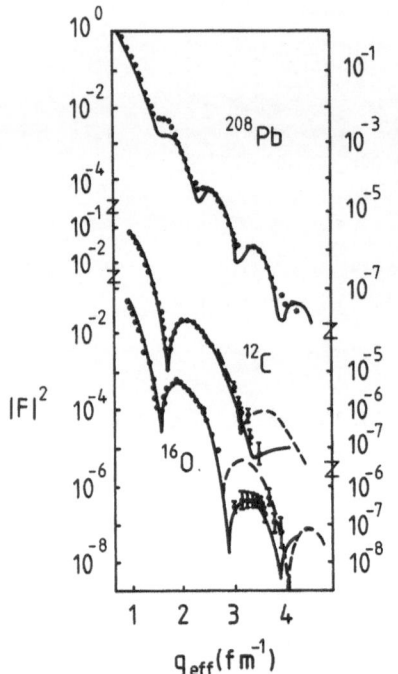

Fig. 9.4. Form factors of electron elastic scattering by ^{12}C and ^{16}O (lefthand scale) and differential cross-section of electron elastic scattering ($E_e = 502$ MeV) on ^{208}Pb (righthand scale). Solid line: the CDFM result in HEA using (9.20) [9.10, 9.11]; dashed line the result of calculations in [9.1] using HEA and the density (9.19). The experimental data are taken from [9.1]

The numerical integration in (9.17) has been carried out taking into account the fact that the CDFM weight function (9.18) is practically equal to zero at $x \lesssim 1$ fm for the nuclei considered.

The analysis of the results obtained for ^{40}Ca shows that the calculations in the HEA within the CDFM using $\rho(r)$ from (9.20) are in good agreement with the experimental data including those in the region of the third maximum of the cross-section at $E_e = 750$ MeV (the solid line in Fig. 9.3). This is not the case when the Fermi-type density distribution (9.19) is used (see the dashed line in the same Figure). The same deviation appears when the phase analysis method with the Fermi-type density ρ_F (9.19) is used. This is the reason why an oscillating term $\Delta\rho$ was added to ρ_F in the work of *Bellicard* et al. [9.14] and *Migdal* [9.15]. The number of the parameters in the density distribution increased to six in this case.

We note also the good agreement between the form factors obtained in the CDFM with $\rho(r)$ from (9.20) for ^{12}C and ^{16}O (the solid lines in Fig. 9.4) and the experimental data. This is in contrast with the results from [9.1] obtained in HEA calculations using the symmetrized Fermi-type distribution

$$\rho_{SF}(r, R, b) = \rho_0 \left\{ \frac{1}{1 + \exp[(r - R)/b]} + \frac{1}{1 + \exp[-(r + R)/b]} - 1 \right\}, \quad (9.21)$$

where

$$\rho_0 = \frac{3}{4\pi R^3 [1 + (\pi b/R)^2]}.$$

The calculations in [9.1] do not describe the experimental form factors at $q_{eff} > 2.5 \, \text{fm}^{-1}$ (see the dashed line in Fig. 9.4).

The results obtained in the CDFM confirm the conclusion in [9.12] that the oscillating density $\Delta\rho$ introduced in [9.14, 9.15] does not necessarily follow from the experimental data on electron–nucleus scattering. The difficulties in describing the cross-sections for ^{12}C, ^{16}O, ^{40}Ca and ^{208}Pb nuclei are overcome in the CDFM using the weight function f corresponding to the density distribution (9.20), but not to the Fermi-density (9.19). The account of the distortion of the electron waves in HEA, introduced in the CDFM, leads to an improvement of the agreement with the experimental data at small angles compared with the results of the calculations in the Born approximation using the same density (9.20) [9.12] and also to the partial filling of the cross-section minima.

An important result obtained in the CDFM is the correct description of the experimental cross-sections and form factors for the elastic electron-nucleus scattering up to transfer momenta: $q \simeq 4 \, \text{fm}^{-1}$ for ^{12}C, ^{16}O and ^{40}Ca and $q \simeq 3 \, \text{fm}^{-1}$ for ^{208}Pb. This fact is related to the effective account of the N–N correlations in the CDFM. As was noted in [9.1], the introduction of short-range N–N correlations in many-particle wave functions gives an improved quantitative agreement with the experimental data at large transfer momenta.

The CDFM results show that the model weight function can be reliably determined from electron elastic scattering on nuclei. This is important because of the fundamental role of this function in the CDFM, and its relation to the one-body density matrix and through it to the basic characteristics of nuclear states and interaction processes.

9.2 Quasielastic Electron Scattering

The cross-section of electron scattering on nuclei is determined by the transition matrix elements of the charge and current operators. This is the reason why the study of such processes gives the possibility of extracting direct information on the nuclear structure. The three-component transfer momentum q and the electron energy loss ω ($\omega = E_{e_i} - E_{e_f}$, E_{e_i} and E_{e_f} being the initial and final electron energy, respectively) can vary satisfying the condition $q^2 = \mathbf{q}^2 - \omega^2 > 0$, where q is the 4-component momentum. Several characteristic regions can be observed in the electron-scattering cross-section $(d^2\sigma/d\Omega_{e_f} dE_{e_f})/\sigma_{\text{Mott}}$ as a function of ω: the elastic scattering peak, peaks corresponding to discrete level excitations ($\leqslant 10$ MeV), the giant resonance region (10–15 MeV), a broad peak between 40 and 140 MeV related to the quasielastic electron scattering and the region $\omega > 140$ MeV in which pion electroproduction takes place.

In this Section we consider the quasielastic electron scattering on nuclei. The term "quasielastic scattering" is related to the fact that at incident electron energies much higher than the nucleon separation energy the nucleons can be considered as free. If the nucleons were unbound and at rest in the nucleus, then the quasielastic peak would be at $\omega_{max} = q^2/2M$, where M is the nucleon mass. Since the nucleons are bound the maximum of the cross-section is shifted and broadened because of the nucleon momentum distribution.

The main characteristics of the quasielastic scattering (e, e') in which only the final electron is detected are discussed by *Moniz* [9.16] using the Fermi-gas model of non-interacting nucleons. In this method the incident electron is scattered elastically by a nucleon from the Fermi sea. Due to the Pauli principle this nucleon must leave the Fermi sea. The model of *Moniz* [9.16] has been applied successfully to the description of the cross-sections of the quasielastic scattering of electrons by a wide range of nuclei at various energies and angles [9.17–9.19]. These analyses enable the values of various nuclear characteristics, such as the Fermi momentum (k_F), the average separation energy $(\bar{\varepsilon})$ as well as the effective mass M^* to be determined. The general behaviour of the quasielastic peak has been reproduced well but some systematic deviations from the experimental cross-sections at small and large values of ω have been noted. In particular the data on high energy $(E_e \gtrsim 2$ GeV) electron scattering cross-sections for ^{6}Li and ^{12}C [9.20] are not reproduced by the model.

The main shortcoming of the Fermi-gas model [9.16] is that the nucleon–nucleon interaction in the bound states and in the continuous spectrum is neglected and it remains "hidden" in the parameters k_F and $\bar{\varepsilon}$. As shown by *Czyż* and *Gottfried* [9.21] the tail of the quasielastic cross-section at large ω contains information about the nucleon–nucleon short-range correlations and they have to be accounted for [9.18, 9.20, 9.22].

A number of theoretical works have been devoted to the inclusion of the nucleon–nucleon interaction effects in the study of the quasielastic electron scattering (QES). The exactly-solvable cases of two- and three-particle systems have been considered by *Fabian* and *Arenhövel* [9.23] and by *Lehman* [9.24] and *Dieperink* et al. [9.25], respectively. Many-body methods have been applied to the case of nuclear matter [9.26]. The Fermi-gas model with a momentum-dependent nuclear potential [9.19], as well as a shell-model potential have been used for the QES cross-section calculations for a number of finite nuclei [9.27–31]. The nuclear interaction has been accounted for in a generalized Fermi-gas model by means of a semiclassical method which is equivalent to the Thomas–Fermi approximation for the bound states and to the eikonal approximation for the continuous spectrum by *Rosenfelder* [9.32]. The relativistic Fermi-gas model in the framework of the relativistic nuclear field theory [9.33, 9.34] has been also considered in [9.32].

As shown in [9.35, 9.36], if only the final electron is detected, any electrodynamic process with one-photon exchange between the nucleus and the electron can be described by means of two form factors $W_{1,2}(q^2, \omega)$. The first studies of the QES as a process of this type has been carried out in [9.36–9.38, 9.21].

The cross-section of the QES (e, e') has the following form in the one-photon-exchange approximation:

$$\left(\frac{d^2\sigma}{d\Omega_{e_f} dE_{e_f}}\right)_{\text{lab.}} = \frac{Z^2}{M_T} \sigma_{\text{Mott}} \left[W_2(q^2, \omega) + 2W_1(q^2, \omega)\tan^2\frac{\theta}{2} \right], \qquad (9.22)$$

where M_T is the target mass, Z is the nuclear charge and

$$\sigma_{\text{Mott}} = \left(\frac{Ze^2}{2E_{e_i}}\right)^2 \frac{\cos^2\theta/2}{\sin^4\theta/2}. \qquad (9.23)$$

Important experimental results on QES have been obtained in the nineteen eighties on the separation of the longitudinal (R_L) and the transverse (R_T) response functions defined by:

$$R_L[q^2, \omega] = \frac{Z^2}{M_T} \frac{q^2}{q^2} \left[-W_1(q^2, \omega) + \frac{q^2}{q^2} W_2(q^2, \omega) \right], \qquad (9.24)$$

$$R_T(q^2, \omega) = \frac{Z^2}{M_T} 2W_1(q^2, \omega). \qquad (9.25)$$

Experimental data for the response functions at various transfer momenta have been obtained for a wide range of nuclei: ^{3}He [9.39], ^{12}C [9.40, 9.41], ^{40}Ca [9.42–9.45], ^{48}Ca [9.43–9.45], ^{56}Fe [9.44–47], ^{238}U [9.48, 9.49]. As is known [9.42, 9.50, 9.51] the longitudinal and the transverse components of the nuclear response are related to the charge and current (magnetic) density distributions in nuclei. The longitudinal response function is determined mainly by the single-nucleon knockout processes and is sensitive to the nucleon wave functions (especially at large transfer momenta when the final state interactions (FSI) of the outgoing nucleons are small [9.39]). The transverse response function contains contributions from the non-nucleonic degrees of freedom, such as those associated with meson exchange currents (MEC), pion and Δ (1236)-resonance electroproduction etc. [9.39, 9.52–55]. This requires the use of single-particle models (for instance, that of the relativistic Fermi-gas [9.56]) to describe the longitudinal response function. The studies, show, however, contrary to expectations, that the Fermi-gas-type models describe successfully the experimentally obtained transverse response function, while the theoretical longitudinal response function is more than twice the experimental data [9.42, 9.43, 9.48, 9.52].

It was shown in [9.57, 9.58] that MEC have significant effects on the transverse response function, while the longitudinal response function is essentially unchanged by them. The MEC taken into account in [9.58] lead to a small decrease of R_L ($\sim 10\%$) and to larger decrease of R_T (~ 20–30%).

As a consequence of this situation, various attempts have been made to clarify the reasons for the failure of the single-particle picture of the scattering to describe the longitudinal response function. Here we mention some of them:

- Conventional non-relativistic considerations, e.g. within the density-dependent Hartree–Fock approximation [9.59], in the Tamm–Dancoff approximation [9.60] and in the random-phase approximation (RPA) [9.61–63].

- An account for the rescattering of the knockout proton in an optical potential [9.60].

- Calculations assuming an increased nucleon charge radius by around 30% [9.52, 9.64] which leads to a correct description of R_L and to the deviation from experimental data of the transverse function. The use of nucleon form factors modified in nuclear medium by *Celenza* et al. [9.65] leads to a satisfactory explanation of the longitudinal response function in ^{12}C [9.66] and ^{40}Ca and ^{56}Fe [9.67] but not of the transverse response function. The analysis of *Mulders* [9.68] of the longitudinal and transverse functions for ^{12}C shows indications of increases of the charge radius by 15% and of the magnetic moment of the nucleons by $15 \pm 5\%$, but not of the rms radius of the magnetic form factor.

The modification of the nucleon properties in the nuclear medium is a common idea in calculations of this type but it is not yet treated in a complete way. Later in this Section we shall consider the influence on the longitudinal response function of the medium-modified electromagnetic form factors and nucleon effective mass obtained in a chiral quark-meson theory [9.69].

- Calculation in the framework of the semi-classical RPA [9.70, 9.71] in which the longitudinal response function is correctly described for ^{12}C but not for the heavier nuclei ^{40}Ca and ^{56}Fe. For them this is achieved [9.72] by an increase of the proton rms radius by 23% for ^{40}Ca and by 21% for ^{56}Fe. The proton and neutron contributions to the charge longitudinal response functions are discussed in [9.73].

- A method taking account of the Fermi-sea depletion due to the nucleon–nucleon short-range correlations (SRC) which leads to a correct description of the longitudinal response function [9.74]. In [9.75] the role of the $\Delta(1236)$-resonance on the behaviour of the transverse response function has been investigated.

The QES data for ^3He and ^4He show the presence of proton–proton correlations [9.76] although they are not accurate enough to allow for a quantitative determination of the size of the repulsive core in the N–N interaction. It is shown also in [9.77] that the SRC effects lead to a reduction of the height of the peak of the longitudinal response function. Furthermore, the SRC, the orthogonality corrections and the d–N interaction in the final state bring the theoretical ^3He longitudinal and transverse and the ^3H transverse response functions into reasonable agreement with the experimental ones [9.78]. The effects of N–N correlations on the response functions are studied also by *Pandharipande* [9.79]. It has been shown by nuclear matter calculations that the SRC have an effect at $|q| \approx 400$ MeV/c [9.80, 9.81].

- A calculation from [9.50, 9.82] in the framework of the extended RPA theory (ERPA) [9.83], which includes 1p–1h and 2p–2h excitations and uses a realistic G-matrix in the local density approximation (LDA). It is shown in [9.50] that the account of the ERPA correlations leads to satisfactory agreement of the longitudinal response function of ^{12}C at least to $|q| \simeq 250$ MeV/c. An important conclusion has been made in [9.82a] that the 2p–2h correlations within the ERPA appear to be equally important for the longitudinal as well as for the transverse response in the quasielastic region.

– The "time of interaction" approximation [9.84a]. It turns out that it is impossible to describe within this approximation the q-dependence of the shift of the quasielastic maxima in the longitudinal and transverse parts of the cross-section for QES on ^{12}C and ^{40}Ca [9.84b].

– The relativistic $\sigma\omega\rho$ model [9.85–9.88] calculations in the RPA show a correct description of the longitudinal response functions for ^{12}C, ^{40,48}Ca but the results for the transverse response functions fall short of the data [9.85]. The effects of the vacuum fluctuations included in this model can be considered as a reason for an effective proton form factor which is reduced in comparison to the free-nucleon one and corresponds to a "swollen proton" [9.87]. The RPA correlations are evaluated in the local density approximation.

– The relativistic Fermi-gas-type models [9.56, 9.57, 9.89]. It is pointed out in [9.89] that the momentum-dependence of the scalar and vector potentials has an essential importance for the behaviour of the response functions. The result obtained in [9.89] is similar to that from the work of *Noble* [9.52] but without the necessity of radical changes in the effective mass.

– The relativistic RPA calculations [9.90, 9.91] which show that the inclusion of the RPA correlations reduces the mean-field theory longitudinal response functions by about 10% at the position of the peak but still overestimates the response on the high-energy side of the peak. It is shown also that the RPA correlations have a small effect on the transverse response; that is a common feature of many single-nucleon models. The high-energy side of the transverse response peak is believed to be dominated by mechanisms like isobar formation and by MEC. There are indications of important changes in the longitudinal response due to the vacuum polarization effects [9.91].

– The relativistic mean-field approach [9.92], in which it is argued that the problem of the simultaneous description of both longitudinal and transverse response functions can be solved by a consistent treatment of the electro-magnetic interaction using the Dirac equation in which the scalar and vector potentials are introduced explicitly. Good agreement is achieved for both functions in the case of ^{12}C ($|q| = 400$ MeV/c), while for ^{40}Ca ($|q| = 410$ MeV/c) the theoretical transverse response function exceeds the experimental one.

– The method of *Shlomo* [9.93], in which it is pointed out that the disagreement between the theoretical results and the experimental response functions for finite nuclei is not due to the effects of the finite size of the nuclear system.

– The concept of y-scaling (see Sect. 7.1) which has been applied to the analyses of the nuclear response functions has the advantage [9.94] that the plot versus q of various sets of data corresponding to different kinematics (but to the same value of y) shows the experimental q-dependence of those effects which break down in the impulse approximation and in the simple Hartree–Fock approximation. The analysis of *Finn* et al. [9.95] shows a large enhancement of the transverse response relatively to the longitudinal response contrary to impulse approximation predictions assuming a free one-body current. Scaling behaviour is recovered when a relativistic effective mass is included in the nucleon current operator.

– The effects of the final state interaction on the nuclear response functions in ^{12}C and ^{40}Ca (at various values of $|q|$) are found [9.96] to be large and important.

The brief review given above shows that considerable theoretical efforts have been made to explain the missing strength in the longitudinal response (often referred as the missing charge problem). Nevertheless as can be seen the problem still remains open.

In the next part of this Section we give the results of the study of the influence on the QES cross-sections and the response functions of the nucleon–nucleon correlations which are included in the coherent density fluctuation model (CDFM) [9.8, 9.9] (see Sect. 4.2) [9.97]. The changes of the nucleon properties in nuclear medium and their effect on the characteristics of the QES have been also considered within the Fermi-gas model and the CDFM in [9.69].

In the Fermi-gas model of *Moniz* [9.16] the nuclear form factors W_1 and W_2 (9.22) have the following form:

$$W_1(q^2, \omega) = \frac{3M_T}{4\pi A k_F^3} \int dk \frac{\delta(\omega + \varepsilon_k - \varepsilon_{k+q})\theta(k_F - k)\theta(|k+q| - k_F)}{\varepsilon_k \cdot \varepsilon_{k+q}}$$

$$\times \left[T_1(q^2) + \frac{1}{2M^2} T_2(q^2)k^2 \sin^2 \tau \right], \tag{9.26}$$

$$W_2(q^2, \omega) = \frac{3M_T}{4\pi A k_F^3} \frac{T_2(q^2)}{M^2} \int dk \frac{\delta(\omega + \varepsilon_k - \varepsilon_{k+q})\Theta(k_F - k)\Theta(|k+q| - k_F)}{\varepsilon_k \cdot \varepsilon_{k+q}}$$

$$\times \left[\left(\varepsilon_k - \frac{\omega}{|q|} k \cos \tau \right)^2 + \frac{1}{2} \frac{q^2}{q^2} k^2 \sin^2 \tau \right], \tag{9.27}$$

where q is chosen as the polar axis, τ is the polar angle, $\Theta(k_F - k)$ is the Fermi momentum distribution, and

$$T_1(q^2) = \tfrac{1}{2} q^2 (F_{1p} + 2MF_{2p})^2 + \tfrac{1}{2} q^2 (F_{1n} + 2MF_{2n})^2, \tag{9.28}$$

$$T_2(q^2) = 2M^2 (F_{1p}^2 + q^2 F_{2p}^2) + 2M^2 (F_{1n}^2 + q^2 F_{2n}^2). \tag{9.29}$$

In (9.28, 9.29) the $F_{1(2), p(n)}$ are the nucleon form factors and M is the nucleon mass.

In the CDFM the form factors (9.26, 9.27) can be generalised by the superpositions [9.97]:

$$W_1(q^2, \omega) = \int_0^\infty dx |f(x)|^2 W_1(q^2, \omega, x), \tag{9.30}$$

$$W_2(q^2,\omega) = \int_0^\omega dx\, |f(x)|^2 W_2(q^2,\omega,x). \tag{9.31}$$

The form factors $W_i(q^2,\omega,x)$ $(i = 1,2)$ have the form of Eqs. (9.26) and (9.27) in which the momentum distribution $\Theta(k_F - k)$ is replaced by $\Theta(k_F(x) - k)$. Here $k_F(x)$ is the Fermi-momentum for a flucton with radius x:

$$k_F(x) = \left(\frac{3\pi^2}{2}\rho_0(x)\right)^{1/3}, \quad \rho_0(x) = 3A/4\pi x^3. \tag{9.32}$$

The relationships (9.30, 9.31) imply that in the CDFM the nuclear form factors $W_1(q^2,\omega)$ and $W_2(q^2,\omega)$ are superpositions of the flucton form factors $W_1(q^2,\omega,x)$ and $W_2(q^2,\omega,x)$ with the participation of the CDFM weight function $|f(x)|^2$. The latter can be obtained by the nuclear local density distribution $\rho(r)$ (in the case of monotonically decreasing densities):

$$|f(x)|^2 = -\frac{1}{\rho_0(x)}\frac{d\rho(r)}{dr}\bigg|_{r=x}. \tag{9.33}$$

The cross-section of the QES have been calculated in the CDFM by means of Eqs. (9.22, 9.23) and (9.26–33). The standard "dipole fit" has been used for the nucleon form factors:

$$G_E^p = \frac{G_M^p}{2.79} = \frac{4M^2}{q^2}\frac{G_E^n}{1.91} = \frac{G_M^n}{-1.91}$$

$$= \left(1 + \frac{q^2}{0.71(\text{GeV/c})^2}\right)^{-2}, \tag{9.34}$$

where

$$G_E = F_1 - \frac{q^2}{2M}F_2, \quad G_M = F_1 + 2MF_2. \tag{9.35}$$

The weight function $|f(x)|^2$ is calculated from (9.33) using the symmetrized Fermi-type distribution (9.21) with parameter values (R and b) obtained from the fit to electron elastic scattering experimental data. Following the prescription given in [9.16], in the CDFM an average separation energy $\bar{\varepsilon}$ and a nucleon effective mass M^* (only in the case of $|q|/k_F < 2$) have been used.

In [9.97] the cross-sections of QES are calculated within the CDFM and are compared with the experimental data and with results of other theoretical methods for the following initial electron energies (E_{e_i}), angles (θ) and nuclei:

(i) $E_{e_i} = 500\,\text{MeV}$; $\theta = 60°$; ^{6}Li, ^{12}C, ^{24}Mg, ^{40}Ca, ^{58}Ni, ^{118}Sn, ^{181}Ta, ^{208}Pb (the experimental data from [9.17]).

(ii) $E_{e_i} = 80.9\,\text{MeV}$, 98 MeV, 148.5 MeV; $\theta = 135°$; ^{12}C (the experimental data from [9.38]).

(iii) $E_{e_i} = 2.0\,\text{GeV}$ and 2.7 GeV; $\theta = 15°$; ^{12}C (the experimental data from [9.98]).

The CDFM calculations have been carried out using the effective mass $M^* = M/1.4$. The values of the average nucleon separation energy (the parameter $\bar{\varepsilon}$) turn out to be in accord with the conclusion of [9.32] that the value $\bar{\varepsilon} \simeq 30$ MeV is almost constant for a wide range of nuclei with the exception of the lightest nuclei. The values of $\bar{\varepsilon}$ used in the CDFM that give the best fit to the data are in agreement (with the only exception of $E_{e_i} = 2.7$ GeV for ^{12}C) with those obtained by the relationship from [9.32]:

$$\bar{\varepsilon}(q) = (q^2 + M^{*2})^{1/2} - M^* - q^2/2M \quad (|q| \geqslant 2k_{\mathrm{F}}) \tag{9.36}$$

for $M^* = M/1.4$ and for the particular value of q^2 which is characteristic of a given experiment. In this way the number of free parameters in the CDFM calculations is actually reduced to only one, namely the effective mass M^*. It has been shown in [9.32] that the value of $M^* \simeq M/1.4$ is, in fact, an average over the nucleus of the local effective mass $M^*(r)$ in the relativistic nuclear field theory from [9.33, 9.34].

The QES cross-sections calculated within the CDFM are in better agreement with the experimental data than those calculated in the Fermi-gas model for large and small energy losses ω (or small and large E_{e_f}, respectively). As an example we present in Fig. 9.5 the cross-section of QES on ^{40}Ca calculated in the CDFM (with added s-wave π-production and isobar excitations contributions calculated in the Fermi-gas model [9.18]). It is seen that the total result (solid line) describes much better the enhancement of the cross section at small E_{e_f} than the Fermi-gas model calculations (large-amplitude dashed curve).

The improvement of the agreement with the experimental data and especially the enhancement of the cross-section in the region of large ω compared with the Fermi-gas model calculations can be considered as a result of the short-range correlations taken into account in the CDFM.

The calculations of the longitudinal R_{L} and the transverse R_{T} response functions for ^{12}C in the CDFM [9.97] using $M^* = M$ show the well-known features of the impulse approximation: a qualitative agreement for R_{T} and larger values than the experimental data for R_{L}. The calculations with $M^* = M/1.4$

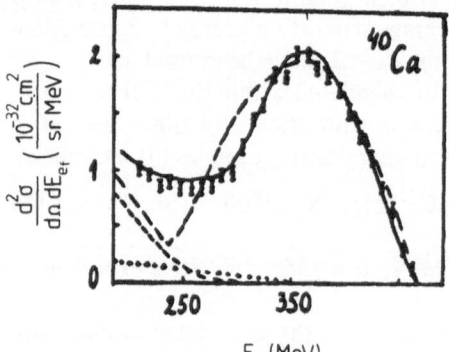

Fig. 9.5. Cross-section of quasielastic scattering of 500 MeV electrons at 60° from ^{40}Ca. The dotted line and the short-dash line are the s-wave π-production and the isobar excitation contributions respectively, the long-dash line is the total result in the Fermi-gas model [9.18]. The solid line is the result of the CDFM calculations [9.97] to which the π-production and isobar excitation contributions from the Fermi-gas model [9.18] have been added

gave an improvement for R_L and a disagreement for R_T. The CDFM results for the longitudinal and transverse response functions [9.97] at $M^* = M/1.4$ are similar to those obtained by the modified impulse approximation [9.74], in which the decrease of R_L at lower energies is related to the reduction of the shell-model orbital occupation probabilities due to the short-range N–N correlations.

The next step of this QES study [9.69] is to incorporate the effects of the nuclear medium on the nucleon properties including the effective mass and the form factors of the nucleons in nuclear medium obtained in the approach from [9.99], within the model of *Nambu-Jona-Lasinio* (*NJL*) [9.100]. In [9.99] was found a decrease of the mass, an increase of the rms radii and a reduction of the form factors of the nucleon in the medium. The application of this result to the shell-model calculations of the longitudinal and transverse response functions for ^{12}C and ^{40}Ca leads to a satisfactory description of the longitudinal response for intermediate values of the transfer momentum [9.101]. In [9.69] the medium-modified nucleon properties are applied to the study of QES within the Fermi-gas (FGM) and the CDFM. The density of the Fermi-gas $\rho_0 = (2/3\pi^2)k_F^3$ has been taken as the medium density at which the calculations in the NJL model have been performed. The effective mass is treated as a consequence of a dynamical effect, modifying both the nucleon current and the kinematical factors. Thus the nucleon binding in the nucleus is also taken into account and it is no longer necessary to introduce an additional parameter $\bar{\varepsilon}$. The values of the Fermi-momenta used in the FGM calculations have been adopted from [9.17] and are given in Table 9.1. For the nuclei not considered in [9.17] (^{48}Ca, ^{56}Fe and ^{238}U) the values of k_F are extrapolated.

As a basis for comparison the longitudinal response function has been calculated in the FGM with the *Gari-Kruempelmann* parametrization [9.102]. The results for R_L in ^{56}Fe are given in Fig. 9.6 since for this nucleus the experimental data over the widest range of momentum transfers from 300 MeV/c to 1.14 GeV/c are available [9.44, 9.47]. Theoretical curves obtained in the FGM with form factors and effective mass at zero medium density and in both FGM and CDFM at finite medium density are compared. For all values of the momentum transfer a significant reduction due to medium effects on the nucleons is observed. In all cases the agreement with the experimental data is considerably improved. At transfer momenta $q < 350$ MeV/c there is still some overestimation and the shape of the curve is not correctly reproduced. This is not surprising since long-range RPA correlations have considerable effect in this region [9.50, 9.80, 9.82]. Final state interactions are also expected to play a more

Table 9.1. Values of the Fermi-momentum k_F used in the Fermi-gas model calculations [9.69]

Nucleus	^{12}C	^{40}Ca	^{48}Ca	^{56}Fe	^{208}Pb	^{238}U
k_F (MeV/c)	221	249	252	253	265	265

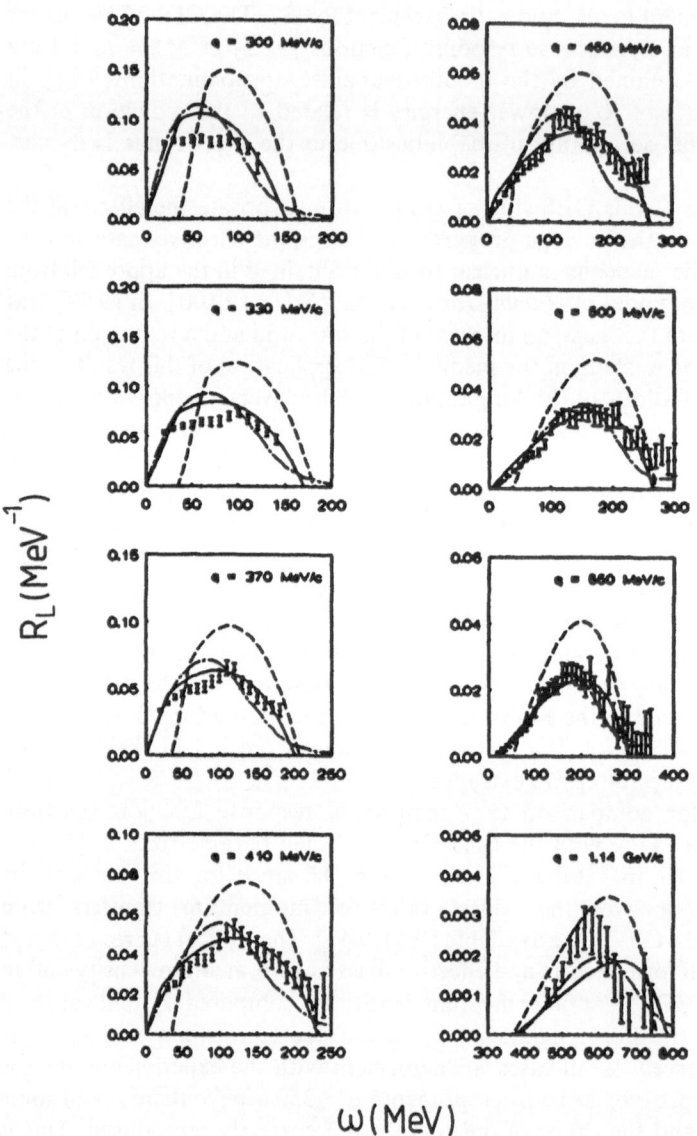

$\omega\,(\mathrm{MeV})$

Fig. 9.6. Longitudinal response functions for ^{56}Fe for different values of the transfer momentum [9.69]. Dashed line: FGM with free nucleon mass and form factors; solid (dash-dotted) lines: FGM (CDFM) with effective nucleon mass and form factors at finite medium density from the chiral theory [9.99]. Experimental data are from [9.44, 9.47]

significant role at such transfer momenta [9.66, 9.67]. At higher values of the transfer momentum (400–600 MeV/c) the effect of long-range correlations is much smaller [9.82] and the longitudinal response can be considered as a pure quasielastic knockout contribution. In this region the agreement of the FGM calculation with effective mass and form factors at finite medium density with the experimental data is reasonable.

As shown in [9.69] the results of the NJL model for the free proton elastic form factor tend to underestimate the experimental data for values of the four-momentum transfer larger than 600 MeV/c. In order to eliminate the effect of this underestimation from the medium modified form factors, we renormalize the form factors from the NJL to the Gari-Kruempelmann fit at zero medium density at high values of the 4-momentum transfer (~ 1 GeV/c). The nucleon form factors at finite density are then obtained by adding the medium correction predicted by the NJL model to the Gari-Kruempelmann parametrization. At the highest value of the transfer momentum now available, 1.14 GeV/c [9.47], our theoretical result tends to underestimate the data at the low energy transfer part of the peak. At such high values of q, however, the single-nucleon knockout could be no longer the dominating process [9.103].

The incorporation into the CDFM of high-momentum components in the nucleon momentum distribution shifts the peak position to lower transfer energy and moves some strength to the "dip" region above the quasielastic peak. This leads to some improvement of the agreement with the experimental data at transfer momenta $q > 500$ MeV/c. For the other nuclei for which the response functions are calculated namely ^{12}C, ^{40}Ca, ^{48}Ca, ^{208}Pb and ^{238}U the situation is very similar to that for ^{56}Fe, with the exception that no experimental data are yet available at transfer momenta near 1 GeV/c.

In Fig. 9.7 are shown the longitudinal response functions for ^{48}Ca and ^{208}Pb at $|q| = 500$ MeV/c calculated using shell-model wave functions in the local density approximation with and without medium effects [9.101]. As for the

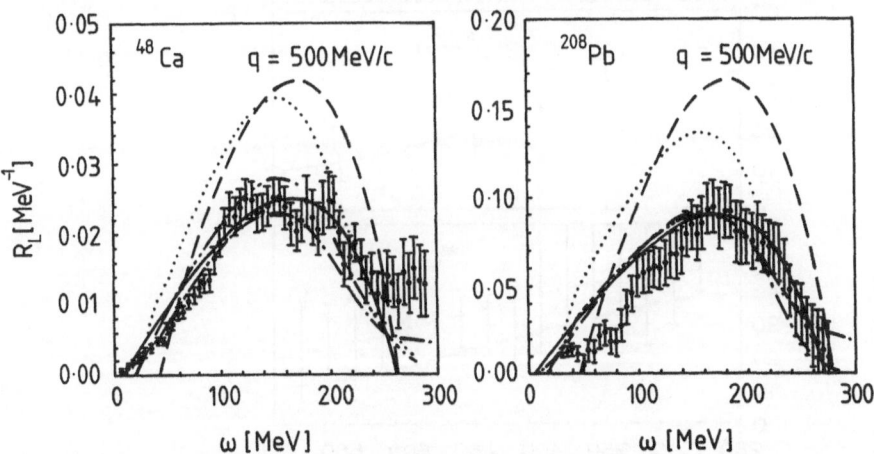

Fig. 9.7. Longitudinal response functions for ^{48}Ca and ^{208}Pb at 500 MeV/c transfer momentum. Dashed, dash-dotted and thin solid lines are similar to those in Fig. 9.6. The thick solid (dotted) lines show the results of shell-model calculations with (without) medium effects from the chiral theory [9.99]

FGM results the inclusion of the medium effects resolves the missing strength problem. The position of the peak and the shape differ slightly from those of the FGM and are very similar to those obtained in the CDFM.

The results can be presented in a more convenient form if the corresponding Coulomb sum rule $C(q)$ is calculated. This quantity is defined as an integral of

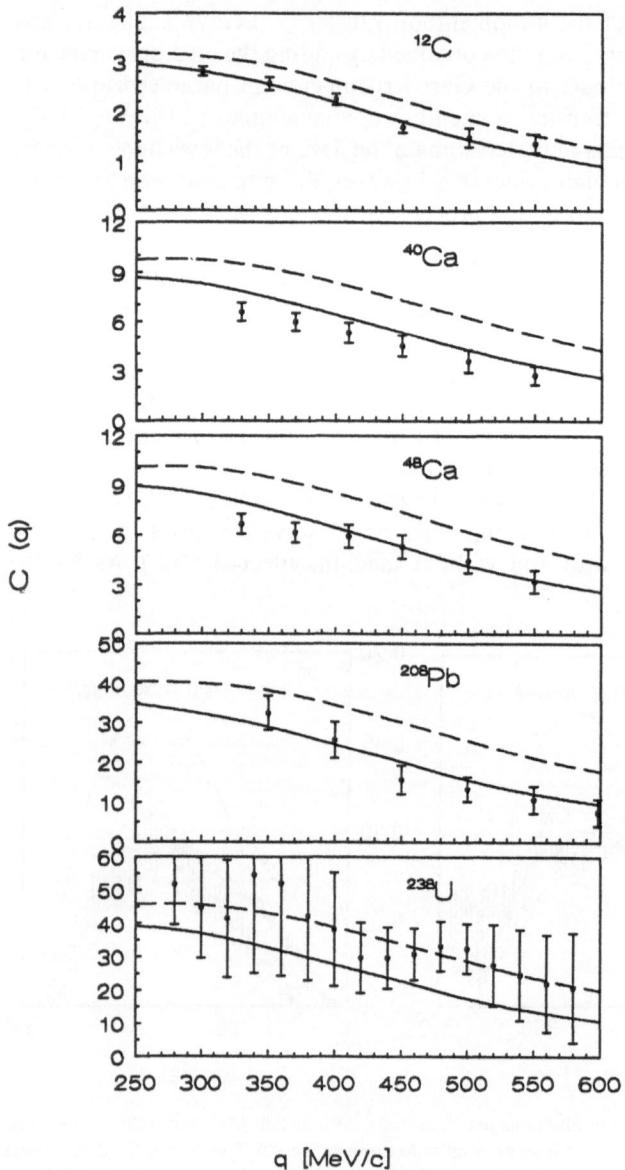

Fig. 9.8. Coulomb sum rule for ^{12}C, ^{40}Ca, ^{48}Ca, ^{208}Pb and ^{238}U. The solid and dashed lines are the same as in Fig. 9.6. The experimental data are from [9.40b, 9.44, 9.49, 9.104]

the longitudinal response function over the transfer energy along the path of constant three-momentum transfer:

$$C(q) = \int_{\omega_{el}^+}^{\omega_{max}} R_L(q, \omega)\, d\omega, \qquad (9.37)$$

where ω_{el}^+ indicates that ω starts just above the elastic peak and ω_{max} is the maximum value of the energy transfer for which the experimental value of R_L is not zero. The results for all nuclei considered are plotted in Fig. 9.8 and Fig. 9.9 and are compared with the experimental data from [9.40b, 9.44, 9.47, 9.49, 9.104]. As can be seen the incorporation of medium effects on the nucleon mass and form factors in the FGM removes the discrepancy between the theory and the experimental data for all the nuclei considered. In particular, for $400 < q < 600\,\text{MeV}/c$, where the data are expected to correspond to a pure quasielastic knockout process, the agreement when medium effects are taken into account is reasonable. It can be seen (Fig. 9.9) that the results obtained using the nucleon form factors in the nuclear medium which are renormalized to the *Gari-Kruempelmann* fit [9.102] at zero density start to deviate from those obtained with the unrenormalized form factors at $q \approx 600\,\text{MeV}/c$. It can be concluded that the renormalization is necessary only at higher values of the transfer momentum.

Concerning the transverse response function, the calculation within the FGM with medium-modified nucleons tends to underestimate the data at energy transfers $\omega > 100\,\text{MeV}$. It should be noted that both pion electroproduction and MEC contributions are essentially transverse [9.56]. Therefore, agreement with experimental data can be achieved only after taking these effects properly into account.

As a result the longitudinal quasielastic response function and the Coulomb sum at momentum transfer larger than $400\,\text{MeV}/c$ are found to be suitable for testing the predictions of the chiral quark-meson theory for the medium modification of nucleon properties. Even the simple Fermi-gas model in which nuclear structure effects are not considered gives reasonable agreement with the experimental data for wide range of momentum transfer and for various nuclei when

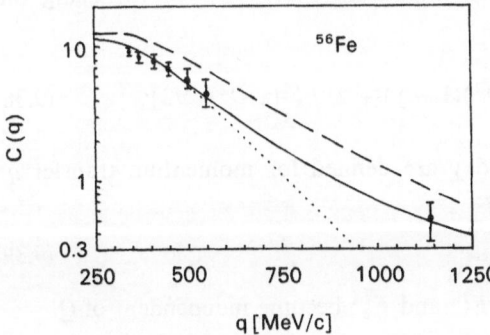

Fig. 9.9. Coulomb sum rule for ^{56}Fe. Dashed and solid lines are the same as in Fig. 9.6; dotted line: FGM with effective nucleon mass and form factors at finite medium density from the NJL model [9.99], but not renormalized to the *Gari-Kruempelmann* [9.102] fit at zero medium density. The experimental data are from [9.44, 9.47]

the medium effects are properly taken into account. In the transfer response Δ-isobar excitations and meson exchange currents appear to be dominating and medium modifications can be studied only after these effects are included in the calculations in a consistent way.

In the end of this Section we note that the theoretical studies of the Coulomb sum (9.37) show the important role of the ground state correlations for the description of this quantity. This is pointed out for light nuclei, such as ^{3}He and ^{4}He by *Schiavilla* et al. [9.76, 9.105] as well as for the ^{40}Ca nucleus by *Takayanagi* [9.106a]. It is shown by *Leidemann* et al. [9.106b] that the short-range correlation effects in heavier nuclei decrease the Coulomb sum rule per proton in the momentum range $0 < |q| < 4 \text{ fm}^{-1}$. The size of the reduction increases with the average proton density reaching its maximum value for medium heavy nuclei ($\sim ^{60}$Ni).

The sensitivity of the Coulomb sum rule to the nucleon–nucleon correlations in the nuclear ground state (short-range and long-range) as well as to the finite size effects is discussed in the review of *Orlandini* and *Traini* [9.106c]. It is pointed out that the effects of short-range correlations are of the order of 10% in the region $|q| = 2\text{--}3 \text{ fm}^{-1}$.

9.3 Deep-inelastic Lepton–nucleus Scattering

The results of the deep-inelastic lepton–nucleus scattering (DILNS) have been among the most interesting ones in experimental nuclear and elementary particle physics in the last decade. The European Muon Collaboration (EMC) obtained data [9.107] on the ratio (R) of the deep-inelastic nucleon structure function measured on iron (F_2^{Fe}) to the nucleon structure function measured in deuterium (F_2^D), establishing that these structure functions are significantly different. This result has been confirmed by the SLAC group observing a similar phenomenon for the Al nucleus [9.108] and later for a variety of nuclei [9.109]. The same effect has been observed in the neutrino experiments [9.110].

The dimensionless structure functions $F_1^{lA}(x, Q^2)$ and $F_2^{lA}(x, Q^2)$ are determined from the deep-inelastic charged lepton scattering ($l = $ e or μ) cross-section in the limit of large lepton laboratory energy $E_l \gg M$ (M being the nucleon mass) [9.111]:

$$\frac{d^2\sigma^{lA}}{dx\,dy} = \frac{4\pi\alpha^2}{Q^4} \frac{2M_A E_l}{A} \left[F_2^{lA}(x, Q^2)(1 - y) + 2xF_1^{lA}(x, Q^2)y^2/2 \right], \tag{9.38}$$

where the scaling variables x and y are defined for momentum transfer q^2 ($= -Q^2$) and energy transfer ν by:

$$x = AQ^2/2M_A\nu, \qquad y = \nu/E_l. \tag{9.39}$$

In the limit $Q \to \infty$ the functions F_1^{lA} and F_2^{lA} become independent of Q.

The main question raised by the DILNS experiments concerns the boundary between their interpretation in terms of quark and that in terms of effective nucleons and meson degrees of freedom [9.112]. Several reviews (e.g. [9.113–115]) have been devoted to the DILNS and the EMC effect and this is the reason that we will note in this Section only some peculiar questions of its interpretation related mainly to the nucleon–nucleon correlations in nuclei.

It turns out that the EMC results cannot be explained using standard methods implying that the momentum distribution of quarks in the nucleus differs from that in the nucleon [9.116]. Various models have been proposed to explain the EMC data [9.116–118] by introducing non-nucleonic degrees of freedom such as nucleon swelling or overlap (see, e.g. [9.119–122]), multi-quark cluster [9.116, 9.123], colour conductivity (see [9.124]), pion and Δ-isobar degrees of freedom [9.125–128].

It was shown, however, that the EMC data in the $0 < x < 1$ region can be explained quantitatively in terms of the nucleonic degrees of freedom taking into account the energy distribution of nucleons in the nucleus (nuclear binding), as well as their momentum distribution (Fermi motion) [9.129–132]. In [9.133] it was concluded that it is not necessary to introduce non-nucleonic degrees of freedom in explaining the EMC effect.

The criticism of *Frankfurt* and *Strikman* [9.134] was related to the fact that when relativistic effects are consistently taken into account by considering the so-called flux factor in the normalization of the relativistic spectral function (see also [9.135–137, 9.114]), the contribution of nucleonic degrees of freedom to the EMC effect should be strongly reduced [9.138]. It has been shown by *Ciofi degli Atti* et al. [9.139, 9.140], however, that taking account of the short-range and tensor correlations resulting from realistic nucleon–nucleon interactions strongly increases the values of the mean removal and kinetic energies of nucleons in nuclei and leads to an enhancement of the calculated EMC effect in accordance with the experimental data in the region $0.2 \leqslant x \leqslant 0.7$ even when the flux factor is considered. The nucleon contributions to the DILNS cross-sections is evaluated assuming the validity of the impulse approximation whose main assumptions are [9.140]:

(i) the nuclear hadronic tensor $W_{\mu\nu}^A$ depends only upon the one-body electromagnetic current $J_\mu^A = \sum_N J_\mu^{(N)}$ with the summation extended to the A nucleons in the nucleus;

(ii) the virtual photon is scattered incoherently from the nucleons, i.e. the interference terms between nucleonic currents do not contribute to the cross-section, and

(iii) the final-state interaction with the residual A-1 system is neglected. Ma noted [9.118, 9.141, 9.142], however, that one of the disadvantages in the conventional nucleonic approaches is that the applicability of the impulse approximation has not been seriously justified. This concerns mainly the role of the final-state interactions. The second disadvantage is related to the existing ambiguites concerning how to identify the off-mass-shell structure functions with the on-mass-shell ones.

Shlomo and *Vagradov* [9.143] have shown that the use of the appropriate experimental data for the energy and the momentum distribution of nucleons in nuclei, which is given by the nuclear spectral function, leads to a correct description of the mass dependence of the EMC effect. This is related in particular to the use of correct nucleon momentum distributions (NMD) which are strongly affected by the short-range correlations (SRC) and tensor nucleon–nucleon correlations in nuclei. It was shown by *Akulinichev* and *Shlomo* [9.112] that a realistic estimate of the high-momentum components of the NMD can account for the enhancement in the DILNS cross-section in the $x > 1$ region for ^{12}C. *Cao* et al. [9.144] noted, however, that these results are sensitive to the value of the parameter β which determines the slope of the high-momentum tail of the NMD and that the value of β used in [9.112] is outside the usually used range of β which approximates the results of the exp(S)-method for the NMD [9.145]. It is shown in [9.144] that the account of the SRC within the Jastrow correlation method [9.146] is not enough to describe the experimental data for the nucleonic structure function in the case of ^{12}C in the $x > 1$ region.

Anisovich et al. [9.147] showed that the ratio of the nuclear structure function to the nucleon structure function $F_2^A(x)/F_2^N(x)$ is very sensitive to the SRC in the region $x \gtrsim 0.3$. Taking account of the SRC and the decrease of the nucleon mass by 65 MeV (related to the existence of a mean meson field inside the nucleus) leads to a satisfactory description of the EMC data.

The realistic high-momentum components of the NMD due to the SRC in nuclei have also been applied in the calculations of DILNS cross-sections by *Araseki* and *Fujita* [9.148]. A realistic treatment of the separation energies for correlated states within the Jastrow method has been employed in the DILNS calculations of *Rozynek* and *Birse* [9.149]. Their results fall considerably below those from [9.144].

In the paper of *Guoju* and *Irvine* [9.111] the use of the NMD from [9.150] and a single-particle energy spectrum derived in [9.151] leads to good agreement of the theoretical calculations with the data for the structure function in ^{12}C obtained in the deep-inelastic muon scattering experiments [9.152].

A more detailed review of the nucleon–nucleon correlation effects on the DILNS cross-sections is given in [9.115]. It is concluded there that while the exact form of the NMD and the single-particle energy spectrum for correlated nucleons in nuclei is not well-known it is likely that they play an important role in obtaining the correct nuclear structure function in both the intermediate x region of the EMC depletion and the $x > 1$ region.

Further in this Section we will present the theoretical calculations concerning the EMC experimental results including the nucleon–nucleon correlations and using the nuclear spectral function obtained in the coherent density fluctuation model (CDFM) [9.153, 9.154].

Firstly, we give here an alternative division of the existing theoretical methods used in the interpretation of the EMC data into two classes [9.136]. The first of them assumes the change of nucleon quark distributions in nuclei at the expense of a possible change of the Q^2-evolution conditions for the nuclear structure functions in the nucleus (Q^2-rescaling) [9.119, 9.155], the nucleon Fermi motion being neglected. In these calculations the nuclear structure functions at a point x and Q^2 are expressed in terms of the nucleon structure function $F_{1,2}^N(x)$ at the same x but at another Q^2, i.e. $F_{1,2}^A(x, Q^2) = F_{1,2}^N(x, \xi Q^2)$. The free parameter ξ can up to now neither be calculated theoretically nor determined from independent experiments. For this reason, in this method ξ is chosen from the condition of the best fit to the experimental data (EMC [9.107, 9.156], BCDMS [9.157], SLAC [9.109]). The second class explains the difference between $F_{1,2}^A(x, Q^2)$ and $F_{1,2}^N(x, Q^2)$ on the basis of the internuclear motion of nucleons taking account of the off-mass-shell effects. The nucleon quark distribution are taken to be the same as for free nucleons. This type of calculation is based on the well-known fact that the properties of the nucleons in the nuclear medium differ from those of the free nucleons. In particular, the bound nucleons have an effective mass depending on the shell energy. This leads to the renormalization of the scaling variable $x \rightarrow xM/M^*$ (x-rescaling) [9.129–131, 9.158]. These methods seem to be preferable as they do not contain free parameters. It has to be emphasized that the proper consideration of the single-particle state characteristics in them leads also to a good description of the (e, e′p) reactions [9.159]. In both cases (DILNS and (e, e′p) reactions) the main idea is to investigate the scattering on the deeply bound nucleons. The characteristics of deep-hole nuclear states, such as spectral functions, widths and centroid energies (see Sect. 5.2), as well as some other nuclear quantities such as NMD and cross-sections of particle and ion-scattering on nuclei have been described [9.9, 9.10, 9.160, 9.161] in the framework of the CDFM suggested in [9.8]. The CDFM has been applied [9.153, 9.154] to the problems of *DILNS* (EMC effect) considering also the behaviour of the nuclear structure function in the region of larger values of x ($x > 0.6$, x being the light-cone variable) where experimental data are already available [9.162]. Of particular interest is the region of x near unity. It is known that for larger values of x ($x \sim 1.2$ for cumulative processes in hadron-nucleus collisions [9.163]) the main contribution to the nuclear structure functions comes from multiquark states [9.164] and the SRC [9.165]. To fix the parameters of such mechanisms, it is very important to know their relative contributions in the boundary region $x \sim 1$, where the role of the Fermi motion of the bound nucleons is still comparable with those of the above-mentioned mechanisms. In this case the nuclear structure function is sensitive to the choice of the spectral function (or of the momentum distribution).

In the impulse approximation the nuclear structure function $F_2^A(x)$ can be connected with the nucleon structure function by the convolution formula (see, e.g. [9.166, 9.134, 9.136, 9.138, 9.139]):

$$F_2^A = \int_{x_N}^A \rho_N(y) F_2^N(x_N/y)\, dy, \tag{9.40}$$

where $x_N = x_A(M_A/M)$, M_A and M being the nuclear and nucleon masses, respectively. The function $\rho_N(y)$ has the meaning of the nucleon distribution in the nucleus where the nucleon is carrying a part y of the total momentum. The explicit expression for $\rho_N(y)$ in the general relativistic case is unknown. This function has to obey two conservation conditions:

$$\int_0^A \rho_N(y)\, dy = A \quad \text{(baryon conservation law)}, \tag{9.41}$$

and

$$\frac{1}{A}\int_0^A \rho_N(y)y\, dy = \langle y \rangle. \tag{9.42}$$

Equation (9.42) is related to the energy conservation law. If the nucleus consists only of nucleons, it is obvious that $\langle y \rangle = 1$. Otherwise, $\langle y \rangle < 1$. In the impulse approximation $\rho_N(y)$ is related to the nuclear spectral function $S(\boldsymbol{k}, \omega)$ which is interpreted as the probability of finding particle with a momentum k in the initial nucleus if after its removal the residual nucleus has an excitation energy ω:

$$\rho_N(y) = \int \frac{d^3k}{(2\pi)^3}\, d\omega\, S(\boldsymbol{k}, \omega)(1 - k_z/M)\delta(y - k_-/M) \tag{9.43}$$

with

$$\int \frac{d^3k}{(2\pi)^3}\, d\omega\, S(\boldsymbol{k}, \omega) = A, \quad k_- = M + \omega - k_z. \tag{9.44}$$

Equations (9.43) and (9.44) involve the effects of bound nucleons through the energy dependence of the spectral function $S(\boldsymbol{k}, \omega)$ and the delta-function. It can be seen that $\rho_N(y)$ from (9.43) satisfies the condition (9.41). As for the condition (9.42) it can be easily found that

$$\langle y \rangle \simeq 1 + \langle \omega \rangle/M, \tag{9.45}$$

where

$$\langle \omega \rangle = \frac{1}{A}\int \frac{d^3k}{(2\pi)^3}\, S(\boldsymbol{k}, \omega)\omega\, d\omega. \tag{9.46}$$

One can see from (9.45) and (9.46) the violation of the energy conservation law; it is obviously due to the binding effects. Since the nucleus can be treated as a system of interacting nucleons and mesons, it is natural that a part of the total nuclear momentum is carried by mesons. This means that Eq. (9.42) is only a part of the total energy conservation law. It should be noted that the expression (9.43) differs from the one in [9.129–131, 9.158] by the flux factor $(1 - k_z/M)$ which in our case appears automatically owing to the non-relativistic reduction of the relevant relativistic expressions [9.167]. The flux factor does not change the general normalization of the spectral function in (9.44) and, consequently does not violate the baryon conservation law (9.41). The inclusion of this factor in our calculations leads to a small deviation in the final results.

To explain the EMC effect, it is necessary to take into account in the spectral function more complicated nuclear excitations as shown in [9.129–131]. For this purpose it is convenient to use the spectral function $S(\mathbf{k}, \omega)$ from [9.160, 9.9] which describes the main characteristics of nuclei, energy and NMD. In this case the NMD is close to the theoretical results obtained by Zabolitzky and Ey [9.145] from microscopic calculations taking account of nucleon–nucleon correlations.

The following expression for the spectral function obtained in the CDFM has been used in the calculations (see Eq. (5.43)):

$$S(\mathbf{k}, \omega) = \frac{16\pi r_0^3}{3} \frac{\alpha}{2|\mathbf{k}|} \frac{|f(r_0)|^2}{[\mu(\omega - E_F)]^{1/2}}, \tag{9.47}$$

where

$$r_0 = \alpha((\omega - E_F)/\mu)^{1/2}/|\mathbf{k}|, \quad \alpha = (9\pi A/8)^{1/3}.$$

The values of the parameters E_F and μ have been taken from [9.160]: this gives $\mu = -50\,\text{MeV}$ and $E_F = -8\,\text{MeV}$. It was shown in the CDFM [9.8] that for monotonically – decreasing density distributions $\rho(r)$ the function $|f(r_0)|^2$ is related to the nuclear density distribution by the expression:

$$|f(r_0)|^2 = -\frac{4\pi r_0^3}{3A} \left(\frac{d\rho(r)}{dr}\right)_{r=r_0}. \tag{9.48}$$

This function can be determined by means of the nuclear density distribution obtained from analyses of the electron scattering from nuclei. In our case the symmetrized Fermi-type distribution (9.21) has been used with values of the parameters R (half-radius) and b (surface thickness) obtained from the experiments on electron–nuclei scattering [9.1]. Then

$$|f(x)|^2 = \frac{4\pi x^3}{3Ab} \rho_0 \left\{ \frac{\exp[-(x+R)/b]}{[1 + \exp[-(x+R)/b]]^2} - \frac{\exp[(x-R)/b]}{[1 + \exp[(x-R)/b]]^2} \right\}, \tag{9.49}$$

where

$$\rho_0 = 3A/\{4\pi R^3 [1 + (\pi b/R)^2]\}.$$

It was shown in [9.160] that the theoretical results of the CDFM are in good agreement with the experimental data for the nuclear hole-state spectral functions extracted from (e, e′p)-reactions [9.159].

The structure functions $F_2^A(x)$ for ^{12}C and ^{56}Fe nuclei have been calculated in the CDFM using Eqs. (9.40, 9.43, 9.44, 9.47–9.49). The following parametrization [9.167] has been used for the $F_2^N(x)$:

$$F_2^N(x) = \tfrac{5}{18}\{x^{0.58}[2.69 + 1.56(1 - x)](1 - x)^{2.7}\}$$
$$+ \tfrac{12}{9} 0.167(1 - x)^7. \tag{9.50}$$

The spectral function (9.47) has been calculated using the following values of the parameters R and b: $R = 2.214$ fm, $b = 0.488$ fm for ^{12}C and $R = 4.054$ fm, $b = 0.600$ fm for ^{56}Fe. The ratio $F_2^A(x)/F_2^N(x)$ in the cases of ^{12}C and ^{56}Fe is shown in Figs. 9.10 and 9.11, respectively. It can be seen that the theoretical results of CDFM (taking account of the flux factor and without it) are in good

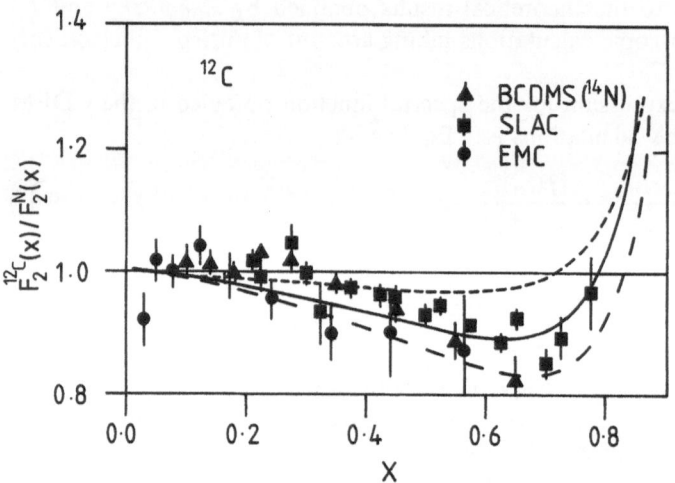

Fig. 9.10. The ratio $F_2^{{}^{12}C}(x) / F_2^N(x)$ calculated in the CDFM. Solid line: taking account of the flux factor; long-dashed line: without the flux factor; short-dashed line: single-particle Hartree approximation with Skyrme forces. The experimental data are taken from [9.107, 9.156, 9.109, 9.157, 9.162]: ▲ BCDMS (^{14}N), ■ SLAC, ● EMC

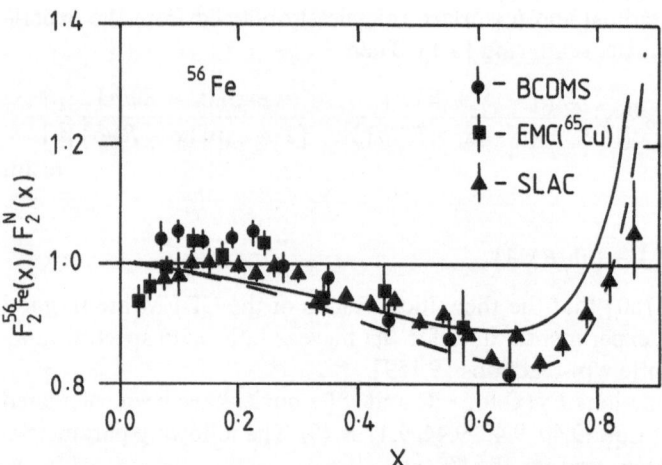

Fig. 9.11. The ratio $F_2^{{}^{56}Fe}(x) / F_2^N(x)$ calculated in the CDFM. Solid line: taking account of the flux factor; long-dashed line: without the flux factor. The experimental data are taken from [9.107, 9.156, 9.109, 9.157, 9.162]: ● BCDMS, ■ EMC (^{66}Cu), ▲ SLAC

agreement with the experimental data in the region $0.3 \leqslant x \leqslant 0.7$. The result for ^{12}C [9.168] obtained by using the spectral function from the Hartree approximation with Skyrme forces is also given in Fig. 9.10. The shapes of the ratio in both cases (CDFM and the Hartree approximation) are close to each other but the depths of the minima are different. The behaviour of the nuclear structure function in the intermediate region of the variable x $(0.3 \leqslant x \leqslant 0.7)$ is essentially determined by the value of $\langle \omega \rangle$ (9.46). In fact, expanding $F_2^N(x/y)$ in (9.40) near $\langle y \rangle$ (the point of maximum $\rho_N(y)$)

$$\frac{1}{A} F_2^A(x) \simeq F_2^N(x/\langle y \rangle)$$

$$+ \frac{1}{2}(\langle y^2 \rangle - \langle y \rangle^2) \frac{\partial^2}{\partial \langle y \rangle^2} F_2^N(x/\langle y \rangle) + \cdots \qquad (9.51)$$

and substituting $\langle y \rangle$ (9.45, 9.46), one can estimate $F_2^A(x)$:

$$F_2^A(x) \approx F_2^N(x)/(1 + \langle \omega \rangle/M). \qquad (9.52)$$

Thus it is clear that the discrepancy between the results of CDFM and of the single-particle approach is due to different values of $\langle \omega \rangle$ used in the two models. The CDFM calculations give $\langle \omega \rangle \simeq -38$ MeV which is in accord with the results from [9.129–9.131, 9.158], whereas in the Hartree approximation $\langle \omega \rangle \approx -20$ MeV.

In Fig. 9.12 the absolute value of the structure functions for a nucleon and for ^{56}Fe obtained within the CDFM and in the calculation with the Fermi

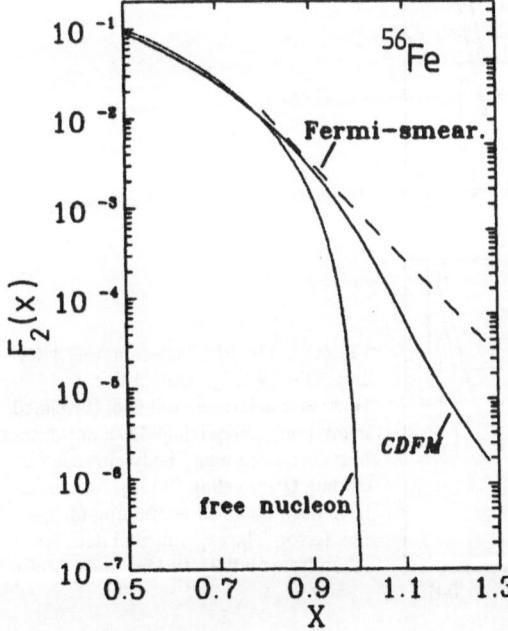

Fig. 9.12. The absolute value of the structure function $F_2(x)$ for the nucleon and for ^{56}Fe calculated in the CDFM and taking account of the Fermi motion without nucleon binding effects (long dashed line)

motion but without nucleon binding effects are given. It is seen that the CDFM curve in the region $0.9 \leqslant x \leqslant 1.2$ is substantially lower than the curve with only the nucleon Fermi motion. This result might be of importance for the future experiments giving the nuclear structure functions in the region $x > 1$.

The result obtained in the CDFM is a further confirmation of the Vagradov hypothesis [9.129–131] about the role of the deep bound nucleons in DILNS. On the other hand, the relative simplicity of the CDFM spectral function expression (9.47) allows one to apply this model to analyse the DILNS experimental data as well as to study the relative contributions of other mechanisms and also to predict the behaviour of the structure functions near $x \sim 1$.

An additional illustration of the role of the N–N correlations included in the correlation approach of *Ciofi degli Atti* et al. [9.139] on the DILNS from ^{12}C,

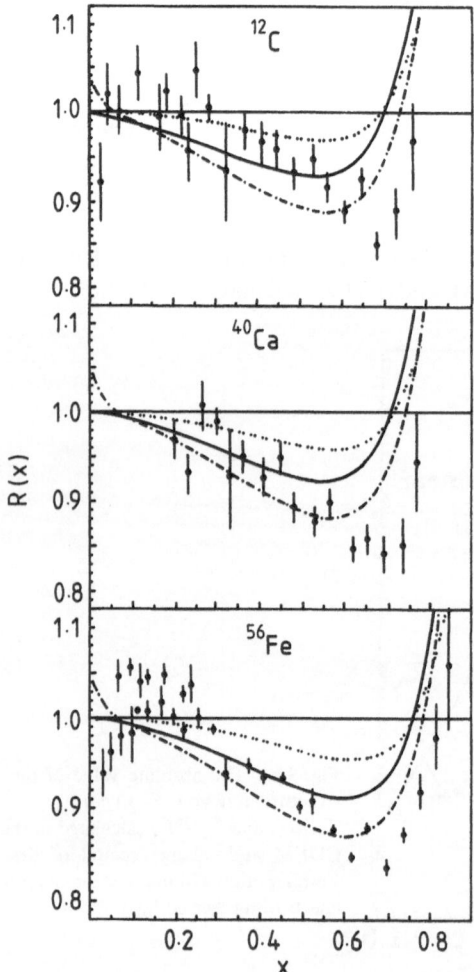

Fig. 9.13. The EMC effect in ^{12}C, ^{40}Ca and ^{56}Fe ([9.139]). Dotted line: Hartree–Fock result; full line: correlated many-body approach [9.139]; dot-dashed line: correlated many-body approach including Q^2-rescaling [9.139]. All curves have been obtained accounting for the flux factor. The experimental data are taken from [9.109], [9.156] and [9.169]

^{40}Ca and ^{56}Fe nuclei is shown in Fig. 9.13. As mentioned above, the nucleon–nucleon correlations resulting from realistic nucleon–nucleon forces lead to a reasonable explanation of the EMC effect in the region $0.2 \leqslant x \leqslant 0.7$ in terms of nucleonic degrees of freedom even when the flux factor is considered. It is pointed out in [9.139], however, that an appreciable discrepancy with the data in the region $0.7 \leqslant x \leqslant 1$ still remains, even if rescaling is considered.

Neglecting non-nucleonic degrees of freedom which turn out to be important only in the small x region, *Nakano* [9.170] confirmed the role of the nucleon–nucleon correlations for the correct description of the EMC data. He emphasized, however, the necessity of developing a realistic vertex function because the non-relativistic spectral function may have only a limited applicability, as far as DILNS is concerned. A detailed relativistic formalism for the nuclear structure function within the conventional nucleon constituents method is derived in [9.171].

The role of the details of nuclear structure (shells, Fermi motion etc.) on the DILNS has been considered by *Kumano* and *Close* [9.172].

A relativistic two-level convolution model for DILNS has been suggested [9.117, 9.118] in which the target nucleus is considered as a baryon–meson composite system. The baryons and mesons are also considered as composite systems of quarks and gluons. The model overcomes both disadvantages of the conventional nucleonic approaches mentioned above. In it the use of the impulse approximation is justified, baryon number conservation is naturally guaranteed and the off-shell ambiguities are avoided. It is confirmed in [9.118] that the EMC data are a signature for non-nucleonic degrees of freedom in nuclei.

A relativistic description of bound fermions is presented within the mesonic model by *Molinari* and *Vagradov* [9.173]. In this model the nucleon–nucleon force in the medium is associated with the exchange of colourless objects corresponding to various mesonic fields. General relationships between the structure functions of the nucleus and of its constituents are established.

Kondratyuk and *Shmatikov* [9.116] described the DILNS cross-sections assuming the existence of 9q- and 12q-bags in nuclei. This is achieved considering the momentum distribution of quarks in multiquarks at $k \gg k_0$ to be $\psi_q^2(k) \sim \exp(-k/k_0)$ with $k_0 \approx 50$–$60 \,\text{MeV}/c$. An 20% admixture of 12q-bags in the case of ^{56}Fe nucleus is necessary to describe the EMC experimental data.

Thomas et al. [9.174] obtained the structure function of an MIT-bag model nucleon bound in self-consistent scalar and vector mean fields corresponding to nuclear matter at an appropriate density. Calculations of a nuclear structure function based upon an explicit quark model of nuclear matter have been performed. The results throw light on the problem of the accuracy of conventional estimates of the nuclear binding and off-mass-shell corrections. The EMC- data for ^{56}Fe nucleus are qualitatively described within this model.

10 Photonuclear Reactions at Intermediate Energies

This chapter is devoted to the study of photonuclear processes as a source of information on the high-momentum components of the nuclear wave function. The (γ, p) and (γ, n) processes are considered in the framework of various theoretical methods and models of the reaction mechanism, such as the single-nucleon model, the quasi-free knockout model, the quasideuteron-type models, the methods taking account of Δ-excitation and ρ-meson exchange, the self-consistent RPA theory, the Jastrow-type models and some others. The role of the nucleon–nucleon correlations is shown in detail. The theoretical results are compared with a variety of experimental data.

In this chapter we consider the photonuclear processes (γ, p) and (γ, n) at photon energies E_γ above the giant-resonance region. These reactions provide a tool for studying the high-momentum components of the nuclear wave function because the kinematics of the process requires the emitted nucleon to have a much higher momentum than the incoming photon [10.1]. According to the mean-field picture, the probability that a single nucleon in the initial state will have a large momentum is very small and so the nucleon–nucleon short-range correlations (SRC), the meson-exchange currents (MEC), and also the Δ-resonance excitations at higher energies will play an important role in these reactions. The sensitivity of the photonuclear reactions to the single-particle aspects of the nuclear many-body problem can be best illustrated by comparing photoreactions, such as (γ, p) with the electronuclear knockout reactions, such as the quasi-free $(e, e'p)$ reaction. The real-photon-induced reaction (γ, p) (Fig. 10.1a) is complementary to a virtual-photon-induced reaction $(e, e'p)$ (Fig. 10.1b) [10.2]. In both diagrams q and ω are the momentum and energy transfer respectively, and p and p' denote the initial and final momentum of a hadron. A real photon, with $|q| = \omega$, cannot be absorbed by a single nucleon, if one excludes reactions such as $\gamma + p \rightarrow p + \pi^0$. A nucleus, or at least one other particle (e.g. $\gamma + d \rightarrow p + n$), must present in order to conserve momentum [10.2]. Whereas the $(e, e'p)$ reaction is usually studied in longitudinal kinematics that mainly probe the low-momentum components (< 300 MeV/c) of the nuclear wave function, the (γ, p) reaction is purely transverse and is associated with higher missing momenta [10.3]. This has the result that two-step processes, such as meson exchange and Δ-excitation, which are relatively unimportant in the $(e, e'p)$ reaction, will play a significant role in reactions with a real photon. The cross-sections of such reactions are determined by an interplay between various processes and there is up to now no well-developed theory for the real photon

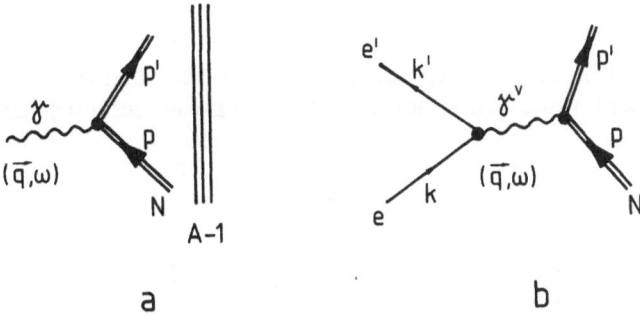

Fig. 10.1. Diagrams for the reactions (γ, N) **(a)** and $(e, e'N)$ **(b)**

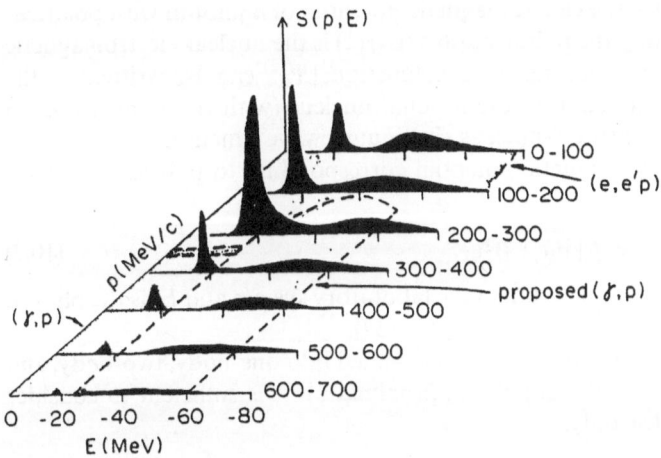

Fig. 10.2. Schematic spectral function for 1p-shell nucleus ([10.2])

processes. As noted by *Matthews* [10.2] if one relates the cross-section to the nuclear spectral function (as is done for the $(e, e'p)$ case, see (5.1) and (5.4)), the validity of this will be questionable in the case of transverse, real photons. One will have additional problems due to the importance of the two-step processes. All these circumstances make it difficult to extract the single-nucleon part of the photo-process and to extend the range of the knowledge of the spectral function $S(p, E)$ (see Chap. 5 and Fig. 10.2).

Comparing the (γ, p) and the $(e, e'p)$ reaction it is noted in [10.4] that there is a limit to the information available from the (γ, p) reaction due purely to the kinematics. The photon brings in relatively little momentum and thus any high-energy proton emitted must have been initially in a high-momentum state. For example [10.4] in the $^{12}C(\gamma, p_0)$ ^{11}B reaction with $E_\gamma = 100$ MeV, the emergent proton has an energy $E_p \geqslant 80$ MeV, and thus a momentum of ≈ 400 MeV/c, only 100 MeV/c of which could have been acquired from the incident photon. It is concluded that the (p, 2p) and (e, e'p) reactions do not probe the same part of the wave function. While small cross-sections and experimental difficulties limit (p, 2p) and (e, e'p) measurements of $q \lesssim 300$ MeV/c, the (γ, p) process (e.g. for $E_\gamma > 60$ MeV) will be restricted kinematically to region of $q \gtrsim 300$ MeV/c. Thus the photoeffect has an almost unique sensitivity to high-momentum components and gives information which is complementary to that from quasi-free scattering and can hardly be obtained any other way [10.4].

Here we give some basic expressions for nuclear photoreactions. The scattering amplitude for the (γ, p) reaction can be written

$$M_{fi}^\lambda = -\langle \Psi_f | \int d\mathbf{x}\, \mathbf{j}(x) \cdot A_\lambda(x) | \Psi_i \rangle, \tag{10.1}$$

where $|\Psi_i\rangle$ is the initial target-nucleus wave function and $|\Psi_f\rangle$ the final state wave function. $A_\lambda(x)$ is the electromagnetic potential of a photon with polarization component λ along the polarization axis. $\mathbf{j}(x)$ is the nuclear electromagnetic current operator. The final state wave function $|\Psi_f\rangle$ can be written as the product of the wave function of the residual nucleus (with $A - 1$ nucleons) in state f and a relative proton–nucleus continuum wave function.

The photon electromagnetic potential corresponding to polarization component λ has the form:

$$A_\lambda(x) = 1/(2E_\gamma)^{1/2} \cdot \exp\left[\mathrm{i}(\mathbf{k}_\gamma \cdot x)\right] \varepsilon_\lambda, \tag{10.2}$$

where $\mathbf{k}_\gamma$ is the photon momentum in the laboratory system and E_γ is the photon energy, $E_\gamma = |\mathbf{k}|$ ($\hbar = c = 1$, $e^2/4\pi = \alpha = 1/137$).

The nuclear current operator can be expanded into one-body, two-body, and higher-order terms. For the impulse approximation it is sufficient to consider the one-body operator only:

$$j_1(x) = -e \sum_{j=1}^{A} \left\{ \frac{e_j}{2\mathrm{i}M} \left[\delta(r_j - x)\nabla\right]_{\text{sym.}} - \delta(r_j - x)\frac{\mu_j}{2M}\, \sigma(j) \times \nabla \right\}, \tag{10.3}$$

where M is the nucleon mass, e_j the charge of the jth nucleon, and μ_j the nucleon magnetic moment (in nuclear magnetons). The notation

$$[\delta(r_j - x)\nabla]_{\text{sym.}} \equiv \delta(r_j - x)(\vec{\nabla}_j - \overleftarrow{\nabla}_j) \tag{10.4}$$

is used in (10.3).

The differential cross-section in the laboratory system, averaged over initial photon polarization and summed over nucleon and residual nucleus polarization is given by

$$d\sigma = 2\pi\delta(E_f - E_i)\tfrac{1}{2}\sum_\lambda \sum_{S_N,\,\alpha} |M_{fi}^\lambda|^2 \frac{d^3 k_N}{(2\pi)^3} \frac{d^3 k_R}{(2\pi)^3}, \tag{10.5}$$

where k_N and k_R are the nucleon and residual nucleus outgoing momenta and S_N are the final spins.

The simplest model for a (γ, N) reaction is one in which a single nucleon bound in a nuclear potential absorbs the photon (Fig. 10.3a). The formalism

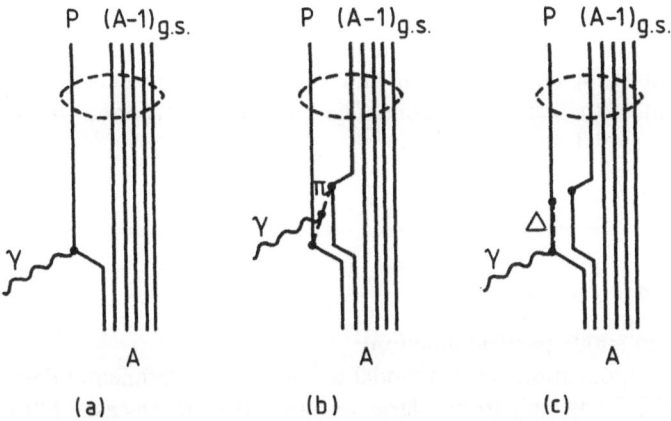

Fig. 10.3. Diagrams for the exclusive (γ, p) reaction. (a) single-nucleon process; (b) and (c) typical two-nucleon processes

based on this model has been suggested by *Shklyarevsky* [10.5] and developed by *Matthews* et al. [10.6]. The photon is considered to interact with the proton charge and magnetic moment.

In the plane-wave impulse approximation (PWIA) the (γ, p) cross-section has the form:

$$\frac{d\sigma}{d\Omega} = C(k_\gamma, q) \frac{1}{(2\pi)^3} \frac{(2j+1)}{(2l+1)} \sum_m |\varphi_{nljm}(q)|^2, \tag{10.6}$$

where $q = k_N - k_\gamma$ and

$$C(k_\gamma, q) = \frac{\pi e^2}{2} \frac{k_N^2}{k_\gamma^2} \frac{dk_N}{dE_f} \left(\frac{k_N^2}{M^2} \sin^2\theta + \frac{k_\gamma^2}{2M^2} \mu_p^2 \right) \tag{10.7}$$

and $\varphi_{nljm}(q)$ is the single-nucleon wave function in momentum space. The (γ, n) cross-section is proportional to the magnetic moment term only. It can be seen from (10.6) that the ratio $(d\sigma/d\Omega)/C$ is just the square of the single-particle wave function, i.e. it is the single-particle momentum density. Thus, in PWIA the (γ, p) cross-section provides directly the possibility of studying the high-momentum behaviour of the single-particle wave function.

It is shown in [10.7] that if the outgoing proton is described by a plane wave and the wave number k is considered to be local inside the nucleus, then the cross-section for the emission of a nucleon with an angular momentum l is:

$$\frac{d\sigma_l}{d\Omega} = \frac{e^2}{4\pi mc^2} \frac{k^2 k_p}{E_\gamma} \left[\sin^2\theta \left| \frac{Z_1}{A} \Phi_l(P_1) - \frac{Z_2}{A-1} \Phi_l(P_2) \right|^2 \right.$$

$$\left. + \frac{k_\gamma^2}{2k^2} |g_1 \Phi_l(P_1) + g_2 \Phi_l(P_2)|^2 \right], \tag{10.8}$$

where $k = E_\gamma/\hbar c$ is the photon momentum, k_p is the asymptotic value of the proton momentum in the centre-of-mass system, Z_1 and g_1 are the charge and the magnetic moment of the emitted proton, and Z_2 and g_2 are the charge and the magnetic moment of the residual nucleus. The momenta transferred to the nucleon and to the residual nucleus are

$$P_1 = k - (A - 1)k_\gamma/A, \qquad P_2 = k_\gamma + k/A, \tag{10.9}$$

respectively. In (10.8)

$$\Phi_l(q) = \int \exp(-i\mathbf{q} \cdot \mathbf{r}) \psi_l(r) \, d\mathbf{r}, \tag{10.10}$$

where ψ_l is the proton single-particle function.

The results of the applications of this model to the early experimental data are reviewed in [10.7]. They vary from a large disagreement in the case of ^{6}Li (γ, p) at $E_\gamma = 100$ MeV [10.6] to a good description of the data for the 1p-proton emission in ^{6,7}Li at $E_\gamma = 60$ MeV (but not of 1s-protons) [10.8] and for ^{12}C (γ, p) reaction [10.5].

In [10.9, 10.10] the importance of including the distortion of the outgoing proton wave function is shown.

The actual nature of the proton removal process remains one of the central problems in the description of the (γ, p) reaction [10.3]. The quasi-free knockout model (QFK), in which the photon interacts with a single nucleon, the other $A - 1$ nucleons being spectators, has been developed by *Boffi* et al. [10.11]. It has been claimed that low energy (γ, p) data can be described to a great extent by direct proton knockout [10.11–10.13, 10.2]. This is shown by *Leitch* et al. [10.12] in the case of ^{16}O (γ, p) reaction for $E_\gamma = 100$–400 MeV. It is noted that the QFK (γ, p) cross-sections are very sensitive to the choice of initial and final state potentials and that within the range of acceptable potentials it is possible to account for the data up to $E_\gamma \approx 100$ MeV using nuclear wave functions which are consistent with (e, e'p) experiments. This is also confirmed by *Ferdinande* et al. [10.14] in the ^{12}C (γ, p) reaction at $E_\gamma = 28$ MeV.

An important test for the single-particle absorption models is the study of the (γ, n)-reaction. As the photon interacts only with the magnetic moment of the neutron, it is estimated in [10.2] that in the plane-wave model the ratio of the (γ, p) and (γ, n) cross-section has the form:

$$\frac{\sigma_p}{\sigma_n} = \frac{\sin^2\theta + g_p^2 q^2/p'^2}{g_n^2 q^2/p'^2}, \tag{10.11}$$

g_p and g_n being the proton and neutron magnetic moments. At angles near $0°$ and $180°$

$$\frac{\sigma_p}{\sigma_n} \approx \left(\frac{g_p}{g_n}\right)^2 \approx 2. \tag{10.12}$$

At intermediate angles the electric ($\sin^2\theta$) term dominates and the result is that the (γ, n) cross-section have to be much smaller than the cross-section for the corresponding (γ, p) reaction. This, however, is not in agreement with the experimental data which show that (γ, n) cross-section [10.15–10.18] are comparable, with an energy and angle dependence similar to the corresponding (γ, p) cross-sections [10.4, 10.19–10.21]. It has been concluded in [10.16] that the results of the (γ, n) and (γ, p) measurements on ^{12}C and ^{16}O combined with each other in the ratio $[d\sigma(\gamma, p)/d\Omega]/[d\sigma(\gamma, n)/d\Omega]$ for $E_\gamma = 60$ MeV favour models in which the absorption of the photon takes place on correlated neutron–proton pairs [10.22–24]. In this way the (γ, N) processes like the (γ, np) ones become potential sources of information concerning nucleon correlations in nuclei.

Studying the ^{16}O (γ, n) reaction in the energy region between the giant resonance and the pion threshold ($60 \leqslant E_\gamma \leqslant 160$ MeV) *Göringer* et al. [10.15] draws attention to the possibility of extracting the high-momentum components of the single-particle proton wave function $\Phi(P)$ from the cross-section within the direct knockout approach [10.25, 10.21]:

$$\frac{d\sigma}{d\Omega}(E_\gamma, \Theta) = C(E_\gamma, \Theta)|\Phi(P)|^2 \tag{10.13}$$

using the experimental (γ, p) data [10.4, 10.19–21]. However, the failure of the single-particle direct model to explain the (γ, n) data mentioned above leads the authors of [10.15] to the assumption that two nucleons are involved in the absorption of photons. The model of *Levinger* [10.26a] has given one of the first accounts of this reaction mechanism (see Fig. 10.4a). In it the (γ, np) cross-section is related to the cross-sections for the photodisintegration on free deuterons. This model explains the results from [10.27–30]. The model has been successfully exploited in [10.26b]. Fig. 10.4b shows (γ, n) and (γ, p) processes as limiting cases of the (γ, np) reaction with one of the nucleons remaining within the nucleus. Various calculations of this type (e.g. [10.22–10.24, 10.31–10.33]) describe both (γ, n) and (γ, p) cross-sections.

A two-step mechanism consisting of a direct knockout (γ, p) process followed by a (p, n) charge-exchange has been proposed [10.34, 10.35] for the description of the (γ, n) reaction at energies below 100 MeV.

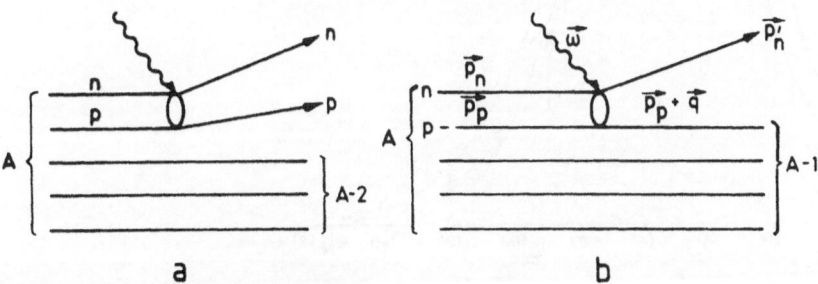

Fig. 10.4. Diagrams for various quasideuteron absorption reaction mechanisms [10.15]: (**a**) quasideuteron (γ, np); (**b**) quasideuteron (γ, N)

The results of *Göringer* et al. [10.15] on ^{16}O (γ, n) ^{15}O at $E_\gamma = 60$ MeV are shown together with other ^{16}O (γ, n) and ^{16}O (γ, p) data in Figs. 10.5 and 10.6. They are compared with the theoretical results of *Gari* and *Hebach* [10.33], of *Saruis* (given in [10.15]) and of the phenomenological model from [10.24] using the reaction mechanism shown in Fig. 10.4b. The comparison of (γ, p) and (γ, n) cross-section shows that they are of similar magnitude. It is concluded in [10.15] that in contrast to the single-particle models (which suppress the (γ, n) cross-section), the models in which the quasideuteron-like absorption mechanism is accepted give results which describe satisfactorily both (γ, p) and (γ, n) cross-sections.

Here we give briefly some more details [10.12] concerning the two-nucleon mechanism of *Schoch* [10.24] and that of *Gari* and *Hebach* [10.33]. The model of *Schoch* [10.24] is based on the diagram from Fig. 10.4b. In this picture the momentum "mismatch"

$$Q = |\boldsymbol{p}' - \boldsymbol{q}| \qquad (10.14)$$

between that of the outgoing nucleon and that of the incoming photon is made up of three contributions, the initial momentum of the knock-out nucleon and both the initial and final momenta of the other nucleon. In the model from

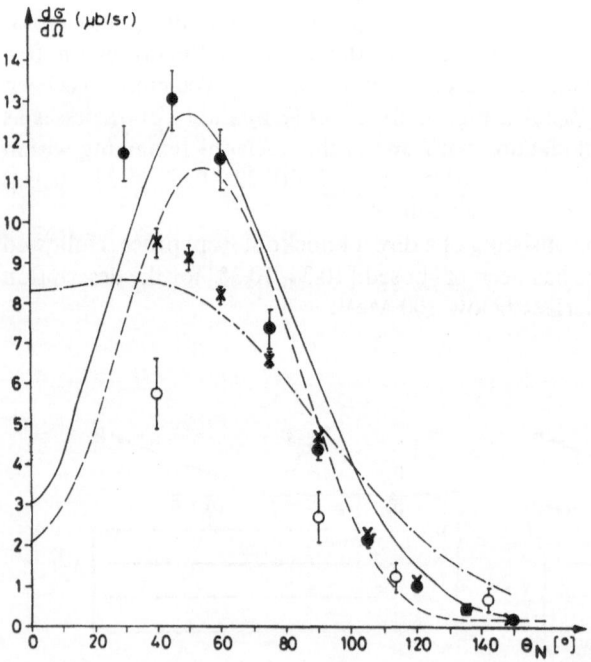

Fig. 10.5. Angular distributions of the differential cross sections for the reactions $^{16}O(\gamma, n_0)$ ^{15}O and $^{16}O(\gamma, p_0)$ ^{15}O, $E_\gamma = 60$ MeV ([10.15]); (γ, n_0): x [10.15]; ○ [10.17]; solid line [10.33]; dashed line (result of *Saruis* from [10.15]); dot-dashed line [10.24]; ● (γ, p_0): [10.21]

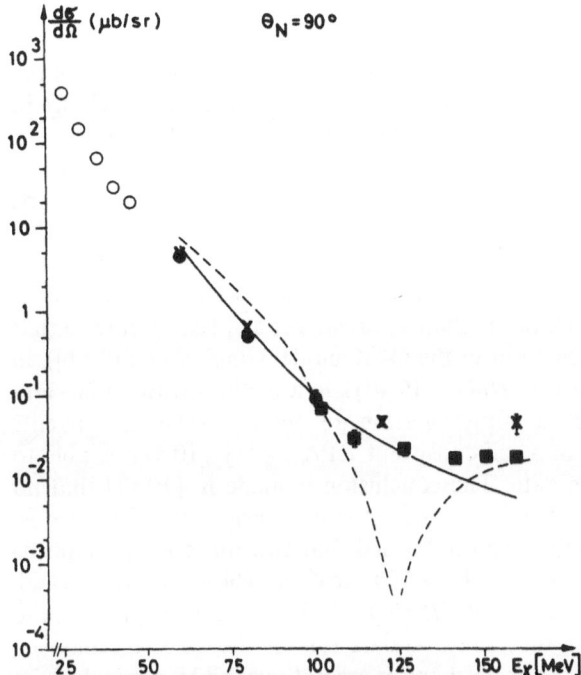

Fig. 10.6. Differential cross section for the reaction $^{16}O(\gamma, n_0)^{15}O$ and $^{16}O(\gamma, p_0)^{15}N$ as a function of photon energy for $\theta = 90°$; (γ, n_0): $\times$ [10.15]; $\bigcirc$ [10.36]; solid line: [10.33]; dashed line [10.24]; (γ, p_0): $\bullet$ [10.21]; $\blacksquare$ [10.20].

[10.24] the latter two contributions are combined into a term $F(q_R)$ which may be identified approximately with the form factor of the residual nucleus. Thus the predicted angular distributions and cross-sections are determined by the shape of the relevant form factor F.

In the calculations of *Gari* and *Hebach* [10.33] the term corresponding to an effective two-body interaction of the photon are not treated individually. Taking into account the contributions due to meson exchange and the initial- and final-state correlations these terms are included by using their relation to the residual nucleon–nucleon interaction. These important terms which are often omitted in many calculations are required by gauge invariance. In the work of [10.33] the gauge terms dominate most of the region studied, while the QFK contributions are small. In this model the meson exchange currents (e.g. Fig. 10.3b) appear in a natural way. The results of the calculations of *Cavinato* et al. [10.37] using a self-consistent RPA theory with Skyrme forces are similar to that from [10.33].

The intermediate Δ-excitation diagram (Fig. 10.3c) has been included in calculations using two-nucleon mechanisms by *Londergan* and *Nixon* [10.38]. Their calculations that account only for the QFK term in addition to Δ-

excitation are claimed to be "parameter-free", since the basic $\gamma N\Delta$ and $\Delta\pi N$ coupling constants are derived from the experimental data. The effects of centre-of-mass corrections, of ρ-meson exchange and of the modification of the Δ width do not alter the qualitative features of the results [10.12] which are in a reasonable agreement with the experimental data for (γ, p) reactions.

It has to be emphasized that the experiments on the ^{16}O (γ, p) ^{15}N reaction for $E_\gamma = 100–400$ MeV [10.12], as well as on ^{7}Li (γ, p) and ^{7}Li (e, p) reactions (Sené et al. [10.1]) and on ^{40}Ca (γ, p) ^{39}K at $E_\gamma = 100–300$ MeV (Leitch et al. [10.39]) essentially extend the range covered by the previous experiments, but the comparison with the available theoretical work still does not permit a definite conclusion about the reaction mechanism. Leitch et al. [10.12] showed that none of the calculations carried out in the QFK model (Boffi et al. [10.11]), in a Jastrow-type model (Weise and Huber [10.40]), in a method with initial- and final-state correlations plus a representation of meson-exchange currents (Hebach et al. [10.22]) and in a self-consistent RPA theory [10.37], is able to reproduce all features of the data. The conclusion is made in [10.12] that no single mechanism dominates the (γ, N) process over a large part of the whole energy and angular range. It is well established that two-nucleon mechanisms provide the major contribution for $E_\gamma > 100$ MeV. As shown by Sené et al. [10.1] the quasi-deuteron-type models (Levinger [10.26a], Schoch [10.24]) are more successful in describing the ^{7}Li (γ, N) data than is the QFK mechanism, though the contributions from this mechanism are not negligible, particularly at backward angles.

The account of the short-range correlations by the Jastrow model gives the possibility considered in [10.40] of using an effective A-body operator of the electromagnetic interaction H_{eff} containing the information about correlations instead of

$$H_{int} = \sum_{j=1}^{A} H_\gamma(j), \tag{10.15}$$

where $H_\gamma(j)$ are one-body operators of interaction between the incoming photon and the particular nucleon. The operators H_{int} and H_{eff} have matrix elements which satisfy the relation:

$$\langle \tilde{\Psi}_f | H_{int} | \tilde{\Psi}_i \rangle = \langle \Psi_f | H_{eff} | \Psi_i \rangle, \tag{10.16}$$

where $\Psi_{i(f)}$ is the initial (final) state nuclear wave function in the independent-particle model and $\tilde{\Psi}_{i(f)}$ is the Jastrow correlated initial (final) state nuclear wave function (see Sect. 3.3, Eq. (3.56)). Neglecting three- and more-particle transitions, H_{eff} is given by:

$$H_{eff} \approx h_1 + h_2, \tag{10.17}$$

where

$$h_1 = \sum_{k=1}^{A} H_\gamma(k) \tag{10.18}$$

and h_2 contains the information about the Jastrow-type N–N short-range correlations:

$$h_2 = -\sum_{k \neq j} \sum \bar{g}(k,j) H_\gamma(k) - \sum_{k' \neq j'} \sum H_\gamma(k') g(k',j'). \tag{10.19}$$

In (10.19) the function g is related to the Jastrow correlation factor (Eq. (3.57)):

$$f(r) \equiv n_c [1 - g(r)]. \tag{10.20}$$

The functions g and $\bar{g}$ are expressed by the correlation factors f and $\bar{f}$ in the initial and final state, respectively.

The results of the ^{16}O (γ, p) reaction at $E_\gamma = 100$–300 MeV (*Matthews* et al. [10.19]) show that the experimental data at $E_\gamma \geqslant 100$ MeV deviate strongly from the predictions that include only single-step photoejection from a simple shell-model orbit. This shows the necessity of including more than one nucleon in the absorption mechanism and also additional high-momentum components in the nuclear wave function. The N–N short-range correlations accounted for by *Małecki* and *Picchi* [10.41] are shown to be capable of enhancing the (γ, p) cross-section above $E_\gamma \simeq 100$ MeV.

The study of the reaction ^{16}O (γ, p) ^{15}N at forward angles $\Theta_p = 5°$–$40°$ for $E_\gamma = 80$ MeV shows that the experimental data can be well described by the method of *Hebach* et al. [10.22], improved by the inclusion of contributions from the spin current. However the modified quasideuteron model of *Schoch* [10.24] fails to describe the data.

Discussing the dynamical aspects of photonuclear reactions *Hebach* et al. [10.22] have noted that the modifications of the shell-model treatment, namely taking into account the Jastrow-type correlations obtained from the Bethe–Goldstone formalism [10.9], are not able to improve the shell-model results for photon energies up to 100 MeV. They emphasized that for these and other models there are two aspects which deserve more attention, namely: (i) the orthogonality of the initial and final state wave functions is not always ensured, and (ii) gauge invariance is not satisfied if the exchange contributions (gauge terms) to the transition matrix are not accounted for. In their method [10.31, 10.44] the nuclear wave functions have been built up by introducing correlations into the shell-model description. They decompose the transition matrix into contributions from (i) the "direct" shell-model transition, (ii) the nucleon–nucleon correlations in the initial and final states and (iii) the gauge (or exchange) contributions where the electromagnetic field is coupled directly to the correlations between a neutron and proton. The approach gives a satisfactory explanation of the (γ, p) and (γ, n) processes on ^{16}O and ^{12}C, of the capture reaction (p, γ) on ^{3}H and for the reaction (γ, pn) on ^{16}O at photon energies between 40 and 140 MeV.

In the method of *Fink* et al. [10.9] mentioned above the two-nucleon correlation function for ^{16}O is calculated using the Bethe-Goldstone equation. The resulting wave function has the correct healing properties and the momentum components introduced are between 400 MeV/c and 1200 MeV/c. The short-range correlation effects are shown to be small ($\approx 10\%$ at $E_\gamma = 100$ MeV). As is noted in [10.4] these effects seem unlikely to be of major significance below 100 MeV. The explanation of this is clear from a simple kinematic estimate [10.9] showing that Fourier components from about 200 MeV/c to 400 MeV/c are needed in the case of (γ, p) and (γ, n) reactions in the energy region $40 < E_\gamma < 100$ MeV. Neither the shell-model nor short-range correlations can provide sufficient momentum components in these momentum regions. The bulk of the momentum components in the method of *Fink* et al. [10.9] lie between 400 and 1200 MeV/c and this is why the short-range correlations are not important for (γ, p) and (γ, n) processes in the energy region considered.

The theoretical predictions of the pure shell-model and of the Jastrow correlation method used by *Weise* and *Huber* [10.40] are compared in [10.4] with the experimental cross-sections of (γ, p) reactions on ^{6}Li, ^{7}Li and ^{12}C at $E_\gamma = 60$–100 MeV and it is found that the shell-model results fall below the experimental measurements. It is concluded that the short-range strong-interaction components of the nuclear force undoubtedly exist so that the wave functions might be parametrized according to the Jastrow model [10.40]. As pointed out by *Ciofi degli Atti* [10.45], however, the correlation parameters cannot be arbitrarily chosen or varied.

Findlay and *Owens* [10.21, 10.25] consider the possibility of using an effective momentum distribution extracted from the (γ, N) data for the photoproton before photoejection. Such a momentum distribution would contain the underlying shell-model momentum distribution together with the Fourier components of the residual interactions. They conclude that the overestimate of the data on the (γ, p) cross-section in the Jastrow correlation method [10.40] and the underestimate when considering hard core effects [10.9] could in this case be interpreted as the addition of residual interactions having unrealistically many and too few Fourier components respectively in the momentum regions studied in these photoreactions.

The simultaneous analyses of (γ, p) and $(e, e'p)$ reactions give the possibility of determining the momentum distribution of a nucleon in a certain shell. The $p_{1/2}$ and $p_{3/2}$-shell nucleon momentum distributions for ^{16}O have been deduced from (γ, p) reactions [10.25] and from $(e, e'p)$ data [10.46]. The 1p-shell nucleon momentum distribution in ^{12}C has been extracted from the (γ, p) reaction [10.47] and from $(e, e'p)$ data [10.48]. The single-particle momentum distributions have been obtained using the so-called "extended plane-wave impulse approximation (PWIA)". This includes an approximate treatment of the distortion of the outgoing proton wave in the optical potential of the residual nucleus, which is applicable to both reactions and preserves the direct PWIA relation between the cross-section and the momentum distribution [10.49].

The proton momentum distribution in ^{4}He has been deduced from the ^{4}He (γ, p) ^{3}H reaction [10.50, 10.51] as well as from the ^{4}He $(e, e'p)$ ^{3}H reaction [10.52]. It was shown that the momentum distribution behaviour at $p_B > 350$ MeV/c supports the distorted-wave impulse approximation (DWIA) calculations corrected for short-range correlation effects [10.52]. The review of

Frullani and *Mougey* [10.53] contains the results for the single-particle momentum distributions in nuclei deduced from (γ, p) and $(e, e'p)$ processes compared with the results of various calculations. We give as an example in Fig. 10.7 the momentum distribution of the $p_{1/2}$-shell protons in ^{16}O. We should mention especially the good agreement of the theoretical calculations from [10.56] in the Jastrow correlation method with the experimental data for the proton momentum distribution at momenta $q \gtrsim 400\,\mathrm{MeV}/c$ (or $\gtrsim 2\,\mathrm{fm}^{-1}$), where the short-range correlation effects are important.

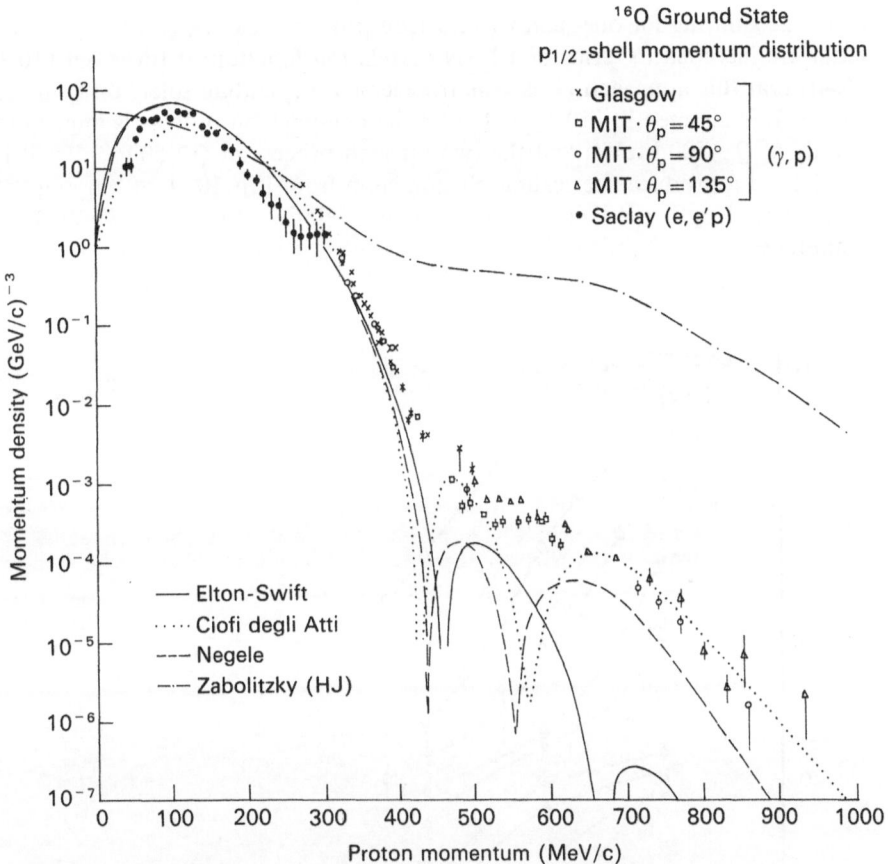

Fig. 10.7. Momentum distribution of $p_{1/2}$-shell protons in ^{16}O deduced from the data of $(e, e'p)$-reaction [10.46] and (γ, p) reaction [10.25, 10.19, 10.20] and presented in the review of *Frullani* and *Mougey* [10.53]. The experimental points deduced from the data through an extended PWIA analysis are compared with several theoretical results: i) calculations using Elton–Swift wavefunctions [10.54] (solid line); ii) density-dependent Hartree–Fock calculations of *Negele* [10.55] (dashed line); iii) calculations of *Ciofi degli Atti* [10.56] using harmonic-oscillator wavefunctions with the Jastrow correlation function (dotted line); iv) the dot-dashed line shows the results of *Zabolitzky* and *Ey* for the entire ^{16}O nucleus [10.57] obtained by the exp (S)-method using the Hamada–Johnston potential

A review of ^{16}O $(\gamma, p)^{15}$N-reaction data has been given by *Matthews* [10.2]. The existing angular distributions from [10.2] are given in Fig. 10.8. It can be seen from Figs. 10.9 and 10.10 that the calculations of *Gari* and *Hebach* [10.33], *Londergan* and *Nixon* [10.38] and *Boffi* et al. [10.11] cannot describe successfully the cross-section data. An interesting fact, namely some kind of scaling phenomenon, has been observed in [10.2] for a large variety of ^{16}O (γ, p_0) cross-section data when they are plotted vs. the momentum mismatch Q (Eq. (10.14)). The data have approximately a common dependence on Q (see Fig. 10.11). It is pointed out in [10.2] that this scaling behaviour is not in contradiction with the two-nucleon mechanism, since a phenomenological model accounting for one- and two-nucleon processes can be constructed. The latter are included in terms of a N–N correlation function. It turns out [10.2, 10.40] that this approach gives a matrix element depending solely on Q in the PWIA. It is claimed in [10.2] that since the one-nucleon process is important mainly at $Q \lesssim 500$ MeV/c and the two-nucleon process at $Q \gtrsim 500$ MeV/c it is possible to reproduce the scaling phenomenon from Fig. 10.11 by varying the correlation length (and correspondingly the relative one- and two-nucleon amplitudes).

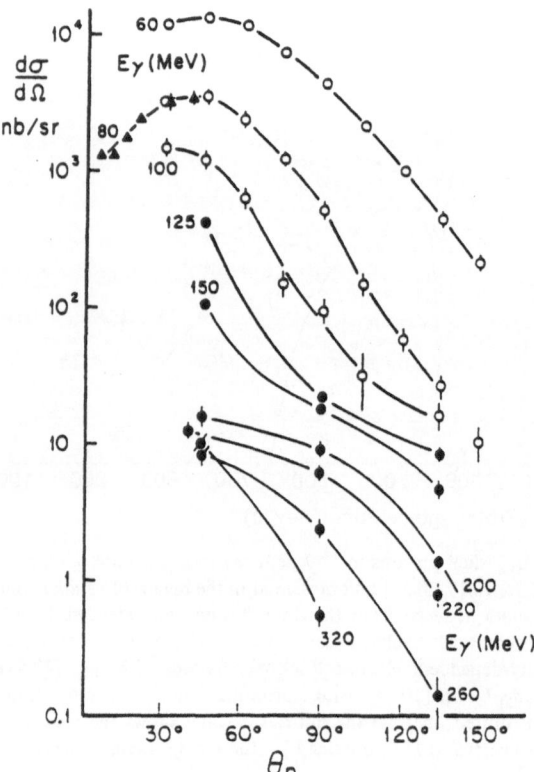

Fig. 10.8. ^{16}O(γ, p_0) angular distributions (taken from *Matthews* [10.2]) $\bigcirc$ – from [10.21]; $\blacktriangle$ – from [10.42]; $\bullet$ – from [10.12] (the curves are drawn to guide the eye)

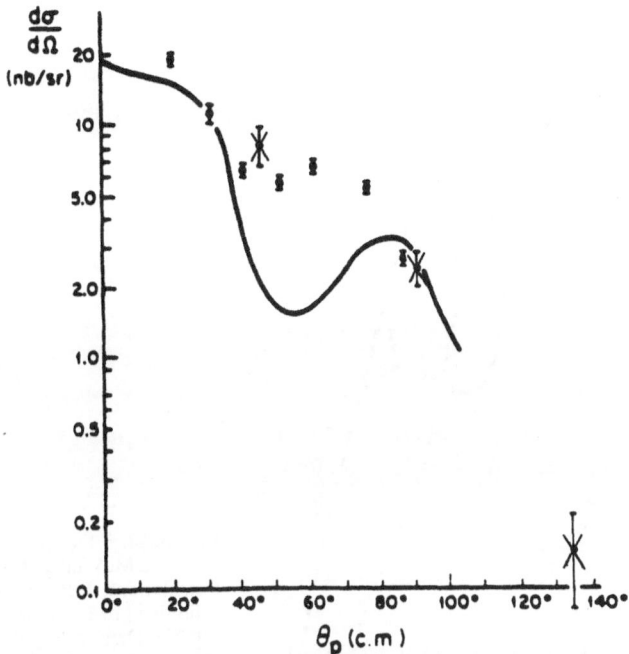

Fig. 10.9. $^{16}O(\gamma, p_0)$ cross-section at $E_\gamma = 260$ MeV taken from [10.2]. Calculations from [10.33] (solid line), data from [10.12] (✖) and from [10.42] (●)

In the paper of *Steenhoven* [10.3] the parameters corresponding to a consistent description of the available ^{12}C (e, e'p) data have been obtained and used for absolute calculations of the ^{12}C (γ, p) cross-section at low energies. This method is based on the fact that the quasi-free DWIA calculations are essentially identical for (e, e'p) and (γ, p) reactions and thus it gives reliable information for the (γ, p) process. The analysis shows that, contrary to what is often stated, even at low energies ($E_\gamma \leqslant 100$ MeV) the (γ, p) reaction cannnot be described by the pure quasi-free knockout model [10.11, 10.58]. It is concluded that either the high-momentum components of the wave function (N–N correlations) have been underestimated, or the contributions from some other mechanisms (e.g. involving exchange currents) cannot be neglected at low energies. It is pointed out in [10.3] that the direct knockout contribution to the cross-section has not been treated in a satisfactory way by *Gari* and *Hebach* [10.33] and *Cavinato* et al. [10.37d]. The (γ, p) and (e, e'p)-reactions on ^{16}O have been studied in a consistent way within the Skyrme–Hartree–Fock-RPA theory by *Ryckebusch* et al. [10.59]. The calculations accounting for long-range N–N correlations, exchange currents and multistep processes in the final step describe well the (γ, p) data over a wide energy range. The role of the velocity-dependent part of the N–N interaction which introduces spatial non-locality in the mean-field description is emphasized.

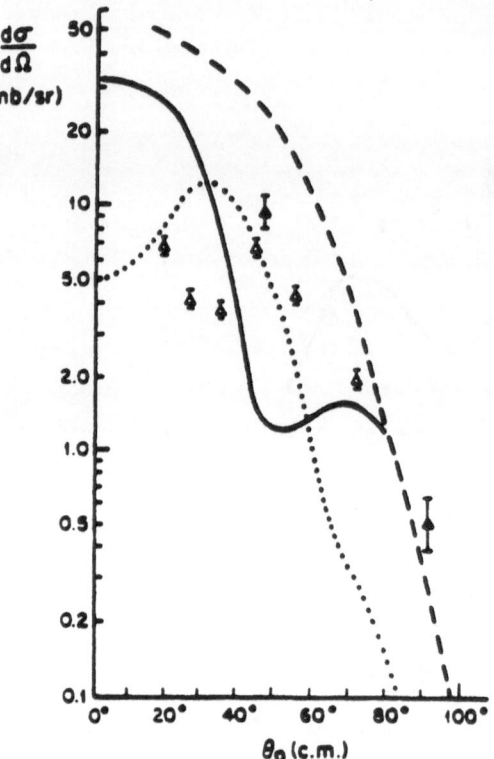

Fig. 10.10. $^{16}O(\gamma, p_0)$ cross-section at $E_\gamma = 312$ MeV taken from [10.2]. Calculations: dotted curve [10.11], solid line [10.33], dashed curve [10.38]. Data from [10.12] (▲) and *Adams* presented in [10.2] (△)

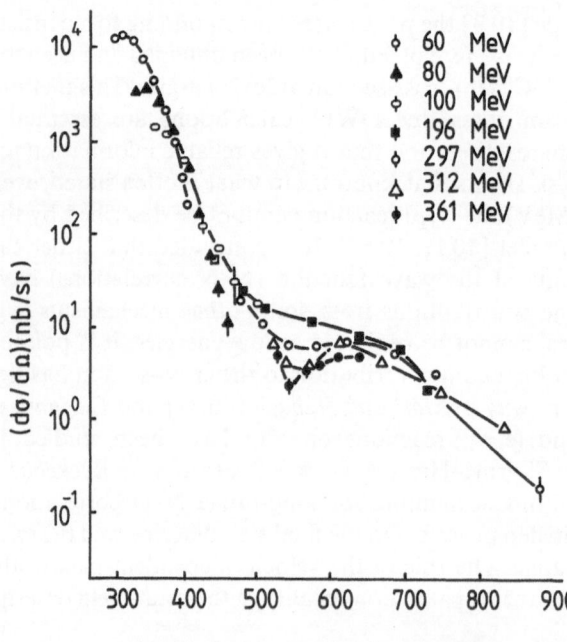

◇   60 MeV
▲   80 MeV
◇ 100 MeV
◼ 196 MeV
◇ 297 MeV
△ 312 MeV
◆ 361 MeV

Fig. 10.11. $^{16}O(\gamma, p_0)$ cross-sections for photon energies between 60 and 360 MeV as a function of momentum mismatch Q (10.14). The values of Q are corrected for the effect of the final-state potential [10.25]

The cross-sections of the ^{16}O (γ, n) reaction at $E_\gamma = 150$, 200 and 250 MeV [10.60] are given in [10.3] and compared with the (γ, p)-ones as a function of the momentum mismatch Q. The (γ, n) cross-section becomes larger than the (γ, p) cross-section as E_γ increases into the region of Δ-production in contrast to the expectation for equivalent cross-sections in a pure Δ-excitation model. The data for (γ, n) cross-section at $E_\gamma = 200$ MeV, as well as those for the (γ, p) cross-section cannot be explained using the models of *Gari* and *Hebach* [10.33] and *Ryckebusch* et al. [10.59]. The method of *McDermott* et al. [10.61] using relativistic forms of the nucleon current operator and four-component nuclear wave functions, gives a better description of the ^{16}O (γ, p) and ^{40}Ca (γ, p) reactions.

The reaction ^{16}O $(\gamma, \pi^- p)$ has been also analysed in [10.3] with the aim of studying the influence of the Δ-isobar contribution to photoreactions in a more direct way. The factorized DWIA calculation results overestimate the data at $E_\gamma = 350$ MeV and $\Theta_\pi = 64°$ and $120°$. The discrepancy between the data and the DWIA result is about twice as large at the forward angle. The differences between the DWIA and PWIA results indicate considerable final-state-interaction effects.

The mechanisms of photon absorption on ^{4}He in the Δ-resonance region has been analysed by *Maruyama* [10.62] on the basis of the reactions ^{4}He (γ, p) and ^{4}He (γ, pn) $(E_\gamma = 170-450$ MeV). It is shown that the contributions of 2N, 3N and 4N absorptions are necessary to reproduce the experimental data. The distribution of the correlation angles between the outgoing protons and neutrons are given in Fig. 10.12 and compared with the calculations of 2N, 3N and 4N absorptions carried out within the independent particle model and the

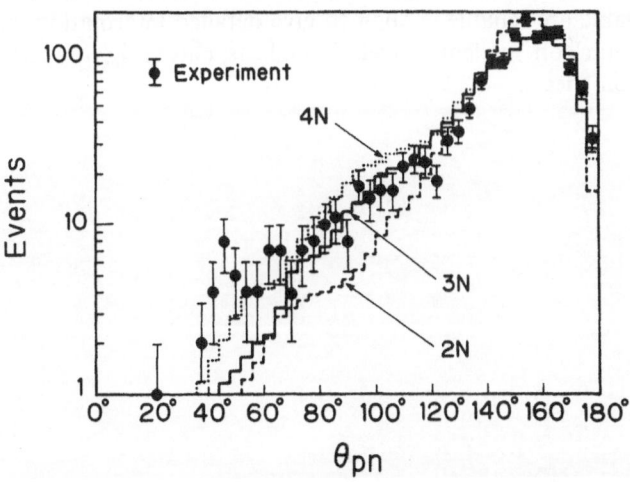

Fig. 10.12. The distribution of correlation angle between the proton and neutron. Calculations for 2N, 3N and 4N absorption mechanisms are shown together [10.62]

impulse approximation with a gaussian form for the nucleon momentum distribution. All three calculations reproduce the data satisfactorily.

A theoretical study of the (γ, np) cross-section in light nuclei at medium energies has been carried out by *Boato* and *Giannini* [10.63] taking into account two-body effects both in the nuclear wave function (correlations) and in the electromagnetic operator (meson exchange currents (MEC)). It is shown that the factorization of the cross-section into two parts in the *Gottfried* formulation [10.64] is also valid in the presence of MEC. The first part of the cross-section is related to the momentum distribution of the nucleon pair in the initial nucleus and the second part is expressed in terms of the Fourier transform of the correlated function. The total cross-section of two-nucleon photoemission is then directly related to the Fourier transform of the nuclear correlation function. Therefore, a proper account for the nucleon–nucleon correlations will be a fundamental ingredient for the description of these particular photonuclear reactions. The total cross-section of the (γ, np) reaction on ^{16}O is evaluated using the ground state wave function from the Jastrow correlation method.

The missing energy spectra for the (γ, np) reaction on ^6Li [10.65, 10.66] $(80 < E_\gamma < 160 \text{ MeV})$ and ^{12}C [10.67] $(83 < E_\gamma < 133 \text{ MeV})$ are given by *Friedrich* in [10.68]. The measured pn-pair initial momentum distribution (for the lp-nucleons) in the ^{12}C (γ, np)-reaction agrees well with the quasideuteron calculations. It is concluded in [10.66] that the quasifree deuteron mechanism seems to be the dominant process in the (γ, np)-reaction considered. The same conclusion has been made by *Homma* et al. [10.69] from their analysis of the Be (γ, p) data at $E_\gamma = 180–420$ MeV. The evidence for such absorption processes makes it very tempting to probe these clusters with (e, e'pn) studies [10.70].

The photonuclear reactions considered in this chapter provide information concerning the separation energies and momentum components in nuclei [10.7]. In practice, however, as was shown, neither the experimental data, nor theoretical models now available are reliable enough to give detailed information on the importance of the nucleon–nucleon correlation effects, clustering in nuclei, or final-state interaction effects.

11 Intermediate Energy Proton–nucleus Scattering

The possible ways of extracting information on nucleon–nucleon correlations in nuclei from elastic proton–nucleus scattering are described in detail in Sect. 11.1. The effects of the nucleon–nucleon correlations on elastic proton–nucleus scattering are considered in the coherent density fluctuation model in Sect. 11.2. The problems of proton production in deep-inelastic intermediate energy proton–nucleus scattering are discussed in relation to the question of the nucleon–nucleon correlations in Sect. 11.3.

11.1 Proton–nucleus Scattering and Nucleon Correlations

The analyses of proton scattering on nuclei at intermediate energies (from several hundred MeV to several GeV) provide information on the scattering mechanism, on nucleon–nucleon scattering in nuclear medium, on the nuclear matter, charge and neutron density distributions and on the nuclear deformation. Proton–nucleus scattering is also informative concerning the dynamical short-range nucleon–nucleon correlations and the correlations leading to the clusterization of the nuclear matter. Such studies therefore allow the range of validity of various nuclear models to be studied.

The scattering of protons with energies of more than several tens MeV has a marked diffraction character. The differential cross-section for elastic scattering has a maximum at small angles provided $kR \gg 1$, where $k = 1/\lambda$ is the wave number of the relative motion of the particles, λ is the wavelength and R is the range of the interaction. The mean free path of the incident proton at these energies is small compared with the nuclear size and in this case the nucleus can be considered as an absolutely black body. The profile function and the differential cross-section then have the form [11.1]:

$$\omega(b) = \begin{cases} 1, & b < R \\ 0, & b > R, \end{cases} \tag{11.1}$$

$$\frac{d\sigma(\Theta)}{d\Omega} = \frac{R^2 J_1^2(k\Theta R)}{\Theta^2}, \tag{11.2}$$

where b is the impact parameter, R the nuclear radius and $J_1(k\Theta R)$ the Bessel function.

At larger incident nucleon energies their mean free path in the nucleus becomes larger than the nuclear size. Since their wavelength is then smaller than the nucleon–nucleon interaction range, the scattering on the nucleus can be considered as a multiple diffractional scattering on the single nucleons. The theory of *Glauber–Sitenko* [11.2, 11.3] (TGS) gives the proton–nucleus scattering amplitude in terms of the amplitudes of scattering on the single nucleons and by structure form factors. Here we give briefly the basic assumptions of this theory [11.1–9]:

1) Small scattering angles $\Theta : \Theta^2 kd \ll 1$, where $d \sim \hbar v / V$ (v is the incident particle velocity and V is the nucleon–nucleon potential depth) is related to the nucleon–nucleon interaction radius.

2) Conditions for high-energy scattering: $V/T_k \ll 1$, $ka \gg 1$, where T_k is the incident nucleon kinetic energy and a is the interaction radius.

These approximations allow the scattering amplitude to be expressed in an eikonal form (the so-called eikonal, or high-energy approximation), so that the trajectory of the incident particle through the nucleus is assumed to be linear.

3) The adiabatic approximation implying that the motions of the nucleons in the nucleus are neglected during the passage of the incident nucleon through the nucleus (the "frozen nucleons" approximation).

4) The dynamical approximation implying that the potentials describing the interaction of the incident particle with the target nucleons are not overlapping. In this case the total phase shift can be expressed by the algebraic sum of the phase shifts due to the scattering on the single nucleons:

$$\chi(\boldsymbol{b}, \boldsymbol{s}_1, \dots, \boldsymbol{s}_A) = \sum_{j=1}^{A} \chi_j(\boldsymbol{b} - \boldsymbol{s}_j), \tag{11.3}$$

where $\boldsymbol{b}$ is the impact parameter and $s_j (j = 1, 2, \dots, A)$ are the projection of the nucleon radius vectors $\boldsymbol{r}_1, \dots, \boldsymbol{r}_A$ on the plane perpendicular to the vector $\boldsymbol{k}$.

The nuclear profile function then has the form:

$$\Gamma(\boldsymbol{b}, \boldsymbol{s}_1, \dots, \boldsymbol{s}_A) = 1 - \exp[i\chi(\boldsymbol{b}, \boldsymbol{s}_1, \dots, \boldsymbol{s}_A)] \tag{11.4}$$

$$= 1 - \prod_{j=1}^{A} \exp[i\chi_j(\boldsymbol{b} - \boldsymbol{s}_j)].$$

The Glauber–Sitenko diffraction amplitude for the process in which the nucleus makes a transition from the initial state $|i\rangle$ to the final state $|f\rangle$

$$F_{fi}(\boldsymbol{q}) = \frac{ik}{2\pi} \int \exp(i\boldsymbol{q} \cdot \boldsymbol{b}) \langle f | \Gamma(\boldsymbol{b}, \boldsymbol{s}_1, \dots, \boldsymbol{s}_A) | i \rangle \, \mathrm{d}^2 \boldsymbol{b} \tag{11.5}$$

($\boldsymbol{q}$ being the transfer momentum) becomes finally:

$$F_{fi}(\boldsymbol{q}) = \frac{ik}{2\pi} \int \mathrm{d}^2 \boldsymbol{b} \exp(i\boldsymbol{q} \cdot \boldsymbol{b}) \int \mathrm{d}\boldsymbol{r}_1 \dots \mathrm{d}\boldsymbol{r}_A \, \Psi_f^*(\boldsymbol{r}_1, \dots, \boldsymbol{r}_A) \delta^{(3)} \left(\frac{1}{A} \sum_j \boldsymbol{r}_j \right)$$

$$\times \left[1 - \prod_{j=1}^{A} \left(1 - \frac{1}{(2\pi ik)} \int e^{-i\boldsymbol{q}' \cdot (\boldsymbol{b} - \boldsymbol{s}_j)} f_j(\boldsymbol{q}') \mathrm{d}^2 \boldsymbol{q}' \right) \right] \Psi_i(\boldsymbol{r}_1, \dots, \boldsymbol{r}_A). \tag{11.6}$$

where $f(q)$ is the nucleon–nucleon scattering amplitude.

The study of the nucleon–nucleon correlations can be made on the basis of the many-particle density expansion in the correlation series [11.8, 11.10]:

$$\rho_A(r_1, \ldots, r_A) = |\Psi_i(r_1, \ldots, r_A)|^2 \delta\left(\sum_{j=1}^{A} r_j\right)$$

$$= \prod_{j=1}^{A} \rho(r_j) + \sum_{j<k} C_2(r_j, r_k) \prod_{\substack{l=1 \\ l \neq k, j}}^{A} \rho(r_l) \tag{11.7}$$

$$+ \sum_{j<k<l} C_3(r_j, r_k, r_l) \prod_{\substack{m=1 \\ m \neq j, k, l}}^{A} \rho(r_m) + \ldots$$

$$+ C_A(r_1, \ldots, r_A),$$

where $\rho(r)$ is the one-body density, $C_2(r_1, r_2)$ is the two-particle correlation function and $C_3(r_1, r_2, r_3), \ldots, C_A$ are the higher order correlation functions. Substitution of (11.7) into (11.6) (at $f = i$) leads to a correlation expansion for the elastic scattering amplitude F_{ii} which is quickly convergent. The use of the first term in (11.7) only means neglecting the nucleon–nucleon correlations (so-called independent-particle model (IPM)).

In the "optical limit" of the Glauber–Sitenko theory [11.7, 11.11] the nucleus is replaced by a single-particle potential giving the same phase shift as the many-particle model. The elastic scattering amplitude then has the form:

$$F_{ii}(q) = \frac{ik}{2\pi} \int \exp(iq \cdot b) \left\{1 - \exp[i\chi_{\text{opt}}(b)]\right\} d^2 b. \tag{11.8}$$

For heavy nuclei $(A \gg 1)$:

$$\exp(i\chi_{\text{opt}}(b)) = \exp\left\{ - A \int dr \rho(r) \Gamma(b - s) + \frac{A^2}{2} \int dr_1 \, dr_2 \left[(A - 1)\rho_2(r_1, r_2)\right.\right.$$

$$\left.\left. - A\rho(r_1)\rho(r_2)\right] \Gamma(b - s_1) \Gamma(b - s_2) + \ldots\right\}, \tag{11.9}$$

where $\rho_2(r_1, r_2)$ is the two-particle density and

$$\Gamma(b - s) = \frac{1}{2\pi i k} \int d^2 q \cdot f_{\text{NN}}(q) \exp[- iq \cdot (b - s)]. \tag{11.9a}$$

In spite of the approximations 1)–4) it turns out that the range of validity of the TGS is much wider, extending to smaller energies and larger angles than expected. Studies of the corrections to the TGS in the framework of the Watson scattering theory [11.12–15] show that the effects related to the deviations from the eikonal approximation, as well as from the "frozen nucleons" approximation, the kinematical effects etc. compensate each other to a large extent [11.12, 11.7, 11.8]. It can be concluded that the omission of any one approximation reduces the accuracy of the method. It is shown [11.16] in the case of ^{40}Ca that taking account of the corrections mentioned leads to a small filling in of the diffractional minima and to changes of several percent in the cross-section at the diffractional maxima. It is concluded that the higher order corrections to the TGS must be considered only after a careful study of the correlation effects.

A basic difficulty in the analysis of proton–nucleus scattering in the TGS is the poor knowledge of the nucleon–nucleon scattering amplitude f_{pN} (especially at $E_p > 600$ MeV) [11.17, 11.18]. While for $q \lesssim 2.5$ fm^{-1} the TGS gives a precise description of the elastic cross-section and polarization, the uncertainties in f_{pN} can lead to substantial changes in the cross-section maxima and minima above the value of the transferred momentum [11.19]. *Alkhazov* [11.18] estimated the possible q-dependence of the ratio Re $f_{pN}(0)/$Im $f_{pN}(0)$ for the correct description of the diffractional minima in elastic and inelastic proton–nucleus scattering.

Another method of analysing proton–nucleus scattering is the optical potential model of *Kerman, McManus* and *Thaler* [11.20] (KMT). The basic equation of this model has the form [11.20, 21, 7]

$$(E - T_k - V_{opt}) \Phi = 0, \tag{11.10}$$

where E is the total energy and T_k the kinetic energy operator. The optical potential V_{opt} is defined by (see also Sect. 2.1):

$$V_{opt} = V_{00} + \sum_{\alpha = 0} V_{0\alpha} G_\alpha V_{\alpha 0} + \sum_{\substack{\alpha \neq 0 \\ \beta \neq 0}} V_{0\alpha} G_\alpha V_{\alpha\beta} G_\beta + \dots, \tag{11.11}$$

where

$$G_\alpha = (E^{(+)} - \varepsilon_\alpha - T_k - V_{\alpha\alpha})^{-1}, \tag{11.12}$$

$$V_{\alpha\beta} = (A - 1)\langle \alpha | \tau | \beta \rangle \tag{11.13}$$

is the effective potential and ε_α the energy of the intermediate nuclear state. The t-matrix of the nucleon–nucleon scattering in the nucleus is determined by solving the equation:

$$\tau = v + v[a/(E^{(+)} - H_n - T_k)] \tau, \tag{11.14}$$

where v is the nucleon–nucleon potential and H_n contains the nuclear potential. The optical potential can be determined [11.22–27] by solving the system of differential equations related by potentials which are proportional to the two-particle and to higher order correlation functions.

In the last two decades 1 GeV proton scattering on various nuclei carried out in Brookhaven, Gatchina, Saclay and other centres have been analysed in the framework of the Glauber–Sitenko and KMT theories. Some of them are reviewed in [11.7, 8]. We should mention also the measurements of proton scattering on the nuclei: $^{144, 150, 152}$Sm (1 GeV) [11.17], ^{9}Be, ^{11}B, $^{12, 13}$C, ^{14}N, ^{16}O (1 GeV) [11.28], $^{40, 48}$Ca, $^{58, 64}$Ni, $^{116, 124}$Sn, ^{208}Pb (800 MeV) [11.29], $^{40, 42, 44, 48}$Ca (800 MeV) [11.30a], $^{24, 26}$Mg (800 MeV) [11.31], ^{16}O, ^{28}Si, ^{32}S, ^{39}K, $^{40, 48}$Ca, ^{90}Zr, ^{208}Pb (1 GeV) [11.32], $^{58, 60, 62, 64}$Ni (1 GeV) [11.33], ^{40}Ca (800 MeV) [11.15, 11.34], $^{40, 48}$Ca, ^{54}Fe (800 MeV) [11.35, 11.36], ^{4}He (500 MeV) [11.37–39], etc. [11.40].

One of the aims in the analysis of these experiments is the study of the matter, charge and neutron density distributions in nuclei. If the charge distribution is taken as known (for instance, from the electron–nucleus scattering experiments) then parametrizing the neutron distribution ρ_n, it is possible to determine the radius and the surface diffuseness of the neutron distribution. This question is considered in TGS for $^{40, 48}$Ca [11.41], $^{40, 42, 44, 48}$Ca, ^{48}Ti [11.42], ^{16}O, ^{28}Si, ^{32}S, ^{39}K, $^{40, 48}$Ca, ^{90}Zr, ^{208}Pb [11.32]. An important result is that for nuclei with $N \approx Z$ the difference between the neutron and proton mean-square radii $(r_n - r_p)$ is rather small ($\lesssim 0.03$ fm). For nuclei with $N > Z$ $r_n - r_p \simeq 0.1$ fm. For ^{48}Ca $r_n - r_p = 0.13$ fm, while for ^{208}Pb $r_n - r_p = 0.06$ fm [11.32]. It is shown in [11.30a] that the neutron "shell" in ^{48}Ca is smaller than is predicted by the Hartree–Fock method. In the earlier studies $r_n - r_p$ for ^{48}Ca was estimated to be 0.21 fm [11.21], 0.19 fm [11.8, 11.30b], see also [11.30c]. The parameters of the nuclear matter distributions in ^{6}Li, ^{9}Be, ^{11}B, $^{12, 13}$C and ^{14}N are obtained in [11.28]. Some difficulties in the study of ρ_n due to the model dependence of the assumed proton distribution have been noted in [11.43].

The analysis of the proton–nucleus experiments serves as a test for the proton and neutron density distributions obtained by various theories. It has been shown by *Auger* and *Lombard* [11.19] that the use of the densities obtained in the Hartree–Fock method taking account of the pairing as well as in the Thomas–Fermi method leads to a correct description of the cross-section for ^{28}Si, $^{32, 34}$S, ^{39}K, $^{40, 42, 44, 48}$Ca, ^{48}Ti, $^{58, 60, 62, 64}$Ni, ^{90}Zr, ^{208}Pb only at angles $\Theta_{c.m.} \leqslant 10°-11°$ ($q \leqslant 1.5-1.7$ fm^{-1}) and for smaller angles in ^{16}O. The reason for this is in the Hartree–Fock method (especially for light nuclei) as well as in the inadequacy of the assumed reaction mechanism, in particular the need to include coupling to the inelastic channels at larger angles. The description of the proton cross-sections is markedly unsuccessful in the cases of ^{48}Ca and ^{48}Ti using the self-consistent density distributions [11.44].

The difficulties of the theory in describing the inelastic proton scattering on ^{12}C (with the excitation of the level 2^+, 4.43 MeV) led the authors of [11.4] to study the effects of the deformation of ^{12}C. These and other studies showed [11.45, 11.46] that it is impossible to describe simultaneously both the inelastic proton cross-section and the charge form factor in ^{12}C taking account of the deformation, as already concluded by [11.47]. A partial filling of the diffractional minima in the cross-sections of ^{9}Be and ^{11}B is considered to be due to the non-spherical component of the density distributions of these nuclei [11.28].

The experiments on intermediate energy proton–nuclei scattering are of particular interest for nuclear structure theory because they give information on the nucleon–nucleon correlations in the nuclear medium. Several types of correlations are considered:

1. Centre-of-mass correlations related to the fact that the centre of the nucleus in its rest system is fixed and therefore the nucleons cannot move arbitrarily (c.m. correlations). In spite of the statement that the c.m. correlations are important for the light nuclei only, it is shown in [11.8, 11.48] that they

cannot be neglected for the medium weight nuclei and even for the ^{208}Pb nucleus.

2. Correlations related to the antisymmetric property of the total wave function (Pauli correlations), which are important for light and for heavy nuclei [11.48].

3. Dynamical short-range correlations reflecting the properties of the nucleon–nucleon forces at small distances.

4. Correlations leading to a clusterization of nuclear matter.

The knowledge of the many-particle density (11.7) gives in principle the possibility of studying the nucleon–nucleon correlations. Since, however, this density is unknown, the two-, three-, etc. particle correlations are studied in the framework of models for $\rho_2(r_1, r_2)$, $\rho_3(r_1, r_2, r_3)$, ... or for the corresponding correlation functions C_2, C_3, etc. For instance, the dynamical correlations are often studied using the assumption for $\rho_2(r_1, r_2)$ [11.8]:

$$\rho_2(r_1, r_2) \simeq \rho(r_1)\rho(r_2)[1 - g_{SR}(r_1 - r_2)], \tag{11.15}$$

where

$$0 \leqslant g_{SR} \leqslant 1 \tag{11.16}$$

and

$$g_{SR}(r) \underset{r \to \infty}{\to} 0.$$

The model investigations of the correlations lead to different and often contradictory conclusions concerning their effects on proton–nucleus scattering. For instance, while in [11.4] it is concluded that the effect of the Jastrow correlations in ^{4}He is negligible, in [11.49] the use of the two-body correlation function for nuclear matter applied to finite nuclei (^{12}C, ^{16}O) leads to substantial changes in the second and third maxima in the cross-sections.

Taking account of the short-range correlations (SRC) by the model of *Gribov* [11.50] shows [11.51] that the correlation effects become more important in the transition from light to heavy nuclei so they cannot be neglected. It is noted by *Starodubsky* [11.52] that the two-particle correlations lead to the increase of the elastic cross-sections in the region of their maxima for ^{208}Pb by 15, 23, 30, 37 and 43%, respectively, for ^{58}Ni by 9, 23 and 33% and for ^{40}Ca by 24 and 35%. The largest changes in the inelastic cross-sections are about 20%. In [11.53] the increase of the maximum in ^{4}He is about 5%. The use of the cluster expansion in the Jastrow correlation method and the TGS leads to an improvement of the fit of the theoretical cross-sections for ^{4}He and ^{16}O to the experimental data compared with the shell-model calculated cross-sections (see Fig. 11.1 for the ^{16}O nucleus).

In contrast to the conclusions made in [11.51, 11.52, 11.54, 11.55], *Alkhazov* [11.56] has shown that the SRC have small effect on the proton cross-section minima (about 10–20% in ^{4}He and 5–10% in ^{16}O and ^{58}Ni).

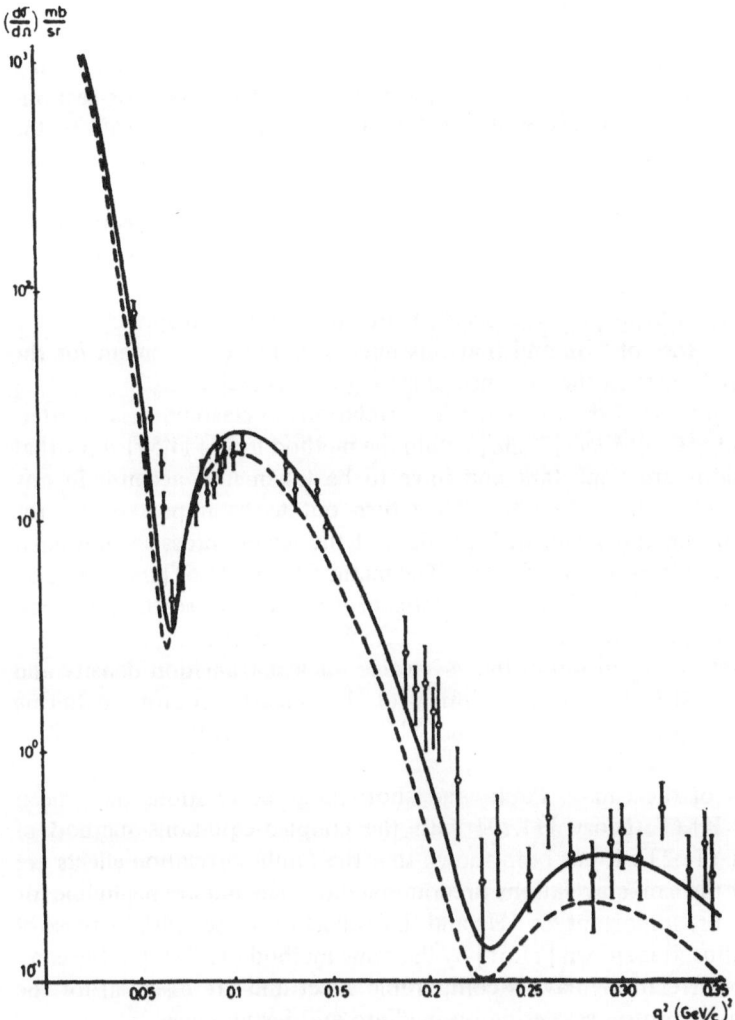

Fig. 11.1. Differential cross section for p + ^{16}O (T_{lab} = 1 GeV) [11.54]. Solid line: Jastrow method result; dashed line: shell model result

It was shown by *Viollier* [11.57] that the inclusion of the dynamical long-range correlations along with c.m.-, Pauli- and short-range correlations in the study of the excitations of 2^+, 0^+ and 3^- states in ^{12}C improves the agreement with the experimental data for the elastic scattering (up to $q^2 = 7$ fm^{-2}) as well as for the inelastic scattering (to the states 0^+, 7.68 MeV and 2^+, 4.43 MeV).

Bleszynski et al. [11.15] showed that the effects of the non-eikonal corrections for large angle elastic and inelastic proton scattering at 800 MeV on ^{40}Ca decrease the oscillation amplitudes without changing their phase, while the SRC and Pauli correlations cause q-dependent changes of the phase at angles $\Theta \geqslant 20°$.

It has been pointed out by *Harrington* and *Varma* [11.16] that taking account of the Pauli-correlations in ^{40}Ca and ^{90}Zr increases the cross-sections by 10–11% at the second maximum, 13% at the third and 21 and 16% at the fourth. These changes are much larger than the non-eikonal-, Fermi motion- and other corrections.

The analysis in [11.9] leads to the conclusion that the SRC have small effect on the proton elastic scattering by ^{12}C and ^{16}O and this makes their study more difficult.

Saudinos and *Wilkin* [11.6] concluded that the Pauli principle effects are negligible in the case of ^{4}He and that this nucleus is most convenient for the study of the SRC even in the presence of strong c.m.-correlations.

The investigations of the two-particle correlations in elastic proton scattering on ^{4}He and ^{12}C by *Khan* [11.58] within the method from [11.59] show that these correlations are important and have to be taken into account in any realistic study of such interactions. They turn out to be important for the improvement of the agreement with the data at the second cross-section maximum for ^{4}He and the third one for ^{12}C. The method from [11.59] has been also applied [11.60] to calculations of inelastic proton–nucleus scattering cross-sections using a spin-dependent nucleon–nucleon interaction.

The correlation expansion of the A-particle nuclear transition density and the role of the SRC for the description of the inelastic proton scattering (800 MeV) on ^{16}O (3^-, 6.13 MeV) and ^{40}Ca (5^-, 4.49 MeV) are studied in [11.61].

The effects of the c.m.-, Pauli- and short-range correlations have been studied using KMT theory [11.20] with the coupled-equations method of *Feshbach* et al. [11.23]. It has been shown that the Pauli-correlation effects are small and that the c.m.-correlations predominate for ^{4}He and are negligible for ^{16}O. The SRC are important for ^{4}He and their effect decreases with increase of the nuclear radius. It is shown [11.61] by the same method [11.23] that the c.m.- and the Pauli-correlations have a comparable effect and are essential for the analyses of elastic proton scattering on medium and heavy nuclei.

A method for constructing a correlation function accounting for dynamical two-particle correlations within the Brueckner-Hartree–Fock approximation has been developed in [11.24]. It has been shown that these correlations are essential in the case of the ^{4}He nucleus. The significant role of the c.m.-, Pauli- and short-range correlations is shown in [11.27]; it is found that they increase the diffraction amplitudes for proton–nucleus scattering cross-sections for ^{40}Ca, ^{58}Ni and ^{208}Pb nuclei.

A detailed review of the effects of correlations on elastic and inelastic proton–nucleus scattering within the Glauber–Sitenko and the KMT theories has been made by *Chaumeaux* et al. [11.21]. As can be seen in Fig. 11.2 the SRC and tensor correlations increase the values of the cross-sections in the regions of the maxima. The same effect has been observed in the case of ^{40}Ca (Fig. 11.3) as

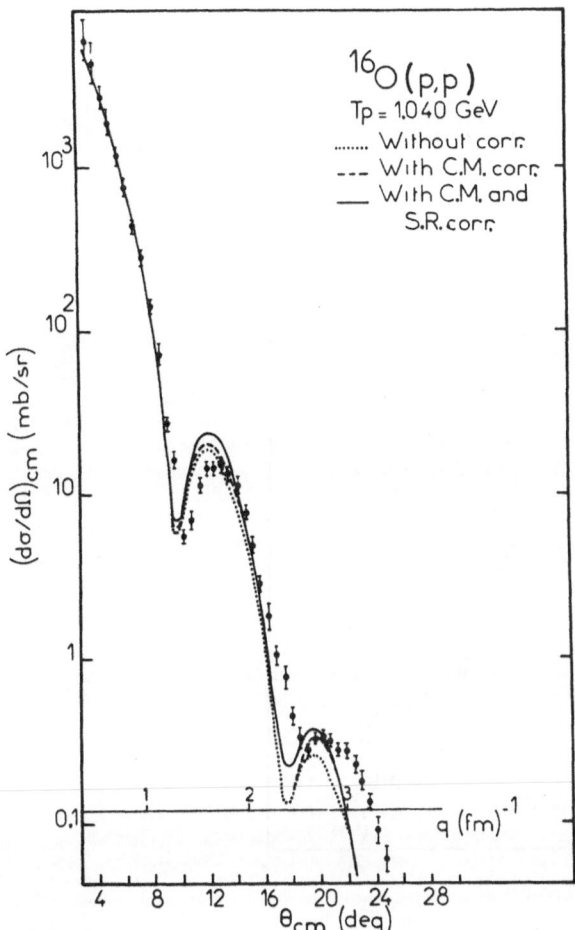

Fig. 11.2. Correlation effects on the differential cross sections of the elastic proton (1.04 GeV) scattering on ^{16}O [11.21]. Dotted line: without correlations; dashed line: with c.m.-correlations; solid line: c.m.-plus SRC. Hartree–Fock–Bogolyubov densities have been used in the calculations

well as for ^{48}Ca, ^{90}Zr and ^{208}Pb. It has to be emphasized that there is systematically unsatisfactory agreement between the theory (without correlations) and the experimental data when Hartree–Fock–Bogolyubov densities are used.

It is noted by *Ray* [11.29] that the correlation effects are small for heavier nuclei and lead to an increase of the diffractional maxima of 10–30%. They have to be taken into account when the neutron density distributions are extracted from the experimental data. It is of interest to compare the relative importance of the various correlations obtained in [11.29]: Pauli-correlations: 80–90%, c.m.: 3–11% and SRC: 10–12%.

The effects of correlations on the total cross-section and the reaction cross-sections for a wide range of nuclei for proton scattering on nuclei from 100 to 2200 MeV are studied by *Ray* [11.62]. The correlation studies using the TGS and KMT [11.63] show that the two- and three-particle correlations are

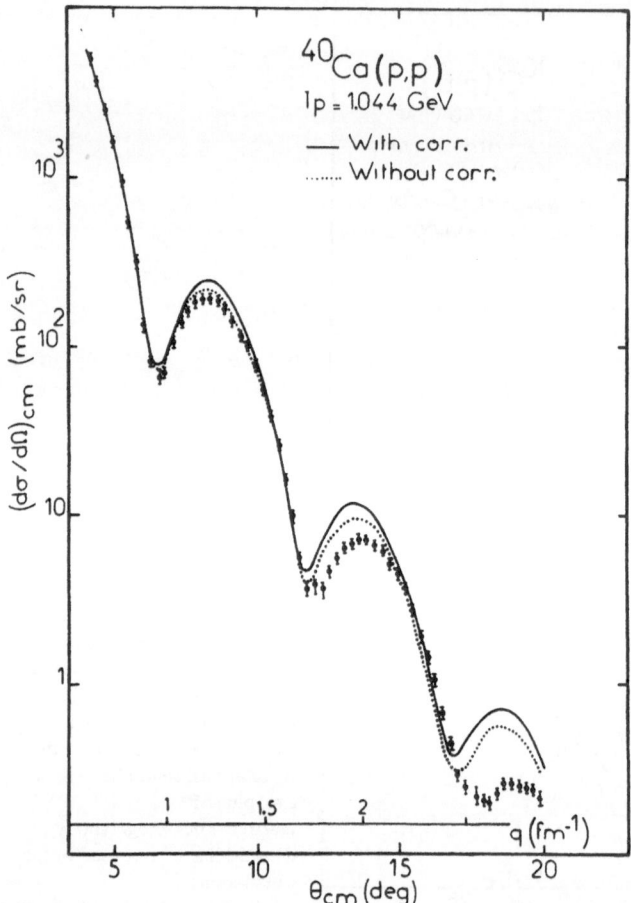

Fig. 11.3. Correlation effects on the differential cross sections of the elastic proton (1.04 GeV) scattering on ^{40}Ca [11.21]. Dotted line: without correlations; solid line: c.m.-plus SRC. Hartree–Fock–Bogolyubov densities have been used in the calculations

important for the ^{4}He nucleus. For heavier nuclei the two-nucleon correlations lead to an increase of the cross–section maxima of 5–20% (see Figs. 11.2, 11.3); the effects decrease with the increasing of the mass number.

It is found by *Lumpe* and *Ray* [11.64] that accounting for correlations within the relativistic impulse approximation does not lead to an improvement of the agreement with the data for the elastic scattering of 500–800 MeV protons by ^{16}O, ^{40}Ca, ^{208}Pb in contrast to the case of the non-relativistic impulse approximation (e.g. [11.21]).

Correlations related to alpha-clustering in nuclear matter play a significant role in the structure of the light nuclei with mass number $A = 4n$ ($n = 1, 2, \ldots$). Correlations of this type have been studied in proton scattering (e.g. [11.65–72,

11.8, 11.56]). It is concluded in [11.8, 11.68, 11.69] that the effects of the clustering and of the effects of the repulsive correlation forces between the clusters, although significant, have different signs, so the resulting effect of the clustering is not large.

The proton elastic scattering on the "alpha-particle" nuclei ^{12}C and ^{16}O has been successfully described with the simple alpha-particle model (APM) in [11.66, 11.67]. A unified description of both proton and electron scattering on ^{12}C has been obtained within the APM in [11.69–70] taking account of the vibrations of the alpha-particles around their equilibrium positions in the ^{12}C nucleus.

The total cross-sections for the process $p(0.15-1\ GeV) + {}^{12}C$ have been described within the APM by *Ahmad* and *Khan* [11.71].

Karmanov [11.72] showed that the data on the elastic and inelastic (3⁻, 6.13 MeV) proton scattering on ^{16}O are not in contradiction with the results of the APM assuming the four alpha-particle structure of the ^{16}O nucleus.

Tabet et al. [11.17] pointed out the necessity of considering processes different from direct scattering, such as intermediate resonance excitations, exchange current effects, etc. for the description of the proton scattering at large transfer momenta. It was shown by *Wallace* [11.73] that the account of the $\Delta(1236)$-isobar excitation in TGS leads to a partial filling of the diffraction minimum in the p + ^{4}He-scattering cross-section. A similar effect has been found in the cross-sections for $p + {}^{12}C$ and $p + {}^{16}O$ scattering where the changes are about 10% [11.74].

There have been attempts to develop the theory of the proton–nucleus scattering taking account of the quark structure of the hadrons (see e.g. [11.75–77]). The amplitude of the proton scattering with inclusion of the hadron quark structure has been obtained in [11.75] and successfully applied to the description of high-energy proton-nucleus scattering on ^{4}He at $T_p = 200\ GeV$ and ^{12}C, ^{27}Al, ^{63}Au, ^{207}Pb at $T_p = 175\ GeV$. The corrections to the amplitude of the elastic proton–deuteron scattering due to the quark structure of the deuteron have been calculated in [11.76]. The inclusion of nucleon structure in the TGS leads [11.77] to an improved agreement with the data for the proton-^{4}He scattering in an accord with the hypothesis of a larger effective size of the nucleon in the nuclear medium.

Of great interest is the study of relativistic dynamical effects in the proton–nucleus scattering using the Dirac equation [11.78–83] along with the study of the nucleon–nucleon correlations in nuclei [11.83].

The experiments on deep-inelastic proton–nucleus scattering (e.g. [11.84–93, 11.7]) are of significant importance for the study of SRC in nuclei. Though model-dependent the treatment of these experiments shows the essential role of the correlations for the nucleon momentum distributions in nuclei and for the reaction mechanism.

These numerous but often contradictory conclusions concerning the effects of correlations on proton–nucleus interactions makes it necessary to study them in more detail, both theoretically and experimentally.

11.2 Elastic Proton–nucleus Scattering within the Coherent Density Fluctuation Model

In the coherent density fluctuation model (CDFM) (see Sect. 4.2) [11.94–98] the amplitude of the elastic particle–nucleus scattering can be written in the form (4.73):

$$A_{00}(q) = \int_0^\infty dx \, |f_0(x)|^2 A_0(x, q), \tag{11.17}$$

where $A_0(x, q)$ is the amplitude of the scattering of the incident particle on the flucton with radius x, q the transfer momentum and $f_0(x)$ the weight function of the model, corresponding to the ground state of the nucleus.

Different variants of the CDFM have been used to calculate the elastic scattering of 1 GeV protons on various nuclei. They are related to the different choice of the proton–flucton scattering amplitude $A_0(x, q)$:

i) The first approximation used for $A_0(x, q)$ is that the fluctons are assumed to scatter as absolutely black bodies. The diffractional amplitude in this case has the form [11.99]:

$$A_0(x, q) = -2\pi \frac{x^2}{L^2} \frac{J_1(qx)}{qx}, \tag{11.18}$$

where $J_1(qx)$ is the first order Bessel function and L is the normalization length. The cross-section is then obtained as

$$\frac{d\sigma}{d\Omega} = |A(q)|^2 \frac{L^4 k^2}{4\pi^2} = \left(\frac{k}{q}\right)^2 \left| \int_0^\infty |f_0(x)|^2 x J_1(xq) \, dx \right|^2, \tag{11.19}$$

where k is the incident proton wave number.

The differential cross sections for 1 GeV proton scattering on ^{12}C, ^{16}O, ^{28}Si, ^{32}S, ^{40}Ca, ^{48}Ca, ^{48}Ti, ^{58}Ni and ^{208}Pb have been calculated in [11.94, 11.100] using (11.19) and

$$|f_0(x)|^2 = -\frac{1}{\rho_0(x)} \frac{d\rho(r)}{dr} \bigg|_{r=x} \tag{11.20}$$

with $\rho_0(x) = 3A/(4\pi x^3)$ and a Fermi-type density distribution $\rho(r)$. It is shown that agreement with the data can be achieved only by using the whole superposition of fluctons in (11.19). The diffractional approximation is reflected mainly in the deviation of the theoretical results from the data in the regions of the cross-section minima.

ii) The second approximation used for $A_0(x, q)$ in the CDFM [11.101] is related to the optical limit of the TGS [11.11]. In this limit the amplitude

$A_0(x, q)$ can be written in the form:

$$A_0(x, q) = \frac{ik}{2\pi} \int d^2 b\, e^{-iq\cdot b} [1 - \exp[i X(b, x)]], \tag{11.21}$$

where the phase X corresponds to the scattering with impact parameter b:

$$i X(b, x) = - \int d^2 s\, \rho_{f1}(s, x)\, \gamma (b - s), \tag{11.22}$$

and ρ_{f1} is the two-dimensional flucton density:

$$\rho_{f1}(s, x) = \int\limits_{-\infty}^{\infty} dz\, \rho_{f1}[(s^2 + z^2)^{1/2}, x]. \tag{11.23}$$

The integrand in (11.23) is:

$$\rho_{f1}[(s^2 + z^2)^{1/2}, x] = \rho_{f1}(r, x) = \frac{3A}{4\pi x^3}\, \Theta (x - r). \tag{11.24}$$

The nuclear profile function γ from (11.22):

$$\gamma(b - s) = \frac{(1 - i\varepsilon)\sigma}{4\pi\beta} \exp[- (b - s)^2/(2\beta)] \tag{11.25}$$

corresponds to the following expression of the nucleon–nucleon amplitude $f_{NN}(q)$:

$$f_{NN}(q) = \frac{(i + \varepsilon)}{4\pi} k\sigma \exp[- \beta q^2/2], \tag{11.26}$$

where σ is the total N–N cross-section and $\varepsilon = \text{Re } f_{NN}(0)/\text{Im } f_{NN}(0)$ is the parameter of the diffractional cone slope. According to (11.23, 24) the flucton density can be presented in the form:

$$\rho_{f1}(s, x) = 2\rho_0(x)(x^2 - s^2)^{1/2}\, \Theta (x - s). \tag{11.27}$$

It can be shown that the scattering phase is:

$$- iX(b, x) = \frac{(1 - i\varepsilon)\sigma\rho_0(x)\exp(- b^2/2\beta)}{\beta}$$

$$\times \int\limits_{0}^{x} ds\cdot s(x^2 - s^2)^{1/2}\, I_0 (bs/\beta)\exp(- s^2/2\beta), \tag{11.28}$$

where I_0 is the zero-order modified Bessel function.

The results for the elastic proton–nucleus cross-sections obtained in [11.101] are in good qualitative agreement with the experimental data with the exception of the region of the cross-section diffractional minima. It is found that the correct description of the latter is strongly dependent on the choice of the value of the N–N amplitude parameter ε, which can vary from -0.25 to -0.60.

iii) It has been shown in [11.102–105] that in the CDFM the Glauber–Sitenko amplitude for the proton elastic scattering on nuclei can be presented in the form (11.17), where the amplitude for the proton scattering on a flucton with radius x is:

$$A_0(x, \boldsymbol{q}) = f_p(q) + ik \int_0^\infty J_0(qb)\{\exp[i\chi_p(b)]$$

$$- G_A(x, b)\exp[i\chi_\rho(x, b)]\} \, b \, db \, . \tag{11.29}$$

In (11.29):

$$f_p(q) = -\left(\frac{2\alpha k}{q^2}\right)\exp[2i \arg \Gamma(1 + i\alpha) - 2i\alpha \, \ln(q/2)] \tag{11.30}$$

is the Coulomb scattering amplitude for a point charge Z. The corresponding point Coulomb phase is

$$X_p(b) = 2\alpha \ln b \, , \tag{11.31}$$

where $\alpha = Ze^2/\hbar V$, V being the velocity of the incoming proton in the laboratory system. $X_\rho(x, b)$ is the Coulomb phase for the proton scattering on a flucton with radius x:

$$\chi_\rho(x, b) = X_p(b) + 2\alpha \int_0^\infty J_0(qb)[1 - S(x, q)]/q \, dq \tag{11.32}$$

$$= X_p(b) + 2\alpha \, \Theta(x - b)\{\ln[x/b + [(x/b)^2 - 1]^{1/2}]$$

$$+ \frac{[(x/b)^2 - 1]^{1/2}}{3(x/b)^3}[1 - 4(x/b)^2]\} \, . \tag{11.33}$$

In (11.32)

$$S(x, q) = 3[\sin(qx) - (qx)\cos(qx)]/(qx)^3 \tag{11.34}$$

is the flucton form factor. The nuclear part of the amplitude (11.29) in [11.102]:

$$G_A(x, b) = \{1 - [G(x, b) + \Delta G(x, b)]\}^A \, , \tag{11.35}$$

where

$$G(x, b) = \frac{1}{ik} \int_0^\infty J_0(qb)f_{pN}(q) \, S(x, q)q \, dq \, . \tag{11.36}$$

The term

$$\Delta G(x, b) = \frac{3(A - 1) \, b^2 \, [G(x, b)]^2}{2(A - 2)\langle r^2 \rangle_x} \tag{11.37}$$

takes into account the centre-of-mass correlations [11.102]. f_{pN} is the proton–nucleon scattering amplitude and $\langle r^2 \rangle_x = 0.6\,x^2$ is the rms radius of the flucton of size x.

In (11.17) the proton–nucleus elastic scattering amplitude is presented as a linear combination of proton–flucton amplitudes ($A_0(x, q)$) weighted by the function $|f_0(x)|^2$. It is shown [11.94–97] that for a monotonically decreasing density distribution $\rho(r)$ the weight function $|f_0(x)|^2$ can be expressed by Eq. (11.20). In order to determine the function $|f_0(x)|^2$ from (11.20) (under the assumption $\rho = 2\rho_p$) one needs the point proton density $\rho_p(r)$ corresponding to the nucleon charge density $\rho_{ch}(r)$:

$$\rho_{ch}(r) = \int \rho_{cp}(r')\,\rho_p(r - r')\,dr'. \tag{11.38}$$

In (11.38) $\rho_{cp}(r)$ is the proton charge distribution. Eq. (11.38) leads to the following expression for the density $\rho_p(r)$:

$$\rho_p(r) = \frac{1}{(2\pi)^3} \int S_{ch}(q)\,S_{cp}^{-1}(q)\exp(iq \cdot r)\,dq, \tag{11.39}$$

where $S_{ch}(q)$ and $S_{cp}(q)$ are form factors corresponding to the densities $\rho_{ch}(r)$ and $\rho_{cp}(r)$, respectively. Normalizing all densities to unity and assuming that the neutron density $\rho_n(r)$ coincides with the proton density one obtains the weight function $|f_0(x)|^2$ in the form:

$$|f_0(x)|^2 = \frac{2x^3}{3\pi} \int\limits_0^\infty j_1(qx)\,S_{ch}(q)\,S_{cp}^{-1}(q)\,q^3\,dq, \tag{11.40}$$

where $j_1(qx)$ is the first-order spherical Bessel function.

In the case of the independent-particle model (IPM) the form of the Glauber–Sitenko amplitude [11.102] coincides with Eq. (11.29), but:

i) The flucton form factor $S(x, q)$ in the Coulomb phase (11.32) is replaced by the charge form factor $S_{ch}(q)$;

ii) The form factor $S(x, q)$ is replaced in the nuclear part (11.35–37) by the expression $S_{ch}(q)\,S_{cp}^{-1}(q)$;

iii) The rms radius of the nucleus is used instead of the flucton rms in Eq. (11.37).

The CDFM zero-motion correlation effects on 1.04 GeV proton elastic scattering on ^{40}Ca have been studied in [11.103]. It was shown that the account of the specific CDFM "flucton" correlations (which are related to the nuclear density zero-motion vibrations) leads to results considerably different from those obtained in the IPM. The use of more realistic charge density distributions for ^{40}Ca [11.106] improves the agreement of the CDFM results with the experimental data in contrast to the case of the IPM. It was pointed out in [11.103], however, that the CDFM results do not fit well the experimental cross-section at angles larger than 18°. It was assumed that this is due to the uncertainties in the proton–nucleon scattering amplitude f_{pN}. Apparently, the poor knowledge of f_{pN}, particularly at energies above 600 MeV can cause uncertainties in the calculated cross-sections at larger transfer momenta.

In [11.105] the study of the flucton correlation effects on the proton-^{40}Ca elastic scattering cross-section has been extended using a new parameterization of the proton–nucleon amplitude obtained by *Golovanova* and *Iskra* [11.107] on the basis of non-eikonal considerations which describe well the experimental proton–proton scattering data:

$$f_{pN}(q) = \frac{ik\sigma}{4\pi} \sum_{n=0}^{\infty} A_{n+1} \left(\frac{\sigma}{4\pi\beta} \right)^n \frac{(1 - i\varepsilon)^{n+1}}{n+1} \exp\left[-\frac{\beta q^2}{2(n+1)} \right]. \tag{11.41}$$

The values of the parameters σ, ε and β in (11.41) are chosen under the conditions that: i) the optical theorem is valid; ii) the ratio Re $f(0)/$Im$f(0)$ is equal to the empirical value, and iii) the experimental elastic proton–nucleon cross-section data are correctly reproduced. In (11.41) A_n are numerical coefficients that satisfy the recurrence relation:

$$A_{n+1} = \frac{A_1}{n(n+1)} + \frac{A_2}{(n-1)n} + \frac{A_3}{(n-2)(n-1)} + \ldots + \frac{A_n}{1.2} \tag{11.42}$$

with $A_1 = 1$.

The Eq. (11.41) is an non-eikonal expression for the proton–nucleon scattering amplitude in which the Fresnel corrections due to the non-linear propagation of the incoming particle are taken into account. It is shown in [11.107] that the amplitude (11.41) describes correctly the experimental data for the elastic proton–proton scattering in the region $22° < \theta° < 90°$ [11.108, 11.109]. The study of the sensitivity of the calculated p–p cross-section with respect to the number of terms in (11.41) shows that at $n \geqslant 50$ the results are practically unchanged. For this reason the proton-^{40}Ca cross-section calculations are made with $n = 50$.

The elastic scattering cross-section of 1.04 GeV protons by ^{40}Ca have been calculated using the relations (11.17, 11.29–40) and the two proton–nucleon amplitudes (11.26) and (11.41). In (11.38) a realistic charge density $\rho_{ch}(r)$ obtained by means of model-independent analysis of experimental data on electron scattering and muonic atoms (type I from [11.106]) is used. For the form factor $S_{ch}(q)$ in (11.40) the dipole formula

$$S_{cp}(q) = (1 + q^2/\lambda)^{-2} \tag{11.43}$$

with $\lambda = 0.71$ (GeV/c)^2 is used. This choice ensures the convergence of the integral in (11.40) in the case of exponential asymptotic behaviour of the charge form factor $S_{ch}(q)$. The values of the parameter set $(\sigma, \beta, \varepsilon)$ in the two-body amplitudes (11.26) and (11.41) are chosen to reproduce as well as possible the available experimental p–p and p–n data. It should be emphasized that the experimental data on 1 GeV p–nucleon scattering are unsufficient to determine them unambiguously. This concerns mainly the proton–neutron scattering data. The calculations show that the following set of average proton and neutron parameter values for σ, β and ε in the amplitude (11.41)

$$\sigma = 3.2128 \, \text{fm}^2, \qquad \beta = 0.30 \, \text{fm}^2, \qquad \varepsilon = -0.18 \tag{11.44}$$

provides an optimal fit to the experimental proton–proton [11.109, 11.110, 11.8] and proton–neutron [11.110] elastic scattering data. The values of the parameters in (11.26):

$$\sigma = 4.4 \, \text{fm}^2, \qquad \beta = 0.24 \, \text{fm}^2, \qquad \varepsilon = -0.28 \tag{11.45}$$

obtained in [11.107] have been used in the calculations. The parameters (11.44, 11.45) satisfy the optical theorem with $\sigma_{tot} = 4.4 \, \text{fm}^2$. In the case (11.44) the ratio $\text{Re} f(0)/\text{Im} f(0) = -0.3426$ is in agreement with the empirical data (within their uncertainties). In Fig. 11.4 the proton–nucleon elastic cross-sections calculated using the amplitudes (11.26) and (11.41) with average parameter sets (11.44, 11.45) are compared with the proton-proton [11.109, 11.110] and proton-neutron [11.110] elastic scattering experimental data for $\theta° > 10°$, where the influence of the Coulomb force is small. As can be seen the cross-section calculated with the non-eikonal amplitude (11.41) is in satisfactory agreement with the p–p and p–n data in the wide region $10° < \theta° < 90°$ which is not the case for the one-term amplitude (11.26). The latter follows well the experimental points up to 50° and deviates significantly from experiment at larger angles.

In Fig. 11.5 we compare the CDFM and IPM results obtained using the amplitude (11.41) and the parameter set (11.44) with the experimental data

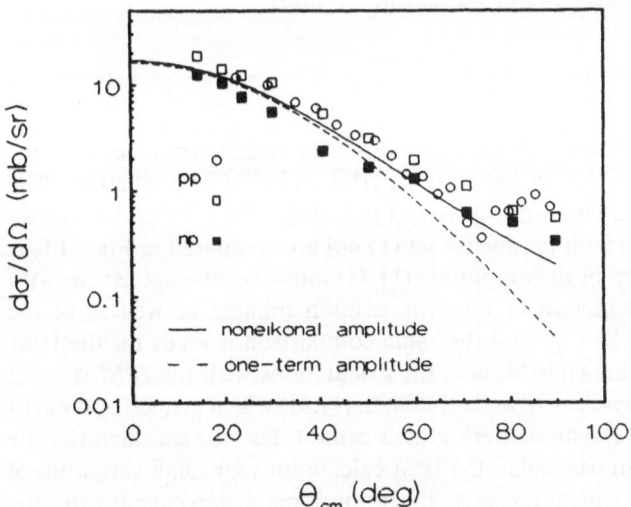

Fig. 11.4. The cross-section of the proton-nucleon elastic scattering. Solid line: calculated using f_{pN} from (11.41), (11.44); dashed line: using f_{pN} from (11.26), (11.45). Experimental data: p–p [11.109, 11.110], p–n [11.110]. The error bars lie within the experimental points

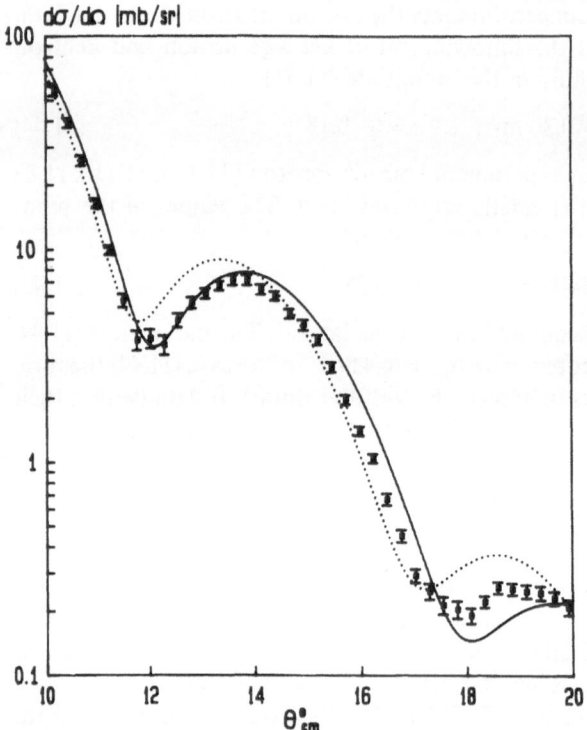

Fig. 11.5. The cross-section of the elastic 1.04 GeV-proton scattering from ^{40}Ca calculated using the two-body amplitude f_{pN} from (11.41) with parameters (11.44). Solid line: the CDFM result; the dotted line: the IPM result. The experimental data are taken from [11.102]

taken from [11.102] for the 1.04 GeV proton-^{40}Ca elastic cross-section. One can see a considerable difference between the IPM and CDFM predictions. The results of the CDFM calculations using $f_{pN}(q)$ from (11.26) with parameter set (11.45) and from (11.41) with parameter set (11.44) are compared in Fig. 11.6. It can be seen that the use of the amplitude (11.41) improves the agreement with the experimental cross-section at the cross-section minima as well as at the angles larger than $18°$. In Fig. 11.7 the same comparison is made for the IPM results. In contrast to the CDFM case, the calculations with the IPM showed that the parameter values for σ, β and ε(11.44) in (11.41) which give an optimal fit to the proton–nucleon elastic scattering data cannot describe satisfactorily the p-^{40}Ca cross-section. In particular, the IPM calculations for small variations of the parameter values which preserve the approximate agreement with the proton–nucleon data give the last minimum in the region of $\theta \simeq 17°$, while the experimental minimum is at $\theta \simeq 18°$. It can be seen from Figs. 11.6, 11.7 that the

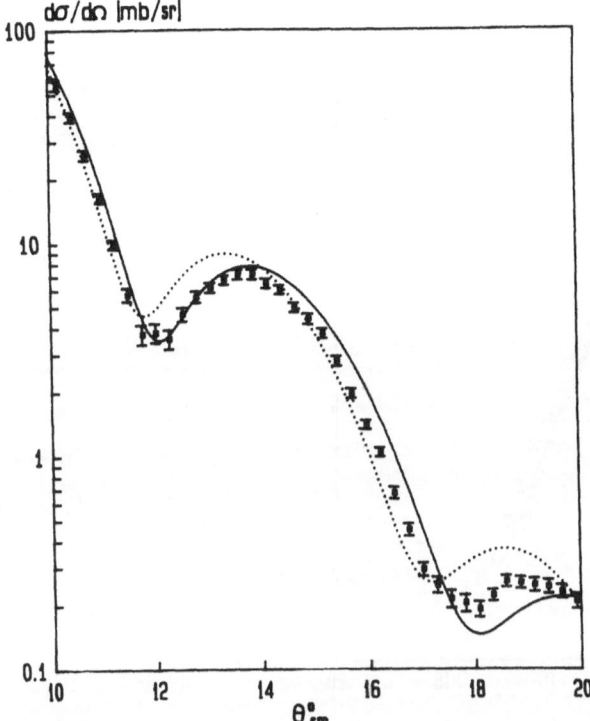

Fig. 11.6. The CDFM results for the elastic 1.04 GeV-proton-^{40}Ca cross-section, Solid line: using f_{pN} from (11.41), (11.44); dotted line: using f_{pN} from (11.26), (11.45). The experimental data as in Fig. 11.5

use of the non-eikonal expression leads to opposite effects in the two models. Contrary to the IPM there is an improved agreement with the experimental data in the case of the CDFM. Apparently, this is due to the zero-motion flucton correlations which are taken into account in the CDFM.

To summarize, the results of the calculations show that:

i) The influence of the zero-motion flucton correlations accounted for in the CDFM is significant.

ii) The CDFM results for the p-^{40}Ca cross-section using a realistic non-eikonal two-body amplitude are in better agreement with the experimental data than the independent-particle model calculations.

iii) The use of the non-eikonal two-body amplitude (11.41) in the CDFM calculations, compared with the use of the amplitude f_{pN} (11.26) improves the agreement with the experimental data which is not the case in the independent-particle model.

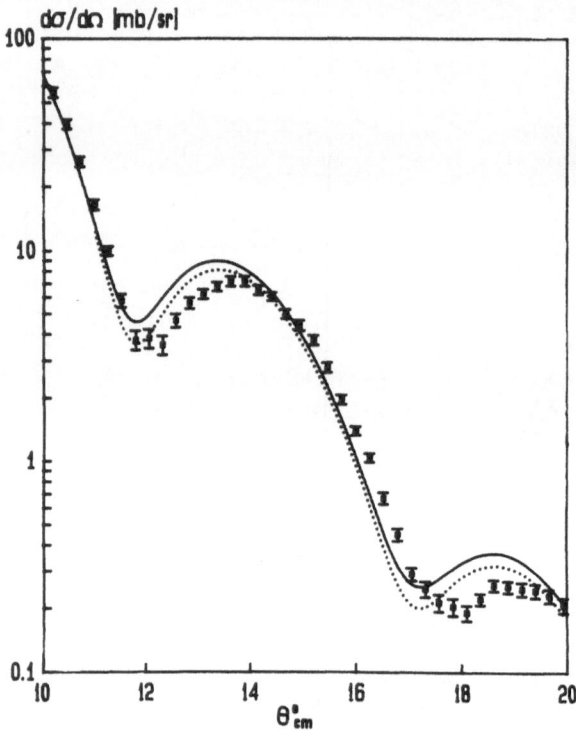

Fig. 11.7. The IPM results for the elastic 1.04 GeV-proton–^{40}Ca cross-section, Solid line: using f_{pN} from (11.41), (11.44); dotted line: using f_{pN} from (11.26), (11.45). The experimental data as in Fig. 11.5

11.3 Deep-inelastic Proton–nucleus Scattering within the Coherent Density Fluctuation Model

The investigations of the high-energy hadron–nucleus interaction show that the energy of the particles detected in the inclusive reactions reaches values which are essentially higher than those determined by the kinematics of the free hadron–nucleon scattering (for instance, in the elementary act $p + N \rightarrow \pi + X$). It has been suggested that such particle energy spectra are due to the short-range correlations in nuclei [11.111–114] and thus to the high-momentum components of the nuclear wave function [11.115]. It can be noted that interest in the SRC effects in nuclei has been renewed by the experiments of *Frankel* et al. [11.84, 11.85] and those from [11.87, 11.88] on the study of the inclusive differential cross-sections for the proton production at large angles for the intermediate energy (600–800 MeV) proton–nucleus scattering. In [11.87, 11.88] the cross-sections of the outgoing protons at various angles have been measured. *Komarov* et al. [11.88] studied the angular dependence of the relativistically-invariant structure function

$$\tilde{f} = \frac{E}{\sigma_t q^2} \frac{d^2\sigma}{d\Omega dq}, \tag{11.46}$$

where q and E are the momentum and the total energy of the detected proton and σ_t is the total proton–nucleus scattering cross-section. Calculations have been carried out in the model of *Weber* and *Miller* [11.116] according to which the emission of the high-energy proton takes place from the ground state of the target nucleus and the incident nucleon interacts with the rest of $A - 1$ nucleons. It is shown in [11.88] that this model cannot give either the absolute values of the observed cross-sections or the relative angular dependence of the structure function $\tilde{f}$ (11.46) (see curve a) from Fig. 11.8).

In the CDFM calculations of these processes [11.94, 11.96a–b, 11.117] another version of the direct interaction has been used, namely the *Amado* and *Woloshyn* [11.115] single-scattering mechanism. In it the incident proton with a momentum p knocks out a target nucleon (with momentum k in the target) which is observed with a momentum q in the final state (see Fig. 11.9).

The differential cross-section of the process is determined by [11.118, 11.115]:

$$\frac{d\sigma}{d^3q} = \frac{M^4}{2(2\pi)^5 p E(q)} \int \frac{dk\, n(k)|m_{NN}|^2}{E(k)\, E(p+k-q)}$$
$$\times \delta(E(p) + M - E(q) - E(p+k-q) - \bar{\varepsilon}), \tag{11.47}$$

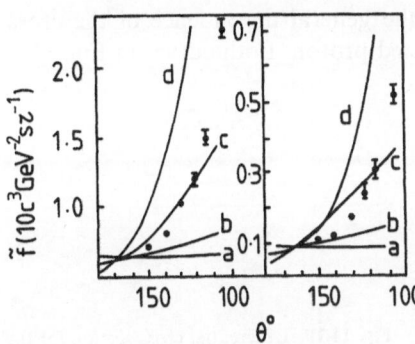

Fig. 11.8. The angular dependence of the structure function (11.46) for two energies of the detected proton, Curve (a): calculations from [11.88] in the *Weber-Miller* model [11.116]; curves (b) and (c): results from [11.88] accounting for three- and two-nucleon clusters; curve (d): the CDFM result [11.117]

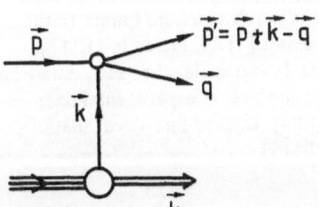

Fig. 11.9. Diagram corresponding to the single-scattering mechanism [11.115]

where $E(q) = (M^2 + q^2)^{1/2}$, M is the nucleon mass, $|m_{NN}|^2$ the square of the invariant N–N scattering amplitude, $n(k)$ the nucleon momentum distribution and $\bar{\varepsilon}$ the average interaction energy of the nucleon in the target, which can be determined from the quasielastic electron–nuclei scattering [11.119] (e.g. $\bar{\varepsilon}(^{12}C) = 25\,\text{MeV}$, $\bar{\varepsilon}(^{181}Ta) = 42\,\text{MeV}$). The following expression for the nucleon–nucleon amplitude has been used:

$$|m_{NN}|^2 = (8\pi)^2 s |f_{NN}|^2,\tag{11.48}$$

where $s = (p + k)^2$, p and k being the four-momenta of the incident proton and target nucleon, respectively,

$$f_{NN}(\delta) = \frac{(i + \varepsilon)}{4\pi} \tilde{p}\, \sigma_t \exp(-\beta\delta^2/2)\tag{11.49}$$

and $\tilde{p}$ the relative momentum. The cross-section (11.47) is calculated by means of the CDFM nucleon momentum distribution (4.65, 4.66). The Fermi momentum $k_F(x)$ for a flucton with radius x is determined from the relativistic normalization condition [11.115]:

$$2 \int \frac{dk}{(2\pi)^3} \frac{M}{E(k)} \tfrac{4}{3}\pi x^3\, \Theta(k_F(x) - |k|) = A.\tag{11.50}$$

The following parameter values taken from [11.120] have been used in the calculations: $\sigma_t = 4.4\,\text{fm}^2$, $\varepsilon = -0.275$, $\beta = 0.212\,\text{fm}^2$.

The cross-sections of the inclusive proton production in the reaction $p + {}^{12}C \rightarrow p + X$ at angles 180°, 160°, 150° and 140° have been calculated within the CDFM in [11.117] and the results for the angles 180° and 160° are given in Figs. 11.10, 11.11. The CDFM results describe satisfactorily the data extrapolated to 180° from [11.87] and the general dependence of the cross-section on the kinetic energy of the detected proton. Both curves in Fig. 11.11

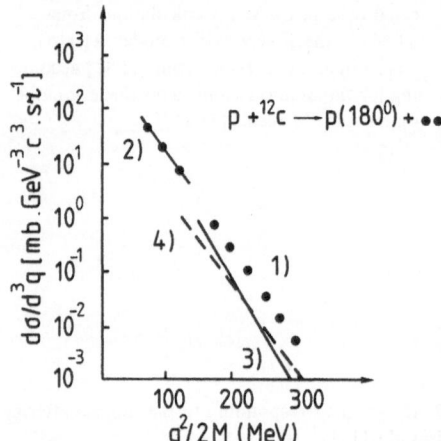

Fig. 11.10. Differential cross-section for the inclusive proton production at 180° in the $p + {}^{12}C \rightarrow p + X$ reaction. Points 1): the CDFM result [11.94, 11.96a,b, 11.117]. Solid line 2): extrapolated to 180° data from [11.87]; solid line 3): experimental data from [11.84]. Dashed line 4): calculations from [11.115]

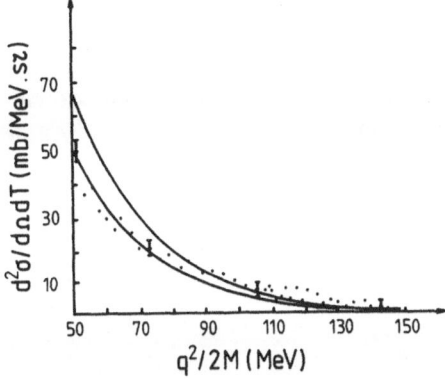

Fig. 11.11. Differential cross-section for the inclusive proton production at 160° in the p + ^{12}C → p + X. Solid lines: CDFM result [11.117] with $\bar{\varepsilon} = 0.13$ fm^{-1} (upper) and $\bar{\varepsilon} = 0.22$ fm^{-1} (lower). Points: experimental data [11.85, 11.86]

show the sensitivity of the cross section at small $q^2/2M$ to the value of the parameter $\bar{\varepsilon}$.

The dependence of the CDFM calculated structure function (Fig. 11.8) on the angle of detecting at fixed energy of the outgoing particles shows a deviation from the data at small angles. The common tendency of a rapid increase of $\tilde{f}$ when the angle decreases is correctly reproduced, however, in contrast to the results obtained by using other scattering mechanisms [11.88] (e.g. curves b) and c) from Fig. 11.8).

To conclude this Section we note that the nucleon momentum distribution obtained within the CDFM describes satisfactorily the experimental results on the inclusive proton production in the intermediate-energy proton scattering. We note the similarity with the results from [11.115]. The slope of $d\sigma/d^3q$ is determined by the behaviour of the nucleon momentum distribution at large momenta ($k \simeq 2$–5 fm^{-1}). The qualitative agreement of the calculated cross-sections with the experimental data strongly depends on the particular form of the high-momentum components of the nucleon momentum distribution.

It has to be emphasized that the problem of the mechanism of the large-angle proton production is far from solution. The existing models, e.g. that of *Amado* and *Woloshyn* [11.115], of *Weber* and *Miller* [11.116], the correlated cluster model of *Fujita* [11.121], the statistical model from [11.122], the model of pion absorption [11.123], the model with a fragmentation of the target [11.124], the model of the correlated two-nucleon clusters [11.125, 11.126] describe successfully one or more properties of the process. We should note that the experimental results on coincidences between the outgoing proton in the back hemisphere and the forward outgoing proton [11.127, 11.128] (in spite of the substantial kinematical limitation in [11.127]) provide a critical test of the range of validity of the various models. It is shown by *Miake* et al. [11.128] that the two-nucleon cluster mechanism can play a more essential role than the single-particle mechanism for the description of the coincidence experiments. *Haneishi* and *Fujita* [11.126] determine the energy region where the single-particle and cluster mechanisms are applicable. Regardless of the differences between the models, however, it is emphasized in [11.125] that the presence of the high-momentum components in the nucleon momentum distribution is confirmed by the analyses of all available experimental data on inclusive proton production in the processes of the proton–nucleus interaction.

12 Elastic Scattering of Complex Nuclear Systems

A brief review of the evidence for nucleon–nucleon correlation effects on the characteristics of alpha-particle (Sect. 12.1) and heavy ion (Sect. 12.2) elastic scattering by nuclei is presented in this chapter. The results of the coherent density fluctuation model are compared with experimental data for alpha-particle ($E_\alpha = 1.37$ GeV) elastic scattering on ^{12}C and ^{40}Ca, as well as for heavy ion elastic scattering.

12.1 Intermediate-energy Alpha-particle Elastic Scattering on Nuclei

The experiments on the intermediate-energy elastic alpha-particle ($E_\alpha = 1.37$ GeV) scattering on ^{12}C [12.1] and on the calcium isotopes ^{40,42,44,48}Ca [12.2] give an essential contribution to the study of the hadron–nucleus and ion–ion interaction. The use of the alpha-particle as the lightest and "the most elementary" heavy ion [12.3] and as a strongly bound system with spin and isospin equal to zero leads to a significant simplification in the investigation of the scattering of complex systems by nuclei. The experiments mentioned give information on the reaction mechanism, on the density distribution in the nuclear surface (due to the strong absorption of the heavy ions in the nuclear interior), as well as on the nucleon–nucleon short-range correlations in the target nucleus and in the incident nuclear system. In this sense the experiments with alpha-particles are complementary to the nuclear structure studies by means of electrons, protons, etc.

The theoretical analysis of the alpha-particle scattering on ^{12}C and ^{40,42,44,48}Ca is carried out mainly in the framework of the *Kerman, McManus* and *Thaler* theory (KMT) [12.4] (see papers [12.1, 12.5]) and of the *Glauber–Sitenko* theory (TGS) [12.6, 12.7] (see e.g. [12.2, 12.3, 12.8–17]). There is a common result obtained that the optical limit [12.18] in both approaches is not able to describe the experimental data even at small transfer momenta [12.16] in the KMT theory: ^{12}C [12.1] and in ^{40}Ca [12.5], in the TGS: ^{12}C [12.8, 12.9] and $^{40-48}$Ca [12.2].

Various modifications of the above-mentioned theories describe to a different extent the data on the alpha-particle elastic scattering on nuclei. The method of the "steady incident particle" [12.2] in which the alpha-particle interacts as a whole with the single nucleons (i.e. it remains in its ground state and is not divided into its constituents during the multiple scattering) describes satisfactorily the data on the scattering on calcium isotopes [12.2, 12.3] but not those on the ^{12}C nucleus [12.3, 12.12]. In the method from [12.15] the roles of the incident particle and the target change places but the results are similar. In these analyses both interacting ions are considered in a non-symmetrical way and no account is taken of all multiple scattering of the nucleons from one of the nuclei on the nucleons from the other [12.16]. A method which is symmetrical with respect to both nuclei is developed in [12.10, 12.11]. In it the phase function is expanded in an infinite series taking account of the correlation effects. In [12.16] an expansion of the nucleus–nucleus scattering S-matrix into a finite series of the so-called effective profile functions is suggested. This gives the possibility of studying the nucleon–nucleon correlations in both nuclei. Both methods describe well the data up to the first diffraction maximum in the cross-section for the alpha-particle elastic scattering on ^{40}Ca [12.10, 12.16], but do not reproduce the results of the scattering on the ^{12}C nucleus [12.11, 12.16, 12.17]. It is shown that the short-range correlation effects are small in the region of the first diffraction maximum and increase for the larger transfer momenta [12.16]. They can be observed in the increase of the cross-section diffraction maxima. Along with this *Ahmad* [12.16] has shown significant difficulties in the nonambiguous recognition of the nucleon–nucleon correlation effects in the nuclei–nuclei scattering.

The use of the TGS and the method similar to that from [12.11] show [12.19, 12.20] that the short-range correlation (SRC) effects are not large and they can be seen in the region of the alpha-particle-calcium isotopes cross-section maxima. This conclusion, however, turns out to be dependent on the choice of the correlation length in the calculations. It is not in accord [12.19] with the result obtained by *Małecki* and *Satta* [12.21] that the SRC effect on the second diffractional maximum in the alpha-particle cross-section (at $q^2 \simeq 0.15$ $(GeV/c)^2$) is very large (up to 350%). It is shown by *Alkhazov* [12.19, 12.20] that the correlations related to the alpha-particle centre-of-mass motion are much larger than the SRC effects.

The use of the t-matrix for the free alpha-particle-nucleus scattering within the KMT theory leads to a better description of alpha-particle scattering on the calcium isotopes [12.15] than that in the optical limit but fails again in the case of the ^{12}C nucleus [12.1]. Also unsuccessful is the description of the data on the alpha $+ {}^{12}$C scattering by the phenomenological method of *Alexander* and *Rinat* [12.22] using the eikonal distorted wave impulse approximation.

Considerably better agreement with the data on the alpha-particle scattering at 1.37 GeV on ^{40}Ca as well as on ^{12}C is achieved by the theoretical approach of *Pak* et al. [12.13, 12.14]. This allows an estimate of the corrections to the scattering amplitude related to the nucleon–nucleon correlations in nuclei, to nuclear excitations and so on.

It should also be mentioned that the phenomenological model of *Choudhury* [12.23] has been successfully applied to the study of the elastic and inelastic

1.37 GeV-alpha-particle scattering on the calcium isotopes. In the framework of the aidabatic approximation and using the partial wave expansion, it is shown that the model is equivalent to the description of the Fraunhofer diffraction of waves on a black sphere with radius equal to the target nucleus radius. The effective equilibrium radius of the target and the effective thickness of the nuclear surface are both parameters of the model. They have larger values than those determined for the Ca-isotopes in other methods [12.2, 12.15] of studying the alpha-particle scattering and from other experiments.

The coherent density fluctuation model (CDFM) [12.24–26] has been applied to the study of the alpha-particle elastic scattering at 1.37 GeV by the ^{12}C and ^{40,42,44,48}Ca nuclei in [12.27, 12.28]. In CDFM [12.27] the amplitude of the elastic scattering of alpha-particle on nuclei (4.73) is expressed approximately as a superposition of the amplitudes of scattering on fluctons with radius x ($A_0(x, q)$). The alpha-particles are considered as point-like particles which are scattered on the absolutely black fluctons (see also Sect. 11.2 for the proton–nucleus scattering in the same approximation). In this case the cross-section can be written in the form (see (11.19)):

$$\frac{d\sigma}{d\Omega} = \frac{L^4 k_0^2}{4\pi^2} |A_{00}(q)|^2$$

$$= \frac{L^4 k_0^2}{4\pi^2} \left| \int\limits_0^\infty dx \, |f(x)|^2 A_0(x, q) \right|^2 \tag{12.1}$$

$$= \left(\frac{k_0}{q}\right)^2 \left| \int\limits_0^\infty dx \, |f(x)|^2 x J_1(qx) \right|^2 ,$$

where k_0 is the wave number of the incident alpha-particle, q is the transfer momentum. $J_1(qx)$ is the first-order Bessel function.

$$A_0(x, q) = -\frac{2\pi}{L^2} \frac{x}{q} J_1(qx) \tag{12.2}$$

is the amplitude of scattering on the absolutely black body (a flucton with radius x in this case) within the diffractional approximation [12.29], and L is the normalizing length.

In the case of monotonically decreasing density distributions the weight function $f(x)$ can be determined by the expression (4.66):

$$|f(x)|^2 = -\frac{1}{\rho_0(x)} \frac{d\rho(r)}{dr}\bigg|_{r=x} , \qquad \rho_0(x) = \frac{3A}{4\pi x^3} . \tag{12.3}$$

The calculations of alpha-particle scattering in the CDFM for ^{12}C and Ca-isotopes have been carried out using the symmetrized Fermi-type density distribution $\rho(r)$ [12.30]. The optimal values of the density parameters, namely the effective radius (R) and the effective thickness of the nuclear surface (b) used

Table 12.1. Effective radius (R) and effective surface thickness (b) used in the CDFM alpha-particle–nucleus cross-section calculations [12.27] and in the work of *Choudhury* [12.23] (in fm)

Nucleus	CDFM [12.27]		Choudhury [12.23]	
	R	b	R	b
^{12}C	2.200	0.540	—	–
^{40}Ca	4.350	0.680	4.959	0.680
^{42}Ca	4.400	0.680	5.040	0.677
^{44}Ca	4.450	0.680	5.120	0.674
^{48}Ca	4.650	0.660	5.086	0.670

in these calculations are given in Table 12.1 and are compared with the corresponding effective parameters of *Choudhury* [12.23].

The cross-sections calculated in the CDFM [12.27] are compared with the experimental data for ^{12}C [12.1] and ^{40}Ca [12.2] in Figs. 12.1, 12.2. The results for alpha-particle scattering on $^{42,44,48}Ca$ are similar to those on ^{40}Ca. We should note the good agreement of the CDFM calculations with the data for the Ca-isotopes, even in the case of the ^{12}C nucleus. For the latter case the agreement is comparable with that from [12.13] and is significantly better than that obtained by other methods [12.1, 12.3, 12.8, 12.11, 12.15, 12.16]. The deviations of the theoretical calculations from the data in the regions of the cross-section minima are due to the black-body diffractional approximation for the amplitude of the alpha-particle–flucton scattering.

As can be seen from Table 12.1, the values of the effective surface thickness parameter b in the CDFM are very close to those from [12.23]. The values of the

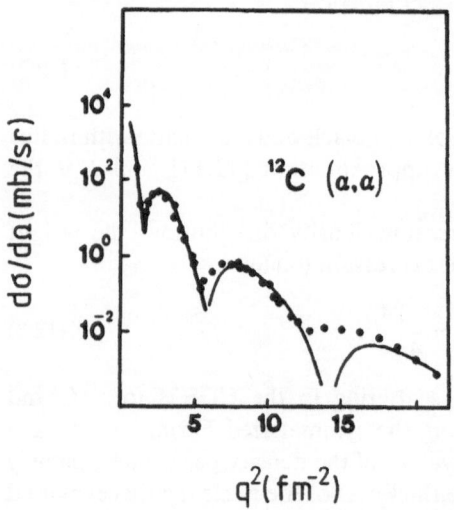

Fig. 12.1. Differential cross-section of elastic scattering of 1.37 GeV-alpha-particles on ^{12}C. Solid line: the CDFM result [12.27], points: experimental data from [12.1]

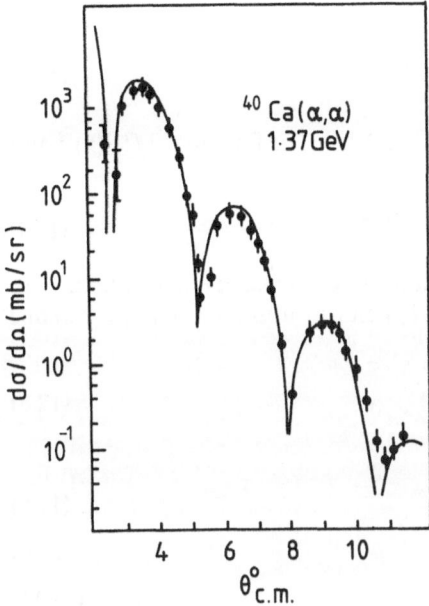

Fig. 12.2. The same as in Fig. 12.1 for ^{40}Ca. The experimental data are taken from [12.2]

parameter R are larger, as in [12.23], than those obtained from the analyses of alpha-particle scattering on Ca-isotopes in [12.2] and [12.15]. In the CDFM the values for R for the Ca-isotopes can be described approximately by $R \simeq 1.27A^{1/3}$ fm. The question of the larger values of the parameters R and b obtained in the CDFM has been discussed in [12.27]. They can be related to the point-like alpha-particle approximation used in the CDFM. Indeed, if the size and the density fluctuations of the incident alpha-particles are accounted for, then the amplitude $A_{00}(q)$ can be written in the form:

$$A_{00}(q) = -\frac{2\pi}{L^2} \int_0^\infty dx_1 |f^{(1)}(x_1)|^2 \int_0^\infty dx_2 |f^{(2)}(x_2)|^2$$

$$\times \frac{(x_1 + x_2)}{q} J_1(q(x_1 + x_2)) , \tag{12.4}$$

where $f^{(1)}(x)$ and $f^{(2)}(x)$ are the weight functions for the ground state of the alpha-particle and the target nucleus, respectively. Both functions $|f^{(1,2)}(x)|^2$ are peaked in the vicinity of the half-radius $R_{1/2}$ of the nuclear density so that the low limits of the integrations over x_1 and x_2 in (12.4) can be approximately extended to $-\infty$. The substitution $x_2 = y - x_1$ leads to the following form of the amplitude (12.4):

$$A_{00}(q) = -\frac{2\pi}{L^2} \int_{-\infty}^\infty dy |\tilde{f}(y)|^2 \frac{y}{q} J_1(qy) , \tag{12.5}$$

where

$$|\tilde{f}(y)|^2 = \int_{-\infty}^{\infty} dx |f^{(1)}(x)|^2 |f^{(2)}(y-x)|^2 \tag{12.6}$$

plays the role of the weight function $|f(x)|^2$ in (12.1). The functions $|f^{(1)}(x)|^2$ and $|f^{(2)}(x)|^2$ may be simplified in the form:

$$|f^{(i)}(x)|^2 = N_i \exp[-\alpha_i(x - R_i)^2], \quad i = 1, 2, \tag{12.7}$$

where R_1 and R_2 are related to the half-radius of both nuclei and α_1 and α_2 – to their surface thicknesses b_1 and b_2 ($\alpha_i \simeq 1/b_i^2$). Substituting (12.7) into (12.6) one gets [12.27, 12.31]:

$$|\tilde{f}(y)|^2 = N \exp[-\alpha(y - R)^2], \tag{12.8}$$

where

$$N = N_1 N_2 (\pi/(\alpha_1 + \alpha_2))^{1/2}, \tag{12.9}$$

$$R = R_1 + R_2, \tag{12.10}$$

$$\alpha = \alpha_1 \alpha_2/(\alpha_1 + \alpha_2). \tag{12.11}$$

From $\alpha = 1/b^2$ it follows that

$$b = (b_1^2 + b_2^2)^{1/2}. \tag{12.12}$$

The parameters R and $b = 1/\alpha^{1/2}$ from (12.8) play the role of the effective radius and surface thickness, respectively. They appear in the CDFM and in the model from [12.23]. It can be seen from (12.12) that if the well-known values of the parameters $b_{^4He} = 0.380$ fm, $b_{^{40}Ca} = 0.578$ fm, $b_{^{48}Ca} = 0.520$ fm extracted from the electron scattering data are substituted in (12.12) one can obtain the effective parameters b for ^{40}Ca and ^{48}Ca to be 0.691 fm and 0.644 fm, respectively. These values are close to the optimal values of b (0.680 fm and 0.660 fm) presented in Table 12.1. The relationship (12.10) gives larger values for the parameter R than those obtained in the CDFM. One of the reasons for this might be the simplified form of the functions $|f^{(i)}(x)|^2$ (12.7) chosen for this qualitative analysis. The given estimation can also explain the increased values of the effective parameters R and b from [12.23].

12.2 Heavy-ion Elastic Scattering

The analysis of elastic heavy-ion scattering allows us to obtain information on the surface structure of complex nuclear systems and of the character of the interaction between them. The basic interest in the study of these processes is focussed on the diffraction picture of the scattering cross-sections, the existence of resonances in the excitation function, as well as on the possibility of nuclear exchange effects (e.g. reactions with elastic particle transfers) [12.32]. Various models and methods are used in the theoretical study of heavy-ion elastic scattering. The models which are used to describe the process give the possibility of relating the measured quantities to the average characteristics of the many-body quantum mechanical system. The optical model which is often applied (e.g. [12.33] and see also Chap. 2) contains free parameters which characterize the nuclear matter distribution and determine the complex optical potential for the interaction between both ions. The sensitivity of the heavy-ion scattering cross-section to the changes in the potential can be used to specify the form and the parameters of the potential itself. A stronger theoretical method for constructing the interaction potential is based on the use of the local density approximation [12.34]. There exists also the approach for the determination of an "effective interaction" of the basis of the density distributions [12.35], as well as various microscopical methods for obtaining of the ion–ion potential (e.g. [12.36]).

Also widely used are the semiclassical approach (e.g. [12.37]), the method of complex angular momentum and the diffractional model based on a phenomenological description of the nuclear absorption and refraction in terms of the S-matrix properties (see e.g. [12.38–41]).

Heavy-ion elastic scattering is also analysed by the multiple-scattering Glauber–Sitenko theory (e.g. [12.32, 12.42–45, 12.11]). There are no free parameters in this approach. The nucleon–nucleon scattering amplitude and the wave functions of the nuclear states are considered as known. It is shown by *Varma* [12.45] that the effects of the SRC as well as of the correlations related to the motion of the centre-of-mass are substantial. They lead to an increase of the diffractional maxima (15–20% for the second and 25–35% for the third maximum in the elastic cross-sections of the processes $^{12}C + {}^{12}C$ at energies 2.1 GeV/N).

Exact equations in the multiple-scattering theory for the case of two nuclei taking complete account of the Pauli principle are derived by *Gurvitz* [12.46].

The heavy-ion elastic scattering has been studied within the coherent density fluctuation model (CDFM) [12.24–26] in the framework of the assumption of diffraction scattering of a flucton in the incident nucleus on an absolutely black flucton from the target nucleus taking account of the superposition of fluctons of both nuclei [12.47, 12.28]. The Coulomb interaction is also taken into account. It is accepted that the fluctons from both ions interact as absolutely black ones at values of the impact parameter $\rho < x_1 + x_2$ and as two charged particles at $\rho > x_1 + x_2$ (x_1 and x_2 being radii of the fluctons from the incident and the target nucleus, respectively). The problem is equivalent to the scattering of a point-like particle with a charge Z_2 on an absolutely black body with radius $x_1 + x_2$ and a charge Z_1. The heavy-ion elastic scattering cross-section has the following form in the CDFM [12.47]:

$$\frac{d\sigma}{d\Omega} = |A_{00}(\theta)|^2, \tag{12.13}$$

where

$$A_{00}(\theta) = \int\limits_{0}^{\infty} dx_1 |f(x_1)|^2 \int\limits_{0}^{\infty} dx_2 |f(x_2)|^2 A_0(\theta, x_1, x_2). \qquad (12.14)$$

The diffractional amplitude of the scattering of a flucton with radius x_1 from the incident nucleus on the absolutely black flucton with radius x_2 from the target nucleus can be expressed [12.48, 12.29] by:

$$A_0(\theta, x_1, x_2) = \frac{i}{k} \left\{ l_0^{2in+1} \frac{J_1(l_0\theta)}{\theta} + 2in\theta^{-2in-2} \times \int\limits_{l_0\theta}^{\infty} J_1(z) z^{2in} \, dz \right\}, \qquad (12.15)$$

where k is the wave number, θ is the scattering angle, J_1 is the first-order Bessel function,

$$l_0 = k(x_1 + x_2), \qquad (12.16)$$

$$n = Z_1 Z_2 e^2 / \hbar v. \qquad (12.17)$$

In (12.17) v is the velocity of the incident nucleus. The functions $|f(x_1)|^2$ and $|f(x_2)|^2$ are the CDFM weight functions and can be determined by means of the density distributions according to (12.3) (for $d\rho/dr \leqslant 0$).

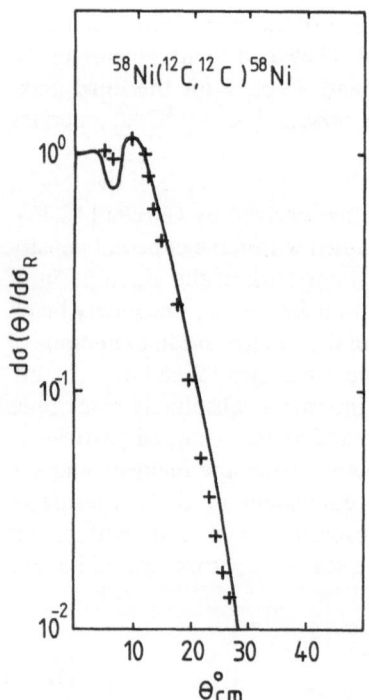

Fig. 12.3. The ratio $\dfrac{d\sigma/d\Omega}{(d\sigma/d\Omega)_R}$ for the elastic scattering ^{58}Ni (^{12}C, ^{12}C)^{58}Ni, $E_{lab} = 124.5$ MeV. Solid line: CDFM result [12.47, 12.28]. Crosses: experimental data taken from [12.3]

This approach is valid for small scattering angles only ($\theta \ll 1$) and under the condition $l_0 \gg 1$. For fluctons with sizes x_1 and x_2, for which the values of $|f(x_1)|^2$ and $|f(x_2)|^2$ are essentially different from zero the latter condition is fulfilled.

The differential cross-sections $d\sigma/d\Omega$ and the ratio $(d\sigma/d\Omega)/(d\sigma/d\Omega)_R$ (where $(d\sigma/d\Omega)_R$ is the Rutherford cross-section) for the elastic scattering:

a) ^{58}Ni (^{12}C, ^{12}C) ^{58}Ni; $E_{\text{lab}} = 124.5$ MeV,

b) ^{209}Bi (^{16}O, ^{16}O) ^{209}Bi; $E_{\text{lab}} = 134$ MeV

are calculated within the CDFM [12.47]. These examples correspond to large Coulomb interaction and to energies E_{lab} higher than the Coulomb barrier. The Fermi density distribution with parameters extracted from the electron–nuclei scattering data have been used in the calculations:

^{12}C: $R = 2.214$ fm, $b = 0.488$ fm,
^{58}Ni: $R = 4.153$ fm, $b = 0.566$ fm,
^{16}O: $R = 2.562$ fm, $b = 0.497$ fm.

For the ^{209}Bi nucleus the parameter values have been chosen to be: $R = 6.557$ fm and $b = 0.515$ fm. The ratio $(d\sigma/d\Omega)/(d\sigma/d\Omega)_R$ for the case ^{12}C + ^{58}Ni is presented in Fig. 12.3 and is compared with the experimental data taken from [12.32]. The good agreement with the experimental data in this case is achieved without use of fitting parameters. The condition for the validity of the approach ($\theta \ll 1$) is satisfied for the scattering ^{12}C + ^{58}Ni ($\theta \lesssim 28°$), while for the case ^{16}O + ^{209}Bi the angular interval is $30° < \theta° < 70°$. Obviously this is the reason for the deviation of the theoretical result from the data in the latter case at angles $\theta° \gtrsim 45°$.

References

Chapter 1

1.1 S. Moshkovsky: In *Handbuch der Physik*, vol. **39** (Ed. by S. Flügge, Springer-Verlag, Berlin-Göttingen-Heidelberg 1957), Part 5

1.2 J.M. Eisenberg, W. Greiner: *Nuclear Theory*, vol. **3** (North-Holland Publishing Company, Amsterdam-London 1972), Chaps 6–7

1.3 K. Huang: *Statistical Mechanics* (Wiley, New York 1963)

1.4 A.G. Sitenko, V.K. Tartakovskiĭ: *Lectures on the Theory of the Nucleus* (Pergamon Press, Oxford 1975); A.N. Antonov, P.E. Hodgson, I.Zh. Petkov: *Nucleon Momentum and Density Distributions in Nuclei* (Clarendon Press, Oxford 1988)

1.5 W. Czyż, K. Gottfried: Nucl. Phys. **21**, 676 (1961)

1.6 A.B. Migdal: ZETP USSR **32**, 399 (1957); Soviet Physics JETP **5**, 333 (1957)

1.7 V.A. Belyakov: ZETP USSR **40**, 1210 (1961); Soviet Physics JETP **15**, 850 (1961)

1.8 R. Sartor, C. Mahaux: Phys. Rev. **C21**, 1546 (1980)

1.9 P. Ring, P. Schuck: *The Nuclear Many-Body Problem* (Springer-Verlag, New York-Heidelberg-Berlin 1980).

1.10 J.P. Elliot, A.M. Lane: In *Handbuch der Physik*, vol. **39** (Ed. by S. Flügge, Springer-Verlag, Berlin-Göttingen-Heidelberg 1957), Part 4, p. 241

1.11 D.R. Hartree: Proc. Camb. Phil. Soc. **24**, 89 (1928)

1.12 V.A. Fock: Z. Phys. **61**, 126 (1930)

1.13 B.I. Bartz, Yu.L. Bolotin, E.V. Inopin, V.Yu. Gonchar: *The Hartree-Fock Method in the Nuclear Theory* (Naukova Dumka, Kiev 1982)

1.14 J.W. Negele: Phys. Rev. **C1**, 1260 (1970)

1.15 J.P. Svenne: Adv. in Nucl. Phys. **11**, 179 (1979)

1.16 T.H.R. Skyrme: Phyl. Mag. **1**, 1043 (1956); Nucl. Phys. **9**, 615 (1959)

1.17 D. Gogny, R. Padjen: Nucl. Phys. **A293**, 365 (1977)

Chapter 2

2.1 H. Feshbach: Ann. Phys. **5**, 357 (1958)

2.2 H. Feshbach: Ann. Phys. **19**, 287 (1962)

2.3 H. Feshbach, J. Hüfner: Ann. Phys. **56**, 268 (1970)

2.4 D.F. Jackson: *Nuclear Reactions* (Methuen & Co LTD, London 1970)

2.5 R.C. Barrett, D.F. Jackson: *Nuclear Sizes and Structure* (Clarendon Press, Oxford 1977)

2.6 I.Zh. Petkov: Particles and Nuclei **2**, 483 (1971); JINR, E4-4901, Dubna, 1970

2.7 P.E. Hodgson: Contemporary Physics **31**, 295 (1990)

2.8 J. Millener, P.E. Hodgson: Nucl. Phys. **A209**, 59 (1973)

2.9 F. Malaguti, P.E. Hodgson: Nucl. Phys. **A215**, 243 (1973)

2.10 F. Malaguti, A. Uguzzoni, E. Verondini, P.E. Hodgson: Nucl. Phys. **A297**, 287 (1978); Nuovo Cim. **A49**, 412 (1979); Nuovo Cim. **A53**, 1 (1979); Rev. Nuovo Cim. **5**, 1 (1982)

2.11 B.A. Brown, S.E. Massen, P.E. Hodgson: Phys. Lett. **89B**, 167 (1979)

2.12 L. Ray, P.E. Hodgson: Phys. Rev. **C20**, 2403 (1979)

2.13   C.M. Perey, F.G. Perey: Atom Data and Nucl. Data Tables **13**, 293 (1974); Ibid. **17**, 1 (1976)

2.14   P.E. Hodgson: *The Optical Model of Elastic Scattering* (Clarendon, Oxford 1963); Nuovo Cim. **A81**, 250 (1984); Rep. Prog. Phys. **47**, 613 (1984)

2.15   B.C. Sinha, V.R.W. Edwards: Phys. Lett. **31B**, 273 (1970); Ibid. **5B**, 391 (1971)

2.16   R.W. Finlay, J.R.M. Annand, J. R. Petler, F.S. Dietrich: Phys. Lett. **155B**, 313 (1985)

2.17   J.R.M. Annand, R.W. Finlay, F.S. Dietrich: Nucl. Phys. **A443**, 249 (1985)

2.18   L.R.B. Elton: Nucl. Phys. **89**, 69 (1966)

2.19   M. Bauer, E. Hernandez-Saldana, P.E. Hodgson, J. Quintanilla: J. Phys. **G3**, 525 (1982)

2.20   S.G. Cooper, P.E. Hodgson: J. Phys. **G6**, L21 (1980)

2.21   J.P. Delaroche, W. Tornow: Phys. Lett. **203B**, 4 (1988)

2.22   P.E. Hodgson: Contemporary Physics **24**, 491 (1983)

2.23   C. Mahaux, H. Ngô, G.R. Satchler: Nucl. Phys. **A449**, 354 (1986)

2.24   G. Passatore: Nucl. Phys. **A110**, 91 (1968)

2.25   C. Mahaux, H. Ngô: Nucl. Phys. **A378**, 205 (1982)

2.26   B. Gyarmati, R.G. Lovas, T. Vertse, P.E. Hodgson: J. Phys. **G7**, L209 (1981)

2.27   I. Ahmad, W. Haider: J. Phys. **G2**, L157 (1976)

2.28   C.H. Johnson, D.J. Horen, C. Mahaux: Phys. Rev. **C36**, 2252 (1987)

2.29   J.-P. Jeukenne, C.H. Johnson, C. Mahaux: Phys. Rev. **C38**, 2573 (1988)

2.30   C.H. Johnson, R.R. Winters: Phys. Rev. **C37**, 2340 (1988)

2.31   C.H. Johnson, C. Mahaux: Phys. Rev. **C38**, 2589 (1988)

2.32   W. Tornow, Z.P. Chen, J.P. Delaroche: Phys. Rev. **C42**, 693 (1990)

2.33   C.H. Johnson, R.F. Carlton, R.R. Winters: Phys. Rev. **C39**, 415 (1989)

2.34   R.W. Finlay, J. Wierzbicki, R.K. Das, F.S. Dietrich: Phys. Rev. **C39**, 804 (1989)

2.35   R.K. Das, R.W. Finlay: Phys. Rev. **C42**, 1013 (1990)

2.36   M.L. Roberts, Z.M. Chen, P.D. Felsher, D.J. Horen, C.R. Howell, K. Murphy, H.G. Pfützer, W. Tornow, R.L. Walter: TUNL Report XXVIII 70 (1988–9)

2.37   Z.P. Chen, W. Tornow: TUNL Report XXVIII 82 (1988–9)

2.38   W. Tornow, J.P. Delaroche: Phys. Lett. **210B**, 26 (1988)

2.39   C. Mahaux, R. Sartor: Nucl. Phys. **A481**, 407 (1988)

2.40   A.B. Smith, P.T. Guenther, R.D. Lawson: Nucl. Phys. **A483**, 50 (1988)

2.41   R.D. Lawson, P.T. Guenther, A.B. Smith: Phys. Rev. **C34**, 1599 (1986); Phys. Rev. **C36**, 1298 (1987)

2.42   S. Chiba, P.T. Guenther, R.D. Lawson, A.B. Smith: Phys. Rev. **C42**, 2487 (1990)

2.43   C. Mahaux, R. Sartor: Phys. Rev. Lett. **57**, 3015 (1986); Nucl. Phys. **A458**, 25 (1986); Ibid. **A460**, 446 (1986); Ibid. **A468**, 193 (1987); Phys. Rev. **C36**, 1777 (1987); Nucl. Phys. **A493**, 157 (1989); Nucl. Phys. **A481**, 381 (1988); Nucl. Phys. **A484**, 205 (1988); Nucl. Phys. **A475**, 247 (1987)

2.44   R.D. Lawson, P.T. Guenther, A.B. Smith: Nucl. Phys. **A493**, 267 (1989)

2.45   J.-P. Jeukenne, L. Lejeune, C. Mahaux: Phys. Rev. **C15**, 10 (1977)

2.46   C. Mahaux, R. Sartor: Nucl. Phys. **A516**, 285 (1990)

2.47   C. Mahaux, R. Sartor: Nucl. Phys. **A528**, 253 (1991)

2.48   C. Mahaux, R. Sartor: Nucl. Phys. **A503**, 525 (1989)

2.49   A.C. Merchant, P.E. Hodgson, H.R. Schelin: Nucl. Sci. Eng. **111**, 132 (1992)

2.50   C. Mahaux, R. Sartor: Advances in Nuclear Physics **20**, 1 (1991)

2.51   G.E. Brown, M. Rho: Nucl. Phys. **A372**, 397 (1981)

2.52   J.-P. Jeukenne, C. Mahaux: Nucl. Phys. **A394**, 445 (1983)

2.53   J.P. Delaroche, Y. Wang, J. Rapaport: Phys. Rev. **C39**, 391 (1989)

2.54   S.E. Hicks, M.T. McEllistrem: Phys. Rev. **C37**, 1787 (1988)

Chapter 3

3.1   P. Ring, P. Schuck: *The Nuclear Many-Body Problem* (Springer-Verlag, New York, Heidelberg, Berlin 1980)

3.2   J.M. Eisenberg, W. Greiner: *Microscopic Theory of the Nucleus* (North-Holland Publishing Company, Amsterdam-London 1972)

3.3   B.I. Bartz, Yu.L. Bolotin, E.V. Inopin, V.Yu. Gonchar: *The Hartree-Fock Method in the Nuclear Theory* (Naukova Dumka, Kiev 1982)

3.4   N.H. March, W.H. Young, S. Sampanthar: *The Many-Body Problem in Quantum Mechanics* (Cambridge University Press, Cambridge 1967)

3.5   J.M. Ziman: *Elements of Advanced Quantum Theory* (Cambridge University Press, Cambridge 1969)

3.6   M. Jaminon, C. Mahaux, H. Ngô: Nucl. Phys. **A473**, 509 (1987)

3.7   B. Frois, J.B. Bellicard, J.M. Cavedon, H. Huet, P. Leconte, P. Ludeau, A. Nakada, P.Z. Ho and I. Sick: Phys. Rev. Lett. **38**, 152 (1977)

3.8   A.E.L. Dieperink, I. Sick: Phys. Lett. **109B**, 1 (1982)

3.9   V.R. Pandharipande, C.N. Papanicolas, J. Wambach: Phys. Rev. Lett. **53**, 1133 (1984)

3.10   P. Grabmayr, S. Klein, H. Clement, K. Reiner, W. Reuter, G.J. Wagner and G. Seegert: Phys. Lett. **164B**, 15 (1985)

3.11   J.W.A. den Herder, J.A. Hendriks, E. Jans, P.H.M. Keizer, G.J. Kramer, L. Lapikás, E.N.M. Quint, P.K.A. de Witt Huberts, H.P. Blok, G. van der Steenhoven: Phys. Rev. Lett. **57**, 1843 (1986)

3.12   E.N.M. Quint, J.F.J. van den Brand, J.W.A. den Herder, E. Jans, P.H.M. Keizer, L. Lapikás, G. van der Steenhoven, P.K.A. de Witt Huberts, S. Klein, P. Grabmayr, G.J. Wagner, H. Nann, B. Frois, D. Goutte: Phys. Rev. Lett. **57**, 186 (1986)

3.13   H. Clement, P. Grabmayr, H. Röhm, G.J. Wagner: Phys. Lett. **183B**, 127 (1987)

3.14   H. Araseki, T. Fujita: Nucl. Phys. **A439**, 681 (1985)

3.15   Y. Haneishi, T. Fujita: Phys. Rev. **C33**, 260 (1986); Phys. Rev. **C35**, 70 (1985)

3.16   T. Fujita: Nucl. Phys. **A457**, 657 (1986)

3.17   M. Avan, A. Baldit, J. Castor, M.El Zoubidi, J. Fargeix, H. Fonvieille, P. Force, J.L. Guelou, B. Harradi, G. Landaud, J.P. Didelez, F. Reide, M. Bernheim, A. Gerard, A. Magnon, C. Marchand, J. Morgenstern, J. Picard, P. Vernin, H. Jackson: Phys. Rev. **C37**, 231 (1988)

3.18   A.N. Antonov, P.E. Hodgson, I.Zh. Petkov: *Nucleon Momentum and Density Distributions in Nuclei* (Clarendon Press, Oxford 1988)

3.19   M. Jaminon, C. Mahaux, H. Ngô: Phys. Lett. **158B**, 103 (1985)

3.20   M. Gaudin, J. Gillespie, G. Ripka: Nucl. Phys. **A176**, 237 (1971)

3.21   O. Bohigas, S. Stringari: Phys. Lett. **95B**, 9 (1980)

3.22   W. Kutzelnigg, V.H. Smith, Jr.: J. Chem. Phys. **41**, 896 (1964)

3.23   D.H. Kobe: J. Chem. Phys. **50**, 5183 (1969)

3.24   P.-O. Löwdin: Phys. Rev. **97**, 1474 (1955)

3.25   E.N.M. Quint, B.M. Barnett, A.M. van den Berg, J.F.J. van den Brand, H. Clement, R. Ent, B. Frois, D. Goutte, P. Grabmayr, J.W.A. den Herder, E. Jans, G.J. Kramer, J.B.J.M. Lanen, L.Lapikás, H. Nann, G. van der Steenhoven, G.J. Wagner, P.K.A. de Witt Huberts: Phys. Rev. Lett. **58**, 1088 (1987)

3.26   A.A. Abrikosov, L.P. Gorkov, I.E. Dzyaloshinsky: *Methods of Quantum Field Theory in Statistical Physics* (Prentice Hall, Englewood Cliffs 1963)

3.27   N.M. Hugenholtz: Physica (Utrecht) **23**, 481 (1957)

3.28   A.B. Migdal: *Theory of Finite Fermi-Systems and Applications to Atomic Nuclei* (Wiley Interscience, New York 1967)

3.29   J.P. Jeukenne, A. Lejeune, C. Mahaux: Phys. Reports **25C**, 83 (1976); C. Mahaux, R. Sartor: Adv. Nucl. Phys. **20**, 1 (1991)

3.30   H. Orland, R. Schaeffer: Nucl. Phys. **A299**, 442 (1978)

3.31   R. Hasse, P. Schuck: Nucl. Phys. **A438**, 157 (1985)

3.32   R. Hasse, P. Schuck: Nucl. Phys. **A445**, 205 (1985)

3.33   M.F. Flynn, J.W. Clark, R.M. Panoff, O. Bohigas, S. Stringari: Nucl. Phys. **A427**, 253 (1984)

3.34   S. Fantoni, V.R. Pandharipande: Nucl. Phys. **A427**, 473 (1984)

3.35   D. Gogny, R. Padjen: Nucl. Phys. **A293**, 365 (1977)

3.36   J. Dechargé, D. Gogny: Phys. Rev. **C21**, 1568 (1980)

3.37   H.A. Bethe: Ann. Rev. Nucl. Sci. **21**, 93 (1971)

3.38   D.W.L. Sprung: Adv. in Nucl. Phys. **5**, 225 (1972)

3.39   H.S. Köhler: Phys. Rep. **18C**, 217 (1975)

3.40   R.C. Barrett, D.F. Jackson: *Nuclear Sizes and Structure* (Clarendon Press, Oxford 1977)

3.41   K.T.R. Davies, R.J. McCarthy, P.U. Sauer: Phys. Rev. **C7**, 943 (1973)

3.42   J.W. van Orden, W. Truex, M.K. Banerjee: Phys. Rev. **C21**, 2628 (1980)

3.43   J.W. Negele: Phys. Rev. **C1**, 1260 (1970)

3.44   J.W. Negele: *Lecture Notes in Physics* (Ed. B.R. Barrett, Springer-Verlag, Berlin 1975)

3.45   J.W. Negele, D. Vautherin: Phys. Rev. **C11**, 1031 (1975)

3.46   X. Campi, D.W. Sprung: Nucl. Phys. **A194**, 401 (1972)

3.47   J. Dechargè, M. Giròd, D. Gogny: in Proc. of the Confer. on Modern Trends in Elastic Electron Scattering (Amsterdam 1978)

3.48   T.H.R. Skyrme: Phil. Mag. **1**, 1043 (1956); Nucl. Phys. **9**, 615 (1959)

3.49   D. Vautherin, D.M. Brink: Phys. Rev. **C5**, 626 (1972)

3.50   M. Beiner, H. Flocard, N. van Giai, P. Quentin: Nucl. Phys. **A238**, 29 (1975)

3.51   M.V. Stoitsov: Bulg. J. Phys. **19**, No. 1–2 (1992)

3.52   J.G. Zabolitzky, W. Ey: Phys. Lett. **76B**, 527 (1978)

3.53   R. Sartor, C. Mahaux: Phys. Rev. **C21**, 1546 (1980)

3.54   W. Czyż, K. Gottfried: Nucl. Phys. **21**, 676 (1961)

3.55   V.A. Belyakov: ZETP USSR **40**, 1210 (1961); Soviet Physics, JETP, **15**, 850 (1961)

3.56   F. Coester: Nucl. Phys. **7**, 421 (1958)

3.57   F. Coester, H. Kümmel: Nucl. Phys. **17**, 477 (1960)

3.58   H. Kümmel: Nucl. Phys. **A176**, 205 (1971)

3.59   H. Kümmel, K.H. Lührman: Nucl. Phys. **A191**, 525 (1972)

3.60   K.H. Lührman, H. Kümmel: Nucl. Phys. **A194**, 225 (1972)

3.61   J.G. Zabolitzky: Phys. Lett. **47B**, 487 (1973)

3.62   H. Kümmel, J.G. Zabolitzky: Phys. Rev. **C7**, 547 (1973)

3.63   J.G. Zabolitzky: Nucl. Phys. **A228**, 272 (1974)

3.64   J.G. Zabolitzky: Nucl. Phys. **A228**, 285 (1974)

3.65   M. Gari, H. Hyuga, J.G. Zabolitzky; Nucl. Phys. **A271**, 365 (1976)

3.66   K. Emrich, K.H. Lührman, J.G. Zabolitzky: Phys. Rev. **C16**, 1650 (1977)

3.67   H. Kümmel, K.H. Lührman, J.G. Zabolitzky: Phys. Rep. **36**, 1 (1978)

3.68   K.A. Brueckner, R.J. Eden, N.C. Francis: Phys. Rev. **98**, 1445 (1955)

3.69   R. Jastrow: Phys. Rev. **98**, 1479 (1955)

3.70   M.L. Ristig, J.W. Clark: Phys. Rev. **B14**, 2875 (1976)

3.71   S. Fantoni: Nuovo Cimento **A44**, 191 (1978)

3.72   S. Fantoni, S. Rosati: Nuovo Cimento **A25**, 593 (1975)

3.73   M. Dal Ri, S. Stringari, O. Bohigas: Nucl. Phys. **A376**, 81 (1982)

3.74   O. Benhar, C. Ciofi degli Atti, S. Fantoni, S. Rosati: Nucl. Phys. **A328**, 127 (1979)

3.75   R.B. Wiringa, V.R. Pandharipande: Nucl. Phys. **A317**, 1 (1979)

3.76   V.R. Pandharipande, R.B. Wiringa: Rev. Mod. Phys. **51**, 821 (1979)

3.77   R. Schiavilla, V.R. Pandharipande, R.B. Wiringa: Nucl. Phys. **A449**, 219 (1986); S.C. Pieper, R.B. Wiringa, V.R. Pandharipande: Phys. Rev. Lett. **64**, 364 (1990)

3.78   J.W. Clark: in *The Many-Body Problem. Jastrow correlations versus Brueckner theory* (Eds. R. Guardiola and J. Ros, Springer-Verlag, Berlin 1981), p. 184

3.79   O. Benhar, C. Ciofi degli Atti, S. Liuti, G. Salmé: Phys. Lett. **177B**, 135 (1986)

3.80   F. Dellagiacoma, G. Orlandini, M. Traini: Nucl. Phys. **A393**, 95 (1983)

3.81   M. Traini, G. Orlandini: Z. Phys. **A321**, 479 (1985)

3.82   J. Da Providencia, C.M. Shakin: Ann. Phys. **30**, 95 (1964)

3.83   A. Małecki, P. Picchi: Riv. Nuovo Cim. **2**, 119 (1970); Phys. Lett. **36B**, 61 (1971); Lett. Nuovo Cim. **8**, 16 (1973)

3.84   A. de Shalit, I. Talmi: *Nuclear Shell Theory* (Academic Press, New York and London 1963)

3.85   A.G. Sitenko, V.K. Tartakovskiï: *Lectures on the theory of the Nucleus* (Pergamon Press, Oxford 1975)

3.86   J.P. Elliot, A.M. Lane: in *Handbuch der Physik*, vol. **39** (Ed. by S. Flügge, Springer-Verlag, Berlin-Göttingen-Heidelberg 1957), Part 4, p. 241

3.87 E.C. Halbert, J.B. McGrory, B.H. Wildenthal, S.P. Pandya: Adv. Nucl. Phys. **4**, 315 (1971)
3.88 R.R. Whitehead: Nucl. Phys. **A182**, 290 (1972)
3.89 R.R. Whitehead, A. Watt, B.I. Cole, I. Morrison: Adv. Nucl. Phys. **9**, 123 (1977)
3.90 N.N. Bogolyubov: Dokl. AN USSR **119**, 52 (1958)
3.91 S. Belyaev: Kgl. Danske vid. selskab. Mat.-fys. medd. **31**, No. 11 (1959)
3.92 V.G. Solovyev: Zh. Eksp. Theor. Phys. **35**, 823 (1958); **36**, 1869 (1959)
3.93 K.W. Schmid, Z. Ren-Rong, F. Grümmer, A. Faessler: Nucl. Phys. **A499**, 63 (1989)

Chapter 4

4.1   D.L. Hill, J.A. Wheeler: Phys. Rev. **89**, 1102 (1953)
4.2   J.J. Griffin, J.A. Wheeler: Phys. Rev. **108**, 311 (1957)
4.3   W. Bauhoff: Ann. Phys. **130**, 307 (1980)
4.4   J.J. Griffin: Phys. Rev. **108**, 328 (1957)
4.5   P.-G. Reinhard, K. Goeke: Rep. Progr. Phys. **50**, 1 (1987)
4.6   K. Wildermuth, Y.C. Tang: *A Unified Theory of the Nucleus* (Vieweg, Braunschweig 1977)
4.7   P.A.M. Dirac: Proc. Cambridge Phil. Soc. **26**, 376 (1930)
4.8   P. Ring, P. Schuck: *The Nuclear Many-Body Problem* (Springer-Verlag, New York, Heidelberg, Berlin 1980)
4.9   T.H.R. Skyrme: Phyl. Mag. **1**, 1034 (1956); Nucl. Phys. **9**, 615 (1959)
4.10 J. Negele, D.Vautherin: Phys. Rev. **C5**, 1472 (1972)
4.11 J. Bartel, P. Quentin, M. Brack, C. Guet, H.-B. Håkansson: Nucl. Phys. **A386**, 79 (1982)
4.12 M.J. Giannoni, P. Quentin: Phys. Rev. **C21**, 2076 (1980)
4.13 B.I. Bartz, Yu.L. Bolotin, E.V. Inopin, Yu.V. Gonchar: *The Hartree-Fock Method in the Nuclear Theory* (Naukova Dumka, Kiev 1982)
4.14 D. Vautherin, D.M. Brink: Phys. Rev. **C5**, 626 (1972)
4.15 M. Beiner, H. Flocard, P. Quentin: Nucl. Phys. **A238**, 29 (1975)
4.16 H. Krivine, J. Treiner, O. Bohigas: Nucl. Phys. **A336**, 155 (1980)
4.17 H. Flocard, D. Vautherin: Phys. Lett. **55B**, 259 (1975)
4.18 H. Flocard, D. Vautherin: Nucl. Phys. **A264**, 197 (1976)
4.19 D.M. Brink: in Proc. Int. School of Physics "E. Fermi", Course 36, 247 (1966)
4.20 A.N. Antonov, Chr. V. Christov, I.Zh. Petkov: Nuovo Cim. **91A**, 119 (1986)
4.21 A.N. Antonov, I.S. Bonev, Chr.V. Christov, I.Zh. Petkov: Nuovo Cim. **100A**, 779 (1988)
4.22 A.N. Antonov, P.E. Hodgson, I.Zh. Petkov: *Nucleon Momentum and Density Distributions in Nuclei* (Clarendon Press, Oxford 1988): A.N. Antonov, I.Zh. Petkov: Particles and Nuclei **22**, 801 (1991)
4.23 A.N. Antonov, I.S. Bonev, I.Zh. Petkov: Rapid Commun. JINR, No.1 [40]-90, p. 35, Dubna, (1990); Bulg. J. Phys. **18**, 169 (1991)
4.24 A.N. Antonov, I.S. Bonev, Chr. V. Christov, E.N. Nikolov, I.Zh. Petkov: Nuovo Cim. **103A**, 1287 (1990)
4.25 A.N. Antonov, V.A. Nikolaev, I.Zh. Petkov: Bulg. J. Phys. **6**, 151 (1979)
4.26 A.N. Antonov, V.A. Nikolaev, I.Zh. Petkov: Preprint ICTP IC/78/152, Trieste, (1978)
4.27 A.N. Antonov, V.A. Nikolaev, I.Zh. Petkov: Z. Phys. **A297**, 257 (1980)
4.28 A.N. Antonov, V.A. Nikolaev, I.Zh. Petkov: Nuovo Cim. **86A**, 23 (1985)
4.29 N.H. March, W.H. Young, S. Sampanthar: *The Many-Body Problem in Quantum Mechanics* (Cambridge University Press, Cambridge 1967)
4.30 R.P. Feynman: *Statistical Mechanics* (W.A. Benjamin, Inc. Advanced Book Program Reading, Massachusetts 1972)
4.31 A.N. Antonov, I.Zh. Petkov, P.E. Hodgson: Bulg. J. Phys. **13**, 110 (1986)
4.32 A.N. Antonov, I.Zh. Petkov: Bulg. J. Phys. **14**, 137 (1987)
4.33 K.A. Brueckner, J.R. Buchler, R.C. Clark, R.J. Lombard: Phys. Rev. **181**, 1543 (1969)
4.34 G.E. Brown, J.W. Durso, M.B. Johnson: Nucl. Phys. **A397**, 447 (1983)
4.35 A.N. Antonov, L.P. Kaptari, V.A. Nikolaev: Comm. JINR, R4-85-396, Dubna, (1985)

4.36 A.N. Antonov, V.A. Nikolaev, I.Zh. Petkov: Bulg. J. Phys. **10**, 42 (1983)
4.37 P.-O. Löwdin: Phys. Rev. **97**, 1474 (1955)
4.38 D.H. Kobe: J. Chem. Phys. **50**, 5183 (1969)
4.39 F. Malaguti, A. Uguzzoni, E. Verondini, P.E. Hodgson: Riv. Nuovo Cim. **5**, 1 (1982)
4.40 F. Malaguti, A. Uguzzoni, E. Verondini, P.E. Hodgson: Nucl. Phys. **A297**, 287 (1978)

Chapter 5

5.1   L.D. Landau: ZETP **30**, 1058 (1956); **32**, 59 (1957); **35**, 97 (1958)
5.2   A.B. Migdal: *Theory of Finite Fermi-Systems and Applications to Atomic Nuclei* (John Wiley and Sons, Inc., New York 1967)
5.3   G.E. Brown: Rev. Mod. Phys. **43**, 1 (1971)
5.4   J. Mougey, M. Bernheim, A. Bussiere De Nercy, A. Gillebert. P.X. Ho, M. Priou, D. Royer, I. Sick, G.J. Wagner: Nucl. Phys. **A262**, 461 (1976); P.K.A. de Witt Huberts: J. Phys. **G16**, 507 (1990); A.E.L. Dieperink, P.K.A. de Witt Huberts: Annu. Rev. Nucl. Part. Sci. **40**, 239 (1990); G. van der Steenhoven: Nucl. Phys. **A527**, 17c (1991)
5.5   K. Nakamura, S. Hiramatsu, T. Kamae, H. Muramatsu, N. Isutsu, Y. Watase: Nucl. Phys. **A268**, 381 (1976); Nucl. Phys. **A271**, 221 (1976)
5.6   A.N. James, P.T. Andrews, P. Butler, N. Cohen, B.G. Lowe: Nucl. Phys. **A133**, 89 (1969)
5.7   A.N. James, P.T. Andrews, P. Kirkby, B.G. Lowe: Nucl. Phys. **A138**, 5 (1969)
5.8   D.H.E. Gross, R. Lipperheide: Nucl. Phys. **A150**, 449 (1970)
5.9   S. Boffi: "Single-Particle Properties of Finite Nuclei", Lecture in Int. School of Physics "E. Fermi", LXXIX Course, Varenna, (1980)
5.10 S. Frullani, J. Mougey: Adv. Nucl. Phys. **14** (Plenum Press, New York 1984), p. 1
5.11 P.E. Hodgson: Rep. on Progr. in Physics **38**, 847 (1975)
5.12 G.W.R. Dean, P.J. Brussard: Z. Phys. **A323**, 351 (1986)
5.13 C. Ciofi degli Atti: Progress in Particle and Nuclear Physics **3**, 163 (1980)
5.14 M. Riou: Revs. Mod. Phys. **37**, 375 (1965)
5.15 J.P. Jeukenne, A. Lejeune, C. Mahaux: Phys. Rep. **C25**, 83 (1976)
5.16 H.S. Köhler: Nucl. Phys. **88**, 529 (1966)
5.17 H. Orland, R. Schaeffer: Nucl. Phys. **A299**, 442 (1978); A. Ramos, A, Polls, W.H. Dickhoff: in *The Nuclear Equation of State*, Part A (Ed. by W. Greiner, H. Stöcker, Plenum Press, New York 1989)
5.18 R. Sartor: Nucl. Phys. **A267**, 29 (1976); Nucl. Phys. **A289**, 329 (1977)
5.19 T.F. Hammann, Q. Ho Kim: Nuovo Cim. **64B**, 356 (1969); **14A**, 638 (1973)
5.20 R. Sartor, C. Mahaux: Phys. Rev. **C21**, 2613 (1980)
5.21 S. Boffi, F. Capuzzi: Nuovo Cim. Lett. **25**, 209 (1979)
5.22 S. Boffi, F. Capuzzi: Nucl. Phys. **A351**, 219 (1981)
5.23 D. van Neck, M. Waroquier, J. Ryckebusch: Phys. Lett. **249B**, 157 (1990)
5.24 A.N. Antonov, V.A. Nikolaev, I.Zh. Petkov: Z. Phys. **A304**, 239 (1982); A.N. Antonov, V.A. Nikolaev, I.Zh. Petkov: Preprint ICTP IC/78/152, Trieste (1978); Bulg. J. Phys. **6**, 151 (1979); Z. Phys. **A297**, 257 (1980)
5.25 L.P. Kadanoff, G. Baym: *Quantum Statistical Mechanics* (W.A. Benjamin, Inc., New York 1962)
5.26 V.V. Burov, Yu.N. Eldyshev, V.K. Lukyanov, Yu.S. Pol': Preprint JINR, E4-8029, Dubna, (1974)
5.27 C.Ciofi degli Atti, S. Liuti: Phys. Rev. **C41**, 1100 (1990); C. Ciofi degli Atti, S. Simula, L.L. Frankfurt, M.I. Strikman: Phys. Rev. **C44**, R7 (1991)
5.28 C. Ciofi degli Atti, E. Pace, G. Salmè: Preprint INFN-ISS 90/8, Roma (1990)
5.29 A.N. Antonov, L.P. Kaptari, V.A. Nikolaev, A. Yu. Umnikov: Rapid Comm. JINR No. 2 [41]-90, Dubna (1990), p. 14; Nuovo Cim. **104A**, 487 (1991)
5.30 D.S. Koltun: Phys. Rev. Lett. **28**, 182 (1972)
5.31 D.S. Koltun: Phys. Rev. **C9**, 484 (1974)

5.32 V.A. Goldshtein, E.L. Kuplennikov, R.I. Jibuti, R. Ya. Kezerashvili: Nucl. Phys. **A355**, 333 (1981)
5.33 R.L. Becker: Phys. Lett. **32B**, 263 (1970)
5.34 C. Ciofi degli Atti, S. Liuti: Phys. Lett. **225B**, 215 (1989); Phys. Rev. **C44**, R1269 (1991)
5.35 J.W. Negele: Phys. Rev. **C1**, 1260 (1970)
5.36 R. Schiavilla, V.R. Pandharipande, R.B. Wiringa: Nucl. Phys. **A449**, 219 (1986)
5.37 G. Orlandini, M. Traini: Rep. Progr. Phys. **54**, 257 (1991)

Chapter 6

6.1   O. Benhar, A. Fabrocini, S. Fantoni: Phys. Rev. **C41**, R24 (1990)
6.2   H. Euteneuer, J. Friedrich, N. Voegler: Nucl. Phys. **A298**, 452 (1978)
6.3   J.M. Cavedon, B. Frois, D. Goutte, M. Huet, Ph. Leconte, C.N. Papanicolas, X.-H. Phan, S.K. Platchkov, S.E. Williamson, W. Boeglin, I. Sick: Phys. Rev. Lett. **49**, 978 (1982)
6.4   B. Frois, C.N. Papanicolas: Ann. Rev. Nucl. Part. Sci. **37**, 133 (1987)
6.5   B. Frois: Rapport DPh-N/Saclay No. 2432 03/1987
6.6   B. Frois, J.M. Cavedon, D. Goutte, M. Huet, Ph. Leconte, C.N. Papanicolas, X.-H. Phan, S.K. Platchkov, S.E. Williamson, W. Boeglin, I. Sick: Nucl. Phys. **A396**, 409c (1983)
6.7   C.N. Papanicolas, L.S. Cardman, J.H. Heisenberg, O. Schwentker, T.E. Milliman, F.W. Hersman, R.S. Hicks, G.A. Peterson, J.S. McCarthy, J. Wise, B. Frois: Phys. Rev. Lett. **58**, 2296 (1987)
6.8   J. Dechargè, D. Gogny: Phys. Rev. **C21**, 1568 (1980)
6.9   T. Suzuki, M. Oka, H. Hyuga, A. Arima: Phys. Rev. **C26**, 750 (1982)
6.10 T. Suzuki, H. Hyuga: Nucl. Phys. **A402**, 491 (1983)
6.11 V.R. Pandharipande, C.N. Papanicolas, J. Wambach: Phys. Rev. Lett. **53**, 1133 (1984)
6.12 C. Mahaux, H. Ngô: Nucl. Phys. **A410**, 271 (1983)
6.13 E.N.M. Quint: Ph. D. thesis, University of Amsterdam (1988)
6.14 P.K.A. de Witt Huberts: in *Electron-Nucleus Scattering* (Ed. by A. Fabrocini et al., World Scientific, Singapore 1989)
6.15 E.N.M. Quint, J.F.J. van den Brand, J.W.A. den Herder, E. Jans, P.H.M. Keizer, L. Lapikás, G. van der Steenhoven, P.K.A. de Witt Huberts, S. Klein, P. Grabmayr, G.J. Wagner, H. Nann, B. Frois: Phys. Rev. Lett. **57**, 186 (1986)
6.16 E.N.M. Quint, B.M. Barnett, A.M. van den Berg, J.F.J. van den Brand, H. Clement, R. Ent, B. Frois, D. Goutte, P. Grabmayr, J.W.A. den Herder, E. Jans, G.J. Kramer, J.B.J.M. Lanen, L. Lapikás, H. Nann, G. van der Steenhoven, G.J. Wagner, P.K.A. de Witt Huberts: Phys. Rev. Lett. **58**, 1088 (1987)
6.17 L. Lapikás: in Proc. 4th Workshop on Perspectives in Nuclear Physics at Intermediate Energies (Trieste, May 8–12, 1989) (World Scientific, Singapore 1989), p. 419; Inst. Phys. Conf. Ser. No. 105, 223 (1990)
6.18 J.W.A. den Herder, P.C. Dunn, E. Jans, P.H.M. Keizer, L. Lapikás, E.N.M. Quint, P.K.A. de Witt Huberts, H.P. Blok, G. van der Steenhoven, Phys. Lett. **161B**, 65 (1985)
6.19 J.W.A. den Herder: Ph. D. thesis, University of Amsterdam (1987)
6.20 P.K.A. de Witt Huberts: Nucl. Phys. **A446**, 301c (1985); J. Phys. **G16**, 507 (1990); A.E.L. Dieperink, P.K.A. de Witt Huberts: Annu. Rev. Nucl. Part. Sci. **40**, 239 (1990); G. van der Steenhoven: Nucl. Phys. **A527**, 17c (1991)
6.21 J. Lanen: Ph. D. thesis, University of Utrecht, (1990)
6.22 G.J. Kramer: Ph. D. thesis, University of Amsterdam, (1990)
6.23 P. Grabmayr, S. Klein, H. Clement, K. Reiner, W. Reuter, G.J. Wagner, G. Seegert: Phys. Lett. **164B**, 15 (1985)
6.24 H. Clement, P. Grabmayr, H. Röhm, G.J. Wagner: Phys. Lett. **183B**, 127 (1987); P. Grabmayr, G.J. Wagner, H. Clement, H. Röhm: Nucl. Phys. **A494**, 244 (1989)
6.25 M.C. Radhakrishna, N.G. Puttaswamy, H. Nann, J.D. Brown, W.W. Jacobs, W.P. Jones, D.W. Miller, P.P. Singh, E.J. Stephenson: Phys. Rev. **C37**, 66 (1988)

6.26 K. Reiner, P. Grabmayr, G.J. Wagner, S.M. Banks, B.G. Lay, V.C. Officer, G.G. Shute, B.M. Spicer, C.W. Glover, W.P. Jones, D.W. Miller, H. Nann and E.J. Stephenson: Nucl. Phys. **A472**, 1 (1987); M. Seeger, Th. Kihm, K.T. Knöpfle, G. Mairle, U. Schmidt-Rohr, J. Hebenstreit, D. Paul, P. von Rossen: Nucl. Phys. **A533**, 1 (1991)

6.27 P. Doll, G.J. Wagner, H. Breuer, K.T. Knöpfle, G. Mairle, H. Riedesel: Phys. Lett. **82B**, 357 (1979)

6.28 H. Langevin-Joliot, E. Gerlic, J. Gnillot, J. van de Wiele: J. Phys. **G10**, 1435 (1984)

6.29 A. Stuirbrink, G.J. Wagner, K.T. Knöpfle, Liu Ken Pao, G. Mairle, H. Riedesel, K. Schindler: Z. Phys. **A297**, 307 (1980)

6.30 G. Seegert, A. Pfeiffer, P. Grabmayr, T. Kihm, K.T. Knöpfle, G. Mairle, G.J. Wagner, V. Bechtold, L. Friedrich: J. de Phys. **45**, C4-85 (1984)

6.31 A. Pfeiffer, G. Mairle, K.T. Knölpfle, T. Kihm. G. Seegert, P. Grabmayr, G.J. Wagner, V. Bechtold, L. Friedrich: Nucl. Phys. **A455**, 381 (1986)

6.32 H. Langevin-Joliot, E. Gerlic, J. Guillot, J. van de Wiele, S.Y. van der Werf, N. Blasi: Nucl. Phys. **A462**, 221 (1987)

6.33 P. Woldt, P. Grabmayr, G. Rau, G.J. Wagner, M.A. Hofstee, J.M. Schippers S.Y. van der Werf, H. Nann: Nucl. Phys. **A518**, 496 (1990)

6.34 F.J. Eckle, H. Lenske, G. Eckle, G. Graw, R. Hertenberger, H. Kader, F. Merz, H. Nann, P. Schiemenz, H.H. Wolter: Phys. Rev. **C39**, 1662 (1989)

6.35 F.J. Eckle, H. Lenske, G. Eckle, G. Graw, R. Hertenberger, H. Kader, H.J. Maier, F. Merz, H. Nann, P. Schiemenz, H.H. Wolter: Nucl. Phys. **A506**, 159 (1990)

6.36 J.B. French, M.H. Macfarlane: Nucl. Phys. **26**, 168 (1961); C.F. Clement, S.M. Perez: Rep. Progr. Phys. **54**, 127 (1991)

6.37 A.J.C. Burghardt: Ph. D. thesis, University of Amsterdam (1989); C. Mahaux, R. Sartor: Adv. Nucl. Phys. **20**, 1 (1991)

6.38 P.K.A. de Witt Huberts: Proc. Int. Nucl. Physics Conf. Harrogate 1986, Inst, of Physics Conference Series No. 86 (Bristol), **2**, p. 61

6.39 C.H. Johnson, C. Mahaux: Phys. Rev. **C38**, 2589 (1988); H. Lenske, J. Wambach: Phys. Lett. **249B**, 377 (1990)

6.40 M. Matoba, H. Ijiri, H. Ohgaki, S. Uehara, T. Fujiki, Y. Uozumi, H. Kugimiya, N. Koori, I. Kumabe, M. Nakano: Phys. Rev. **C39**, 1658 (1989)

6.41 Z.Y. Ma, J. Wambach: Nucl. Phys. **A402**, 275 (1983); C. Mahaux, R. Sartor: Nucl. Phys. **A484**, 205 (1988)

6.42 J.W.A. den Herder, H.P. Blok, E. Jans, P.H.M. Keizer, L. Lapikás, E.N.M. Quint, G. van der Steenhoven, P.K.A. de Witt Huberts: Nucl. Phys. **A490**, 507 (1988); D. van Neck, M. Waroquier, J. Ryckebusch: Phys. Lett. **249B**, 157 (1990)

6.43 O. Benhar: in Proc. of 3rd Workshop on Perspectives in Nuclear Physics at Intermediate Energies (Trieste, May 18–22, 1987) (World Scientific, Singapore 1988), p. 403

6.44 J. Dechargé, L. Šips, D. Gogny: Phys. Lett. **98B**, 229 (1981); J. Dechargé, L. Šips: Nucl. Phys. **A407**, 1 (1983)

6.45 M. Jaminon, C. Mahaux, H. Ngô: Phys. Lett. **158B**, 103 (1985)

6.46 M. Jaminon, C. Mahaux, H. Ngô: Nucl. Phys. **A440**, 228 (1985)

6.47 M. Jaminon, C. Mahaux, H. Ngô: Nucl. Phys. **A452**, 445 (1986)

6.48 C. Mahaux, P.F. Bortignon, R.A. Broglia, C.H. Dasso: Phys. Rep. **120C**, 1 (1985)

6.49 L. Bennour, P.-H. Heenen, P. Bonche, J. Dobaczewski, H. Flocard: Phys. Rev. **C40**, 2834 (1989)

6.50 L.S. Celenza, A. Harindranath, C.M. Shakin: Preprint B. C. I. N. T. 85/051/139, City Univ. of New York (1985)

6.51 C. Mahaux, R. Sartor: Nucl. Phys. **A493**, 157 (1989); Z.Y. Ma, J. Wambach: Phys. Lett. **256B**, 1 (1991)

6.52 P.-O. Löwdin: Phys. Rev. **97**, 1474 (1955)

6.53 F. Malaguti, A. Uguzzoni, E. Verondini, P.E. Hodgson: Riv. Nuovo Cim. **5**, 1 (1982)

6.54 S. Boffi, F.D. Pacati: Nucl. Phys. **A204**, 485 (1973)

6.55 I.S. Gulkarov: Yad. Fiz. **51**, 97 (1990)

6.56 M. Jaminon, C. Mahaux, H. Ngô: Nucl. Phys. **A473**, 509 (1987)
6.57 J.G. Zabolitzky, W. Ey: Phys. Lett. **76B**, 527 (1978)
6.58 C. Ciofi degli Atti, E. Pace, G. Salmé: Phys. Lett. **141B**, 14 (1984)
6.59 J.W. van Orden, W. Truex, M.K. Banerjee: Phys. Rev. **C21**, 2628 (1980)
6.60 S. Fantoni, V.R. Pandharipande: Nucl. Phys. **A427**, 473 (1984)
6.61 M. Gaudin, J. Gillespie, G. Ripka: Nucl. Phys. **A176**, 237 (1971); M.V. Stoitsov, A.N. Antonov, S.S. Dimitzova: Phys. Rev. **C47**, No. 2 (1993)
6.62 A.N. Antonov, Chr. V. Christov, E.N. Nikolov, I.Zh. Petkov, P.E. Hodgson: Nuovo Cim. **102A**, 1701 (1989)
6.63 V.V. Burov, Yu.N. Eldyshev, V.K. Lukyanov, Yu.S. Pol': Preprint JINR, E4-8029, Dubna (1974)
6.64 I. Sick: Phys. Lett. **53B**, 15 (1974)
6.65 A.N. Antonov, P.E. Hodgson, I.Zh. Petkov: Nuovo Cim. **97A**, 117 (1987)
6.66 O. Benhar, C. Ciofi degli Atti, S. Liuti, G. Salmé: Phys. Lett. **177B**, 135 (1986)
6.67 F. Dellagiacoma, G. Orlandini, M. Traini: Nucl. Phys. **A393**, 95 (1983); M. Traini, G. Orlandini: Z. Phys. **A321**, 479 (1985)
6.68 A.N. Antonov, V.A. Nikolaev, I.Zh. Petkov: Z. Phys. **A297**, 257 (1980)
6.69 A.N. Antonov, Chr. V. Christov, I.Zh. Petkov: Nuovo Cim. **A91**, 119 (1986)
6.70 A.N. Antonov, I.S. Bonev, Chr.V. Christov, I.Zh. Petkov: Nuovo Cim. **A100**, 779 (1988)
6.71 A.N. Antonov, I.S. Bonev, Chr.V. Christov, E.N. Nikolov, I.Zh. Petkov: Nuovo Cim. **A103**, 1287 (1990)
6.72 J. Bartel, P. Quentin, M. Brack, C. Guet, H.-B. Håkansson: Nucl. Phys. **A386**, 79 (1982)
6.73 R. Muthukrishnan, M. Baranger: Phys. Lett. **18**, 160 (1965)
6.74 H. Orland, R. Schaeffer: Nucl. Phys. **A299**, 442 (1978); B. Harradi: Master's thesis, University of Clermont-Ferrand II (1986)
6.75 M. Avan, J. Baldit, M. Bernheim, J. Fargeix, H. Fonvielle, P. Force, A. Gérard, J.L. Guelou, B. Harradi, H. Jackson, A. Magnon, C. Marchand, J. Morgenstern, J. Picard, P. Vernin: Annual Saclay Report 1985–1986, p. 17 (CEN Saclay ISSN 0750-6678)

Chapter 7

7.1   M. Jaminon, C. Mahaux, H. Ngô: Phys. Lett. **158B**, 103 (1985)
7.2   A.N. Antonov, P.E. Hodgson, I.Zh. Petkov: *Nucleon Momentum and Density Distributions in Nuclei* (Clarendon Press, Oxford 1988)
7.3   L.H. Thomas: Proc. Cambridge Phil. Soc. **23**. 542 (1927)
7.4   E. Fermi: Z. Phys. **48**, 73 (1928)
7.5   D.J. Thouless: *The Quantum Mechanics of Many-Body Systems* (2nd edn., Academic Press, New York 1972)
7.6   H.A. Bethe: Phys. Rev. **167**, 879 (1968)
7.7   P.J. Siemens: Phys. Rev. **C1**, 98 (1970)
7.8   K.A. Brueckner, J.R. Buchler, R.C. Clark, R.J. Lombard: Phys. Rev. **181**, 1543 (1969)
7.9   R.J. Lombard: Ann. Phys. **77**, 380 (1973)
7.10 K.A. Brueckner, J.R. Buchler, S. Jorna, R.J. Lombard: Phys. Rev. **171**, 1188 (1968)
7.11 K.A. Brueckner, W.-F. Lin, R.J. Lombard: Phys. Rev. **181**, 1506 (1969)
7.12 K.A. Brueckner, W.-F. Lin, R.J. Lombard, R.C. Clark: Phys. Rev. **C1**, 249 (1970)
7.13 O. Bohigas, X. Campi, H. Krivine, J. Treiner: Phys. Lett. **64B**, 381 (1976)
7.14 C. Guet, M. Brack: Z. Phys. **A297**, 247 (1980)
7.15 M. Brack, C. Guet, H.-B. Håkansson: Phys. Rep. **123**, 275 (1985)
7.16 D.A. Kirzhnitz: *Field Theoretical Methods in Many-Body Systems* (Pergamon Press, Oxford 1967)
7.17 P. Ring, P. Schuck: *The Nuclear Many-Body Problem* (Springer-Verlag, New York, Heidelberg, Berlin 1980)
7.18 J. Bartel, M. Brack, M. Durand: Nucl. Phys. **A445**, 263 (1985)
7.19 P. Hohenberg, W. Kohn: Phys. Rev. **136**, B864 (1964)

7.20   W. Kohn, L.J. Sham: Phys. Rev. **140**, A1133 (1965)
7.21   A.N. Antonov, I.Zh. Petkov: Nuovo Cimento **94A**, 68 (1986)
7.22   J. Hüfner, M.C. Nemes: Phys. Rev. **C23**, 2538 (1981)
7.23   A.N. Antonov, V.A. Nikolaev, I.Zh. Petkov: Bulg. J. Phys. **6**, 151 (1979)
7.24   A.N. Antonov, V.A. Nikolaev, I.Zh. Petkov: Z. Phys. **A297**, 257 (1980)
7.25   K. Gottfried: Ann. Phys. **21**, 29 (1963)
7.26   R.D. Amado: Phys. Rev. **C14**, 1264 (1976)
7.27   R.D. Amado, R.M. Woloshyn: Phys. Lett. **62B**, 253 (1976)
7.28   R.D. Amado, R.M. Woloshyn: Phys. Rev. **C15**, 2200 (1977)
7.29   J.B. McGuire: J. Math. Phys. **5**, 622 (1964)
7.30   F. Calogero, A. Degasperis: Phys. Rev. **A11**, 265 (1975)
7.31   I.M. Narodetsky, Yu.A. Simonov: Phys. Lett. **58B**, 125 (1975)
7.32   S. Frankel, W. Frati, O. van Dyck, R. Werbeck, V. Highland: Phys. Rev. Lett. **36**, 642 (1976)
7.33   R.D. Amado, R.M. Woloshyn: Phys. Rev. Lett. **36**, 1435 (1976)
7.34   J.G. Zabolitzky and W. Ey; Phys. Lett. **76B**, 527 (1978)
7.35   A.N. Antonov, I.S. Bonev, Chr.V. Christov, I.Zh. Petkov: Nuovo Cim. **A100**, 779 (1988)
7.36   A.N. Antonov, I.Zh. Petkov, P.E. Hodgson: Bulg. J. Phys. **13**, 110 (1986)
7.37   R. Muthukrishnan, M. Baranger: Phys. Lett. **18**, 160 (1965)
7.38   C. Ciofi degli Atti, E. Pace, G. Salmé; Preprint INFN-1SS 87/7, Roma (1987); Nucl. Phys. **A508**, 349c (1990)
7.39   F. Dellagiacoma, G. Orlandini, M. Traini:Nucl. Phys **A393**, 95 (1983)
7.40   M. Traini, G. Orlandini: Z. Phys **A321**, 479 (1985)
7.41   O. Benhar, C. Ciofi degli Atti, S. Liuti, G Salmé: Phys. Lett. **177B**, 135 (1986)
7.42   Y. Akaishi: Nucl. Phys. **A416**, 409c (1984)
7.43   O. Bohigas. S. Stringari: Phys. Lett. **95B**, 9 (1980)
7.44   D.B .Day, J.S. McCarthy, Z.E. Meziani, R. Minehart, R. Sealock, S.T. Thornton, J. Jourdan, I. Sick, B.W. Filippone, R.D. McKeown, R.G. Milner, D.H. Potterveld, Z. Szalata: Phys. Rev. Lett. **59**, 427 (1987)
7.45   C. Ciofi degli Atti, E. Pace, G. Salmé: Nucl. Phys. **A497**, 361c (1989)
7.46   C. Ciofi degli Atti, E. Pace, G. Salmé: Preprint INFN-ISS 90/8, Roma (1990)
7.47   R. Schiavilla, V.R. Pandharipande, R.B. Wiringa: Nucl. Phys. **A449**, 219 (1986); R. Schiavilla, V.R. Pandharipande, D.O. Riska: Phys. Rev. **C41**, 309 (1990); R.B. Wiringa: Phys. Rev. **C43**, 1585 (1991); S.C. Pieper, R.B. Wiringa, V.R. Pandharipande: Phys. Rev. Lett. **64**, 364 (1990); Phys. Rev. **C46**, 1741 (1992)
7.48   R.L. Hatch, S.E. Koonin: Phys. Lett. **81B**, 1 (1979)
7.49   M. di Toro: *Competition Between One- and Two-Body Dynamics in Medium Energy*, High-Energy Ion Collisions, Dubna School, September (1986)
7.50   R.Ya. Zulkarneev, R.H. Kutuev. H. Murtazaev: Yad. Fiz. **32**, 889 (1980)
7.51   M. Lifshitz, P. Singer: Phys. Rev. **C22**, 2135 (1980)
7.52   A.N. Antonov, V.A. Nikolaev, I.Zh. Petkov, P.E. Hodgson: Bulg. J. Phys. **10**, 590 (1983)
7.53   I. Sick: Phys. Lett. **53B**, 15 (1974).
7.54   F. Malaguti, A. Uguzzoni, E. Verondini, P.E. Hodgson: Riv. Nuovo Cim. **5**, 1 (1982)
7.55   A.N. Antonov, P.E. Hodgson, I.Zh. Petkov: Nuovo Cim. **97A**, 117 (1987)
7.56   M. Jaminon, C. Mahaux, H. Ngô: Nucl. Phys. **A452**, 445 (1986)
7.57   J. Dechargé, L. Šips: Nucl. Phys. **A407**, 1 (1983)
7.58   V.R. Pandharipande, C.N. Papanicolas, J. Wambach: Phys. Rev. Lett. **53**, 1133 (1984)
7.59   A.N. Antonov, I.Zh. Petkov: Bulg. J. Phys. **14**, 137 (1987)
7.60   B.A. Brown, S.E. Massen, J.I. Escudero, P.E. Hodgson, G. Madurga, J.Viñas: J. Phys. **G9**, 423 (1983)
7.61   M. Jaminon, C. Mahaux, H. Ngô: Nucl. Phys. **A473**, 509 (1987)
7.62   B. Harradi: Master's thesis, University of Clermont-Ferrand II (1986)
7.63   M. Avan, J. Baldit, M. Bernheim, J. Fargeix, H. Fonvielle, P. Force, A. Gerard, J.L. Guelou, B. Harradi, H. Jackson, A. Magnon, C. Marchand, J. Morgenstern, J. Picard, P. Vernin: Annual Saclay Report 1985-1986, p.17 (CEN Saclay, ISSN 0750-6678)
7.64   A. Małecki, A.N. Antonov, I.Zh. Petkov, P.E. Hodgson: Z. Phys. **A328**, 393 (1987)

7.65   A.N. Antonov, J. Kanev, I.Zh. Petkov, M.V. Stoitsov: Nuovo Cim. **101A**, 525 (1989)

7.66   S. Song, M.F. Rivet, R. Bimbot, B. Borderie, I. Forest, J. Galin, D. Gardes, B. Gatty, M. Lefort, H. Oeschler, B. Tamain, X. Tarrago: Phys. Lett. **130B**, 14 (1983); A.D. Panagiotu, M.W. Curtin, P.J. Siemens: Phys. Rev. Lett. **52**, 52 (1984)

7.67   L.D. Landau, E.M. Lifshitz: *Statistical Physics* (Pergamon Press, London-Paris 1959)

7.68   M.V. Stoitsov, I.Zh. Petkov, E.S. Kryachko: Compt. rend. Bulg. acad. Sci. **40**, 45 (1987); M.V. Stoitsov: Nuovo Cim. **98A**, 725 (1987)

7.69   M. Casas, J. Martorell, E. Moya de Guerra: Phys. Lett. **167B**, 263 (1986)

7.70   M. Casas, J. Martorell, E. Moya de Guerra, J. Treiner: Nucl. Phys. **A473**, 429 (1987)

7.71   H. Krivine: Nucl. Phys. **A457**, 125 (1986)

7.72   M.V. Zverev, E.E. Saperstein: Yad. Fiz. **43**, 304 (1986)

7.73   V.A. Khodel, E.E. Saperstein: Phys. Rep. **92**, 183 (1982)

7.74   J.A. Caballero, E. Moya de Guerra: Nucl. Phys **A509**, 117 (1990): E. Moya de Guerra, J.A. Caballero, P. Sarriguren: Nucl. Phys. **A477**, 445 (1988); E. Moya de Guerra, P. Sarriguren, J.A. Caballero, M. Casas, D.W.L. Sprung: Nucl. Phys. **A529**, 68 (1991)

7.75   J. Lanen: Ph.D. thesis, Utrecht (1990)

7.76   C. Alexandrou, J. Myczkowsky, J.W. Negele: Phys. Rev. **C39**, 1076 (1989)

7.77   C. Alexandrou: Preprint Paul Scherrer Institute PR-89-29. Villingen (1989)

7.78   S.E. Massen, C.P. Panos: J. Phys. **G15**, 311 (1989)

7.79   M. Grypeos, K. Ypsilantis: J. Phys. **G15**, 1397 (1989)

7.80   H. Morita, Y. Akaishi, H. Tanaka: Progr. Theor. Phys. **79**, 863 (1988)

7.81   S. Tadokoro, T. Katayama, Y. Akaishi, H. Tanaka: Progr. Theor. Phys. **78**, 732 (1987)

7.82   Y. Akaishi, M. Sakai, J. Hiura, H. Tanaka: Progr. Theor. Phys. Suppl. **56**, 6 (1974)

7.83   J. Lomnitz-Adler, V.R. Pandharipande, R.A. Smith: Nucl. Phys. **A361**, 399 (1981)

7.84   M. Fabre de la Ripelle: Ann. Phys. **147**, 281 (1983); Lecture Notes in Phys. **273**, 283 (1986)

7.85   J.L. Ballot: Phys. Lett. **127B**, 399 (1983); Few Body Systems, Suppl. **1**, 140 (1986)

7.86   Y. Akaishi: Few Body Systems, Suppl. **2**, 64 (1987)

7.87   Y. Akaishi: Lecture Notes in Phys. **273**, 324 (1986)

7.88   H. Morita, Y. Akaishi, O. Endo, H. Tanaka: Progr. Theor. Phys. **78**, 1117 (1987)

7.89   S. Stringari, M. Traini, O. Bohigas: Nucl. Phys. **A516**, 33 (1990)

7.90   M.F. Flynn, J.W. Clark, R.M. Panoff, O. Bohigas, S. Stringari: Nucl. Phys. **A427**, 253 (1984); S. Fantoni, V.R. Pandharipande: Nucl. Phys. **A427**, 473 (1984)

7.91   M. Baldo, I. Bombachi, G. Giansiracusa, U. Lombardo, C. Mahaux, R. Sartor: Phys. Rev. **C41**, 1748 (1990)

7.92   X. Viñas, A. Guirao: Nucl. Phys. **A464**, 326 (1987)

7.93   H.J. Weber, L.D. Miller: Phys. Rev. **C16**, 726 (1977)

7.94   V.V. Burov, V.K. Lukyanov, A.I. Titov: Phys. Lett. **67B**, 46 (1977)

7.95   T. Fujita: Phys. Rev. Lett. **39**, 174 (1977)

7.96   R.D. Amado, R.M. Woloshyn: Phys. Lett. **69B**, 400 (1977)

7.97   R.D. Amado: Phys. Rev. **A16**, 1725 (1977)

7.98   J.V. Noble: Phys. Rev. **C17**, 2151 (1978)

7.99   J.M. Eisenberg, J.V. Noble, H.J. Weber: Phys. Rev. **C19**, 276 (1979)

7.100   S. Boffi, F. Cannata, C. Giusti, F.D. Pacati: Nucl. Phys. **A379**, 509 (1982)

7.101   S. Frankel: Phys. Rev. Lett. **38**, 1338 (1977); Phys. Rev. **C17**, 691 (1978)

7.102   C. Ciofi degli Atti, E. Pace, G. Salmé: Few Body Systems Suppl. **1**, 280 (1986)

7.103   C. Ciofi delgi Atti, E. Pace, G. Salmé: Phys. Rev. **C36**, 1208 (1987)

7.104   C. Ciofi degli Atti, E. Pace, G. Salmé: Phys. Rev. **C39**, 259 (1989)

7.105   C. Ciofi degli Atti: Nucl. Phys. **A463**, 127c (1987)

7.106   G.B. West: Phys. Rep. **18**, 263 (1975)

7.107   D. Day, J.S. McCarthy, I. Sick, R.G. Arnold, B.T. Chertok, S. Rock, Z.M. Szalata, F. Martin, B.A. Mecking, G. Tamas: Phys. Rev. Lett, **43**, 1143 (1979)

7.108   S. Rock, R.G. Arnold, B.T. Chertok, Z.M. Szalata, D. Day, J.S. McCarthy, F. Martin, B.A. Mecking, I. Sick, G. Tamas: Phys. Rev. **C26**, 1592 (1982)

7.109   D. Day, J.S. McCarthy, Z.E. Meziani, R.C. Minehart, R.M. Sealock, S.T. Thornton, J. Jourdan, I. Sick, B.W. Filippone, R.D. McKeown, R.G. Milner, D.H. Potterveld, Z. Szalata: Phys. Rev. **C40**, 1011 (1989)

7.110   C. Ciofi degli Atti, L. Frankfurt, S. Simula, M. Strikman: in Proc. "4th Workshop on Prespectives in Nuclear Physics at Intermediate Energies" (Trieste, May 8–12, 1989) (Ed. S. Boffi, C. Ciofi degli Atti and M. Giannini, World Scientific, Singapore 1989), p. 312

7.111   C. Ciofi degli Atti, S. Liuti, S. Simula: Phys. Rev. **C41**, R2474 (1990)

7.112   R.V. Reid, Jr.: Ann. Phys. **50**, 411 (1968)

7.113   C. Ciofi degli Atti, E. Pace, G. Salmé: Phys. Lett. **141B**, 14 (1984)

7.114   S. Turck-Chieze, P. Barreau, M. Bernheim, P. Bradu, Z.E. Meziani, J. Morgenstern, A. Bussière, G.P. Capitani, E. de Sanctis, S. Frullani, F. Garibaldi, J. Mougey: Phys. Lett. **142B**, 145 (1984)

7.115   E. Jans, P. Barreau, M. Bernheim, J.M. Finn, J. Morgenstern, J. Mougey, D. Tarnowski, S. Turck-Chieze, S. Frullani, F. Garibaldi, G.P. Capitani, E. de Sanctis, M.K. Brussel, I. Sick: Phys. Rev. Lett. **49**, 974 (1982)

7.116   E. Jans, M. Bernheim, M.K. Brussel, G.P. Capitani, E. de Sanctis, S. Frullani, F. Garibaldi, J. Morgenstern, J. Mougey, I. Sick, S. Turck-Chieze: Nucl. Phys. **A475**, 687 (1987)

7.117   C. Marchand, M. Bernheim, P.C. Dunn, A. Gérard, J.M. Laget, A. Magnon, J. Morgenstern, J. Mougey, J. Picard, D. Reffay-Pikeroen, S. Turck-Chieze, P. Vernin, M.K. Brussel, G.P. Capitani, E. de Sanctis, S. Frullani, F. Garibaldi: Phys. Rev. Lett. **60**, 1703 (1988)

7.118   J.F.J. van den Brand, H.P. Blok, R. Ent, E. Jans, G.J. Kramer, J.B.J.M. Lanen, L. Lapikás, E.N.M. Quint, G. van der Steehoven, P.K.A. de Witt Huberts: Phys. Rev. Lett. **60**, 2006 (1988)

7.119   J.M. Legoff, A. Magnon, M. Bernheim, M.K. Brussel, G.P. Capitani, E. de Sanctis, S. Frullani, A. Gérard, H.E. Jackson, C. Marchand, Z.E. Meziani, J. Morgenstern, J. Picard, D. Reffay, S. Turck-Chieze, P. Vernin, A. Zghiche; in Proc. "4th Workshop on Perspectives in Nuclear Physics at Intermediate Energies" (Trieste, May 8–12, 1989) (Ed. S. Boffi, C. Ciofi delgi Atti and M. Giannini, World Scientific, Singapore 1989), p. 376

7.120   A. Magnon, M. Bernheim, M.K. Brussel, G.P. Capitani, E. de Sanctis, S. Frullani, F. Garibaldi, A. Gerard, H.E. Jackson, J.M. Legoff, C. Marchand, Z.E. Meziani, J. Morgenstern, J. Picard, D. Reffay, S. Turck-Chieze, P. Vernin, A. Zghiche: Phys. Lett. **222B**, 352 (1989)

7.121   S. Frullani, J. Mougey: Adv. Nucl. Phys. **14**, 1 (1984)

7.122   X. Ji, R.D. McKeown: Phys. Lett. **236B**, 130 (1990)

7.123   V.A. Goldstein, Eh.L. Kuplennikov, R.I. Jibuti, R. Ya Kezerashvili: Nucl. Phys. **A355**, 333 (1981)

7.124   A. Bodek, J.L. Ritchie: Phys. Rev. **D23**, 1070 (1981)

7.125   M. Bernheim, A. Bussiere, J. Mougey, D. Royer, D. Tarnowski, S. Turck-Chieze, S. Frullani, G.P. Capitani, E. de Sanctis, E. Jans: Nucl. Phys. **A365**, 349 (1981)

7.126   S.I. Nagornyi, Yu.A. Kasatkin, I.K. Kirichenko, E.V. Inopin: Yad. Fiz. **42**, 870 (1985)

7.127   E. Jans: Nucl. Phys. **A508**, 433c (1990); P.K.A. de Witt Huberts: J. Phys. **G16**, 507 (1990); A.E.L. Dieperink, P.K.A. de Witt Huberts: Annu. Rep. Nucl. Part. Sci. **40**, 239 (1990); G. van der Steenhoven: Nucl. Phys. **A527**, 17c (1991)

7.128   S. Frankel, W. Frati, G. Blanpied, G.W. Hoffmann, T. Kozlowski, C. Morris, H.A. Thiessen, O. van Dyck, R. Ridge, C. Whitten: Phys. Rev. **C18**, 1375 (1978)

7.129   S. Frankel, W. Frati, R.M. Woloshyn, D. Yang: Phys. Rev. **C18**, 1379 (1978)

7.130   V.I. Komarov, G.E. Kosarev, H. Müller, D. Netzband, T. Stiehler: Phys. Lett. **69B**, 37 (1977)

7.131   K.R. Cordell, S.T. Thornton, L.C. Dennis, R.R. Doering, R.L. Parks, T.C. Schweizer: Nucl. Phys. **A352**, 485 (1981)

7.132   G. Roy, L.G. Greeniaus, G.A. Moss, D.A. Hutcheon, R.L. Liljestrand, R.M. Woloshyn, D.H. Boal, A.W. Stetz, K. Aniol, A. Willis, N. Willis, R. Camis: Phys. Rev. **C23**, 1671 (1981)

7.133   V.I. Komarov, G.E. Kosarev, H. Müller, D. Netzband, T. Stiehler, S. Tesch: Comm. JINR, E1-11354, Dubna (1978); E1-12393, Dubna (1979)

264   References

7.134 M. Avan, A. Baldit, J. Castor, G. Chaigne, A. Devaux, J. Fargeix, P. Force, G. Landaud,
       G. Roche, J. Vicente, J.P. Didelez, F. Reide, S.A. Gurvitz: Phys. Rev. **C30**, 521 (1984)
7.135 M. Avan, A. Baldit, J. Castor, M. El Zoubidi, J. Fargeix, H. Fonvieille, P. Force, J.L. Guelou,
       B. Harradi, G. Landaud, J.P. Didelez, F. Reide, M. Bernheim, A. Gérard, A. Magnon, C.
       Marchand, J. Morgenstern, J. Picard, P. Vernin, H. Jackson: Phys. Rev. **C37**, 231 (1988)
7.136 K.R. Cordell, S.T. Thornton, L.C. Dennis, R.R. Doering, R.L. Parks, T.C. Schweizer: Nucl.
       Phys. **A362**, 431 (1981)
7.137 L.S. Azhgirey, I.K. Vzorov, V.N. Zhmyrov, V.V. Ivanov, M.A. Ignatenko, A.S. Kuznetzov,
       M.G. Mescheryakov, S.V. Razin, G.D. Stoletov: JINR, I-10842, Dubna (1977)
7.138 D.H. Boal, R.M. Woloshyn: Phys. Rev. **C20**, 1878 (1979)
7.139 R.H. Landau: Phys. Rev. **C17**, 2144 (1978)
7.140 A.F. Grashin, Ya.Ya. Shalamov: Yad. Fiz. **29**, 625 (1979)
7.141 T. Fujita, J. Hüfner: Nucl. Phys. **A343**, 493 (1980)
7.142 G.F. Bertsch, Phys. Rev. Lett. **46**, 472 (1981)
7.143 B. Hiller, J. Hüfner: Nucl. Phys. **A382**, 542 (1982)
7.144 H. Araseki, T. Fujita: Nucl. Phys. **A439**, 681 (1985)
7.145 D.H. Boal: Phys. Rev. **C21**, 1913 (1980)
7.146 M. Lifshitz, P. Singer: Phys. Rev. **41**, 18 (1978)
7.147 T. Fujita: Nucl. Phys. **A457**, 657 (1986)
7.148 R.C. Barrett, D.F. Jackson: *Nuclear Sizes and Structure* (Clarendon Press, Oxford 1977)
7.149 S. Boffi, F. Cappuzzi: Nucl. Phys. **A351**, 219 (1981)
7.150 S. Boffi, C. Giusti, F.D. Pacati: Nucl. Phys. **A359**, 81 (1981)
7.151 A.N. Gorbunov: Phys. Lett. **27B**, 436 (1968)
7.152 S.E. Kiergan, A.O. Hansen, L.J. Koester, Jr.: Phys. Rev. **C8**, 431 (1973)
7.153 D.J.S. Findlay, R.O. Owens: Nucl. Phys. **A292**, 53 (1977)
7.154 J.L. Matthews, D.J.S. Findlay, S.N. Gardiner, R.O. Owens: Nucl. Phys. **A267**, 51 (1976)
7.155 M.J. Leitch: Ph.D. Thesis, MIT (1979)
7.156 D.J.S. Findlay, R.O. Owens, M.J. Leitch, J.L. Matthews, C.A. Peridier, B.L. Roberts, C.P.
       Sargent: Phys. Lett. **74B**, 305 (1978)
7.157 M.J. Leitch, J.L. Matthews, W.W. Sapp, C.P. Sargent, S.A. Wood, D.J.S. Findlay, R.O.
       Owens, B.L. Roberts: Phys. Rev. **C31**, 1633 (1985)
7.158 M.R. Sené, I. Anthony, D. Brandford, A.G. Flowers, A.C. Shotter, C.H. Zimmerman, J.C.
       McGeorge, R.O. Owens, P.J. Thorley: Nucl. Phys. **A442**, 215 (1985)
7.159 M.J. Leitch, F.C. Lin, J.L. Matthews, W.W. Sapp, C.P. Sargent, D.J.S. Findlay, R.O. Owens,
       B.L. Roberts: Phys. Rev. **C33**, 1511 (1986)
7.160 J.L. Matthews, W. Bertozzi, M.J. Leitch, C.A. Peridier, B.L. Roberts, C.P. Sargent, W.
       Turchinetz, D.J.S. Findlay, R.O. Owens: Phys. Rev. Lett. **38**, 8 (1977)
7.161 D.J.S. Findlay, R.O. Owens: Nucl. Phys. **A279**, 385 (1977)
7.162 W. Weise, M.G. Huber: Nucl. Phys. **A162**, 330 (1971)
7.163 H. Hebach, A. Wortberg, M. Gari: Nucl. Phys. **A267**, 425 (1976)
7.164 J.S. Levinger: Phys. Rev. **84**, 43 (1951)
7.165 B. Schoch: Phys. Rev. Lett. **41**, 809 (1978)
7.166 H.Göringer, B. Schoch, G. Luhrs: Nucl. Phys. **A384**, 414 (1982)
7.167 H. Göringer, B. Schoch: Phys. Lett. **97B**, 41 (1980)
7.168 H. Schier, B. Schoch: Nucl. Phys. **A229**, 93 (1974)
7.169 J.-S. Tsai, W.V. Prestwich, T.J. Kennet: Z. Phys. **A322**, 597 (1985)
7.170 S.N. Belyaev, O.V. Vasiliev, A.B. Kozin, A.A. Nechkin, V.A. Semyonov: Izv. AN USSR, ser.
       fiz. **48**, 1940 (1984)
7.171 H.G. Miller, W. Buss, J.A. Rawlins: Nucl. Phys. **A163**, 637 (1971)
7.172 H. Schier, B. Schoch: Lett. Nuovo Cim. **12**, 334 (1975)
7.173 S.R. Cotanch: Phys. Lett. **76B**, 19 (1978)
7.174 S. Boffi, C. Giusti, F.D. Pacati: Nucl. Phys. **A386**, 599 (1982)
7.175 J. Mougey, M. Bernheim, A. Bussiere de Nercy, A. Gillebert, P.X. Ho, M. Priou, D. Royer,
       I. Sick, G.J. Wagner: Nucl. Phys, **A262**, 461 (1976)

7.176   M. Bernheim, A. Bussiere, J. Mougey, D. Royer, D. Tarnowski, S. Turck-Chieze, S. Frullani, S. Boffi, C. Giusti, F.D. Pacati, G.P. Capitani, E. de Sanctis, G.J. Wagner: Nucl. Phys. A375, 381 (1982)

7.177   E.N.M. Quint: Ph.D. thesis, University of Amsterdam (1988)

7.178   L. Lapikás: in Proc. "4th Workshop on Perspectives in Nuclear Physics at Intermediate Energies" (Trieste, May 8–12, 1989) (Ed. S. Boffi, C. Ciofi degli Atti and M. Giannini, World Scientific, Singapore 1989), p. 419; Inst. Phys. Conf. ser. No. 105, 223 (1990)

7.179   J.W. Lightbody Jr.: in Proc. "3rd Workshop on Perspectives in Nuclear Physics at Intermediate Energies" (Trieste, May 18–22, 1987) (Ed. S. Boffi, C. Ciofi degli Atti and M. Giannini, World Scientific, Singapore 1988), p. 455

7.180   C. Ciofi degli Atti: in Proc. "3rd Workshop on Perspectives in Nuclear Physics at Intermediate Energies" (Trieste, May 18–22, 1987) (Ed. S. Boffi, C. Ciofi degli Atti and M. Giannini, World Scientific, Singapore 1988), p.367

7.181   J.Y. Mougey: in Proc. "4th Workshop on Perspectives in Nuclear Physics at Intermediate Energies" (Trieste, May 8–12, 1989) (Ed. S. Boffi, C. Ciofi degli Atti and M. Giannini, World Scientific, Singapore 1989), p. 237

7.182   Y. Haneishi, T. Fujita: Phys. Rev. C33, 260 (1986)

7.183   Y. Haneishi, T. Fujita: Phys. Rev. C35, 70 (1987)

7.184   Y. Miake, H. Hamagaki, S. Kadota, S. Nagamiya, S. Schnetzer, Y. Shida, H. Steiner, I. Tanihata: Phys. Rev. C31, 2168 (1985)

7.185   A.N. Antonov, I.S. Bonev, Chr.V. Christov, I.Zh. Petkov: Nuovo Cim. A101, 639 (1989)

7.186   A.N.Antonov, I.S. Bonev, I.Zh. Petkov: Rapid Comm. JINR, No.1 [40]-90, Dubna (1990), p. 35; Bulg. J. Phys. 18, 169 (1991)

7.187   P.-O. Löwdin: Phys. Rev. 97, 1474 (1955); Phys. Rev. 97, 1490 (1955); D.L. Hill, J.A. Wheeler: Phys. Rev. 89, 1102 (1953); J.J. Griffin, J.A. Wheeler: Phys. Rev. 108, 311 (1957); C.W. Wong: Phys. Rep. 15, 283 (1975)

7.188   J. Bartel, P. Quentin, M. Brack, C. Guet, H.-B. Håkansson: Nucl. Phys. A386, 79 (1982)

7.189   P.E. Hodgson: in Proc. Fifth Int. Conf. Clustering Aspects in Nuclear and Subnuclear Systems, Kyoto, (1988), J. Phys. Soc. Jpn. 58, 755 (1989); D. Brink, J.J. Castro: Nucl. Phys. A216, 109 (1973); A.N. James, H.G. Pugh: Nucl. Phys. 42, 441 (1963)

7.190   J.W. Watson, H.G. Pugh, P.G. Roos, D.A. Goldberg, R.A. Riddle, D.I. Bonbright: Nucl. Phys. A172, 513 (1971); M. Jain, P.G. Roos, H.G. Pugh and H.D. Holmgren: Nucl. Phys. A153, 49 (1970)

7.191   N. Chirapatpimol, J.C. Fong, M.M. Gazzaly, G. Igo, A.D. Liberman, R.J. Ridge, S.L. Verbeck, C.A. Whitten Jr., D.G. Kovar, V. Perez-Mendez, N.S. Chant, P.G. Roos: Nucl. Phys. A264, 379 (1976)

7.192   W.E. Dollhopf, C.F. Perdrisat, P. Kitching, W.C. Olsen: Nucl. Phys. A316, 350 (1979)

7.193   T.A. Garey, P.G. Ross, N.S. Chant, A. Nadasen, H.L. Chen: Phys. Rev. C23, 576 (1981)

7.194   C.W. Wang, N.S. Chant, P.G. Roos, A. Nadasen, T.A. Garey: Phys. Rev. C21, 1705 (1980)

7.195   N.S. Chant, P.G. Roos: Phys. Rev. C15, 57 (1977)

7.196   N.S. Wall, J.R. Wu, C.C. Chang, H.D. Holmgren: Phys. Rev. C20, 1079 (1979)

7.197   R. Bonetti, F. Crespi, K.-I. Kubo: Nucl. Phys. A499, 381 (1989)

7.198   R. Bonetti, R. Colombo, K.-I. Kubo: Nucl. Phys. A420, 109 (1984)

7.199   J.P. Genin, J. Julien, M. Rambaut, C. Samour, A. Palmeri and Vinciguerra: Phys. Lett. 52B, 46 (1974)

7.200   Th. Kihm, K.T. Knöpfle, H.Riedesel, P. Voruganti, H.J. Emrich, G. Fricke, R. Neuhausen, R.K.M. Schneider: Phys. Rev. Lett. 56, 2789 (1986)

7.201   V.F. Dmitriev, D.M. Nikolenko, S.G. Popov, I.A. Rachek, D.K. Toporkov, E.P. Tsentalovich, B.B. Woitsekhowski, V.G. Zelevinsky: Nucl. Phys. A464, 237 (1987)

7.202   I.G. Evseev, V.P. Lichachev, V.L. Agranovich, Yu.V. Vladimirov, S.A. Paschuk, G.A. Savitzkyi, V.B. Shostak: Ukr. Fiz. Zh. 35, 839 (1990)

7.203   H.P. Blok: In "Proc. 5th Int. Conf. Clustering Aspects in Nuclear and Subnuclear Systems", Kyoto (1989), J. Phys. Soc. Jpn. 58, 409 Suppl. (1989)

7.204   J.J. Murphy, II, D.N. Skopik, J. Asai, J. Uegaki: Phys. Rev. C18, 736 (1978)

7.205   A.G. Flowers, A.C. Shotter, D. Brandford, J.C. McGeorge, R.O. Owens; Phys. Rev. Lett. **40**, 709 (1978)
7.206   T.A. Griffi, R.J. Oakes, H.M. Schwartz: Nucl. Phys. **86**, 313 (1966)
7.207   A.N. Antonov, E.N. Nikolov, I.Zh. Petkov, P.E. Hodgson, G.A. Lalazissis: Preprint OUNP-91-32, Oxford (1991); Bulg. J. Phys. **19**, No 1–2 (1992)
7.208   P. Beregi, N.S. Zelenskaya, V.N. Neudatchin, Yu.F. Smirnov: Nucl. Phys. **66**, 513 (1965)
7.209   K. Wildermuth, Y.C. Tang: *An Unified Theory of the Nucleus* (Academic Press, New York 1977)
7.210   V.V. Balashov, A.N. Boyarkina, I. Rotter: Nucl. Phys. **59**, 417 (1964)
7.211   J.M. Laget: Phys. Lett. **151B**, 325 (1985)
7.212   V.G. Ableev, S.V. Dshemuchadse, Ch. Dimitrov, A.P. Kobushkin, B. Naumann, L. Naumann, A.A. Nomofilov, L. Penchev, N.M. Piskunov, V.I. Sharov, I.M. Sitnik, E.A. Strokovsky, L.N. Strunov, S. Tesch, S.A. Zaporozhets: Few-Body Syst. **8**, 137 (1990)
7.213   J.F.J. van den Brand, H.P. Blok, R. Ent, E. Jans, J.M. Laget, L. Lapikás, C. de Vries, P.K.A. de Witt Huberts: Nucl. Phys. **A534**, 637 (1991)
7.214   H.H. Gan, S.J. Lee, S.Das Gupta, J. Barrette: Phys. Lett. **234B**, 4 (1990)

Chapter 8

8.1   A.N. Antonov, V.A. Nikolaev, I.Zh. Petkov: Bulg. J. Phys. **6**, 151 (1979)
8.2   A.N. Antonov, V.A. Nikolaev, I.Zh. Petkov: Nuovo Cim. **86A**, 23 (1985)
8.3   A.N. Antonov, P.E. Hodgson, I.Zh. Petkov: *Nucleon Momentum and Density Distributions in Nuclei* (Clarendon Press, Oxford 1988)
8.4   K.A. Brueckner, J.R. Buchler, S. Jorna, R.J. Lombard: Phys. Rev. **171**, 1188 (1968)
8.5   K.A. Brueckner, J.R. Buchler, R.C. Clark, R.J. Lombard: Phys. Rev. **181**, 1543 (1969)
8.6   P.M. Morse: Phys. Rev. **34**, 57 (1929)
8.7   V.A. Kravtsov: *Masses of Atoms and Nuclear Binding Energies* (Atomizdat, Moscow 1974)
8.8   G.E. Brown, J.W. Durso, M.B. Johnson: Nucl. Phys. **A397**, 447 (1983)
8.9   S. Flügge: *Practical Quantum Mechanics*, vol. I, (Springer-Verlag, Berlin-Heidelberg-New York 1971)
8.10   L.D. Landau, E.M. Lifshitz: *Quantum Mechanics* (3rd edn, Pergamon, Oxford 1977)
8.11   N.L. Balazs, H.C. Pauli: Z. Phys. **A281**, 395 (1977)
8.12   D.M. Brink, J.J. Castro: Nucl. Phys. **A216**, 109 (1973)
8.13   D.H. Wilkinson: in Proc. 3rd Int. Conf. on Nuclei far from Stability, Cargese, Corsica (France), Geneva (1976), p. 71
8.14   F.W. Schlepütz, J.C. Comiso, T.C. Meyer, K.O.H. Ziock: Phys. Rev. **C19**, 135 (1979)
8.15   A.I. Yavin: Nucl. Phys. **A374**, 297c (1982)
8.16   Yu.A. Chestnov, B.L. Gorshkov, A.I. Iljin, B.Yu. Sokolovsky, G.E. Solyakin: in Abstr. of the Contrib. Papers in 9 ICOHEPANS, Versailles, France (1981), p. 572
8.17   B.L. Gorshkov, A.I. Iljin, B.Yu. Sokolovsky, G.E. Solyakin, Yu.A. Chestnov: Pis'ma ZETP **37**, 60 (1983)
8.18   Yu.A. Chestnov, B.L. Gorshkov, A.I. Iljin, A.V. Kravtsov, A.M. Nikitin, B.Yu. Sokolovsky, G.E. Solyakin: Preprint LNPI, No. 941, Leningrad (1984)
8.19   B.D. Wilkins, S.B. Kaufman, E.P. Steinberg, J.A. Urban, D.J. Henderson: Phys. Rev. Lett. **43**, 1080 (1979)
8.20   S. Pandian, N.T. Porile: Phys. Rev. **C23**, 427 (1981)
8.21   H. Sauvageon, S. Regnier, G.N. Simonoff: Phys. Rev. **C25**, 466 (1982)
8.22   D.H. Youngblood, P. Bogucki, J.D. Bronson, U. Garg, Y.-W. Lui, C.M. Rosza: Phys. Rev. **C23**, 1997 (1981)
8.23   J. Speth, A. van der Woude: Rep. Progr. Phys. **44**, 719 (1981)
8.24   N.A. Voinova-Eliseeva, I.A. Mitropolsky: Preprint LINP, No. 1104, Leningrad (1985)
8.25   Y.-W. Lui, J.D. Bronson, D.H. Youngblood, Y. Toba, U. Garg: Phys. Rev. **C31**, 1643 (1985)

8.26 H.Y. Lu, S. Brandenburg, R. de Leo, M.N. Harakeh, T.D. Poelhekken, A. van der Woude: Phys. Rev. **C33**, 1116 (1986)

8.27 S. Brandenburg, W.T.A. Borghols, A.G. Drentje, L.P. Ekström, M.N. Harakeh, A. van der Woude, A. Håkansson, L. Nilsson, N. Olsson, M. Pignanelli, R. de Leo: Nucl. Phys. **A466**, 29 (1987)

8.28 M.M. Sharma, W.T.A. Borghols, S. Brandenburg, S. Crona, A. van der Woude, M.N. Harakeh: Phys. Rev. **C38**, 2562 (1988)

8.29 W.T.A. Borghols, S. Brandenburg, J.H. Meier, J.M. Schippers, M.M. Sharma, A. van der Woude, M.N. Harakeh, A. Lindholm, L. Nilsson, S. Orona, A. Håkansson, L.P. Ekström, N. Olsson, R. de Leo: Nucl. Phys. **A504**, 231 (1989)

8.30 K.A. Brueckner, M.J. Giannoni, R.J. Lombard: Phys. Lett. **31B**, 97 (1970)

8.31 V.R. Pandharipande: Phys. Lett. **31B**, 635 (1970)

8.32 J. Treiner, H. Krivine, O. Bohigas, J. Martorell: Nucl. Phys. **A371**, 253 (1981)

8.33 N.A. Voinova-Eliseeva, I.A. Mitropolsky: Preprint LINP, No. 1095, Leningrad (1985)

8.34 M.M. Sharma, W. Stocker, P. Gleissl, M. Brack: Nucl. Phys. **A504**, 337 (1989)

8.35 A.N. Antonov, V.A. Nikolaev, I.Zh. Petkov: Bulg. J. Phys. **18**, 107 (1991)

8.36 K.A. Bruekkner, J.L. Gammel: Phys. Rev. **109**, 1023 (1958)

8.37 D.M. Brink, G.F. Nash: Nucl. Phys. **40**, 608 (1963)

8.38 J. Cerny, R.H. Pehl, G.T. Garvey: Phys. Lett. **12**, 234 (1964)

8.39 D.L. Hill, J.A. Wheeler: Phys. Rev. **89**, 1102 (1953); J.J. Griffin, J.A. Wheeler: Phys. Rev. **108**, 311 (1957)

8.40 A.N. Antonov, I.S. Bonev, Chr.V. Christov, I.Zh. Petkov: Nuovo Cim. **100A**, 779 (1988)

8.41 J. Bartel, P. Quentin, M. Brack, C. Guet, H.-B. Håkansson: Nucl. Phys. **A386**, 79 (1982)

8.42 H. Flocard, D. Vautherin: Phys. Lett. **55B**, 259 (1975); Nucl. Phys. **A264**, 197 (1976)

8.43 I. Sick: Phys. Lett. **53B**, 15 (1974)

Chapter 9

9.1   V.K. Lukyanov, Yu.S. Pol': Particles and Nuclei **5**, 955 (1974)

9.2   I.Zh. Petkov, V.K. Lukyanov, Yu.S. Pol': Yad. Fiz. **4**, 57 (1966)

9.3   L.J. Schiff: Phys. Rev. **103**, 443 (1956)

9.4   D. Saxon, L.J. Schiff: Nuovo Cim. **6**, 614 (1957)

9.5   A. Baker: Phys. Rev. **134**, B240 (1964)

9.6   D.R. Yennie, F.L. Boos, D.C. Ravenhall: Phys. Rev. **137**, B882 (1965)

9.7   V.K. Lukyanov, I.Zh. Petkov, Yu.S. Pol': Yad. Fiz. **9**, 349 (1969)

9.8   A.N. Antonov, V.A. Nikolaev, I.Zh. Petkov: Preprint ICTP IC/78/152, Trieste (1978); Bulg. J. Phys. **6**, 151 (1979); Z.Phys. **A297**, 257 (1980)

9.9   A.N. Antonov, P.E. Hodgson, I.Zh. Petkov: *Nucleon Momentum and Density Distributions in Nuclei* (Clarendon Press, Oxford 1988)

9.10 A.N. Antonov, V.A. Nikolaev, I.Zh. Petkov: Izv. AN USSR, ser. fiz. **47**, 134 (1983)

9.11 A.N. Antonov, V.A. Nikolaev, I.Zh. Petkov: Compt. rend. Acad. bulg. Sci. **34**, 19 (1981)

9.12 A.N. Antonov, V.P. Garistov, I.Zh. Petkov: Phys. Lett. **68B**, 305 (1977)

9.13 R.F. Frosch, R. Hofstadter, J.S. McCarthy, G.K. Nödleke, K.J. van Oostrum, M.R. Yearian, B.C. Clark, R. Herman, D.G. Ravenhall: Phys. Rev. **174**, 1380 (1968)

9.14 J.B. Bellicard, P. Bounin, R.F. Frosch, R.Hofstadter, J.S. McCarthy, F.J. Uhrhane, M.R. Yearian, B.C. Clark, R. Herman, D.G. Ravenhall: Phys. Rev. Lett. **19**, 527 (1967)

9.15 A.B. Migdal: Pis'ma JETP **19**, 539 (1974)

9.16 E.J. Moniz: Phys. Rev. **184**, 1154 (1969)

9.17 E.J. Moniz, I. Sick, R.R. Whitney, J.R. Ficenec, R.D. Kephart, W.P. Trower: Phys. Rev. Lett. **26**, 445 (1971)

9.18 R.R. Whitney, I. Sick, J.R. Ficenec, R.D. Kephart, W.P. Trower: Phys. Rev. **C9**, 2230 (1974)

9.19 F.A. Brieva, A. Dellafiore: Nucl. Phys. **A292**, 445 (1977)
9.20 F.H. Heimlich, M. Köbberling, J. Moritz, K.H. Schmidt, D. Wegener, D. Zeller, J.K. Bienlein, J. Bleckwenn, H. Dinter: Nucl. Phys. **A231**, 509 (1974)
9.21 W. Czyż and K. Gottfried: Ann. Phys. **21**, 47 (1963)
9.22 S.V. Dementji, N.G. Afanasiev, I.M. Arkatov, V.G. Vlasenko, V.A. Goldstein, E.L. Kuplennikov: Yad. Fiz. **11**, 19 (1970)
9.23 W. Fabian, H. Arenhövel: Nucl. Phys. **A314**, 253 (1979)
9.24 D.R. Lehman: Phys. Rev. **C3**, 1827 (1971)
9.25 A.E.L. Dieperink, T.De Forest, I. Sick, R.A. Brandenburg: Phys. Lett. **63B**, 261 (1976)
9.26 W.M. Alberico, R. Cenni, A. Molinari: Riv. Nuovo Cim. **4**, 1 (1978)
9.27 T. De Forest: Nucl. Phys. **A132**, 305 (1969)
9.28 T.W. Donnelly: Nucl. Phys. **A150**, 393 (1970)
9.29 S. Klawansky, H.W. Kendall, A.K. Kerman: Phys. Rev. **C7**, 795 (1973)
9.30 Y. Kawazoe, G. Takeda, H. Matsuzaki: Progr. Theor. Phys. **54**, 1394 (1975)
9.31 W.C. Haxton: Nucl. Phys. **A306**, 429 (1978)
9.32 R. Rosenfelder: Ann. Phys. **128**, 188 (1980)
9.33 J.D. Walecka: Ann. Phys. **83**, 491 (1974)
9.34 F.E. Serr, J.D. Walecka: Phys. Lett. **79B**, 10 (1978)
9.35 S.D. Drell, J.D. Walecka: Ann. Phys. **28**, 18 (1964)
9.36 T. De Forest, Jr., J.D. Walecka: Adv. Phys. **15**, 1 (1966)
9.37 A.G. Sitenko, V.N. Gur'ev: ZETP **39**, 1260 (1960)
9.38 W. Czyż: Phys. Rev. **131**, 2141 (1963)
9.39 C. Marchand, P. Barreau, M. Bernheim, P. Bradu, G. Fournier, Z.E. Meziani, J. Miller, J. Morgenstern, J. Picard, B. Saghai, S. Turck-Chieze, P. Vernin, M.K. Brussel: Phys. Lett. **153B**, 29 (1985)
9.40 P. Barreau, M. Bernheim, M. Brussel, G.P. Capitani, J. Duclos, J.M. Finn, S. Frullani, F. Garibaldi, D. Isabelle, E. Jans, J. Morgenstern, J. Mougey, D. Royer, B. Saghai, E. de Sanctis, I. Sick, D. Tarnowsky, S. Turck-Chieze, P.D. Zimmerman: Nucl. Phys. **A358**, 287 (1981); P. Barreau, M. Bernheim, J. Duclos, J.M. Finn, Z. Meziani, J. Morgenstern, J. Mougey, D. Royer, B. Saghai, D. Tarnowski, S. Turck-Chieze, M. Brussel, G.P. Capitani, E. de Sanctis, S. Frullani, F. Garibaldi, D.B. Isabelle, E. Jans, I. Sick, P.D. Zimmerman: Nucl. Phys. **A402**, 515 (1983)
9.41 A.Yu. Buki, N.G. Shevchenko, N.G. Afanasiev, V.N. Polishchuk, A.A. Homich, B.V. Mazanko: Ukr. Phys. Zh. **28**, 1654 (1983)
9.42 M. Deady, C.F. Williamson, J. Wong, P.D. Zimmerman, C. Blatchley, J.M. Finn, J. Le Rose, P. Sioshansi, R. Altemus, J.S. McCarthy, R.R. Whitney: Phys. Rev. **C28**, 631 (1983)
9.43 M. Deady, C.F. Williamson, P.D. Zimmerman, R. Altemus, R.R. Whitney: Phys. Rev. **C33**, 1897 (1986)
9.44 Z.E. Meziani, P. Barreau, M. Bernheim, J. Morgenstern, S. Turck-Chieze, R. Altemus, J. McCarthy, L.J. Orphans, R.R. Whitney, G.P. Capitani, E. de Sanctis, S. Frullani, F. Garibaldi: Phys. Rev. Lett. **52**, 2130 (1984)
9.45 Z.E. Meziani, P. Barreau, M. Bernheim, J. Morgenstern, S. Turck-Chieze, R. Altemus, J. McCarthy, L.J. Orphans, R.R. Whitney, G.P. Capitani, E. de Sanctis, S. Frullani, F. Garibaldi: Phys. Rev. Lett. **54**, 1233 (1985)
9.46 R. Altemus, A. Cafolla, D. Day, J.S. McCarthy, R.R. Whitney, J.E. Wise: Phys. Rev. Lett. **44**, 965 (1980)
9.47 J.P. Chen, Z.E. Meziani, D. Beck, G. Boyd, L.M. Chinitz, D.B. Day, L.C. Dennis, G. Dodge, B.W. Filippone, K.L. Giovanetti, J. Jourdan, K.W. Kemper, T. Koh, W. Lorenzon, J.S. McCarthy, R.D. McKeown, R.G. Milner, R.C. Minehart, J. Morgenstern, J. Mougey, D.H. Potterveld, O.A. Rondon-Aramayo, R.M. Sealock, L.C. Smith, S.T. Thornton, R.C. Walker, C. Woodward: Phys. Rev. Lett. **66**, 1283 (1991)
9.48 C. Blatchley: Ph.D. Thesis, Lousiana State University (1985)
9.49 C.C. Blatchley, J.J. Le Rose, O.E. Pruet, P.D. Zimmerman, C.F. Williamson, M. Deady: Phys. Rev. **C34**, 1243 (1986)
9.50 S. Drożdż, G.Co', J. Wambach, J. Speth: Phys. Lett. **185B**, 287 (1987)

9.51   S. Boffi, C. Giusti, F.D. Pacati: Nucl. Phys. **A386**, 599 (1982)
9.52   J.V. Noble: Phys. Rev. Lett. **46**, 412 (1981)
9.53   T.W. Donnelly, J.W. van Orden, T. De Forest, W.C. Hermans: Phys. Lett. **76B**, 393 (1978)
9.54   J.M. Laget: Phys. Rep. **69**, 1 (1981)
9.55   J.M. Laget: Nucl. Phys. **A358**, 275c (1981)
9.56   J.W. van Orden, T.W. Donnely: Ann. Phys. **131**, 451 (1980)
9.57   P.G. Blunden, M.N. Butler: Phys. Lett. **219B**, 151 (1989)
9.58   M. Kohno, N. Ohtsuka: Phys. Lett. **98B**, 335 (1981)
9.59   M. Kohno: Nucl. Phys. **A410**, 349 (1983)
9.60   A. Dellafiore, F. Lenz, F.A. Brieva: Phys. Rev. **C31**, 1088 (1985)
9.61   M. Cavinato, D. Drechsel, E. Fein, M. Marangoni, A.M. Saruis: Nucl. Phys. **A423**, 376 (1984)
9.62   M. Cavinato, M. Marangoni, A.M. Saruis: Nucl. Phys. **A496**, 108 (1989)
9.63   M. Cavinato, M. Marangoni, A.M. Saruis: Phys. Lett. **235B**, 15 (1990)
9.64   P.D. Zimmerman: Phys. Rev. **C26**, 265 (1982)
9.65   L.S. Celenza, A. Rosental, C.M. Shakin: Phys. Rev. **C31**, 232 (1985)
9.66   L.S. Celenza, A. Harindranath, C.M. Shakin: Phys. Rev. **C32**, 248 (1985)
9.67   L.S. Celenza, A. Harindranath, C.M. Shakin: Phys. Rev. **C32**, 650 (1985)
9.68   P.J. Mulders: Phys. Rev. Lett. **54**, 2560 (1985)
9.69   E.N. Nikolov, M. Bergmann, C. Christov, A.N. Antonov, K. Goeke: RUB-TP II (28 June 1991, Bochum); in "Proc. 5th Workshop on Persp. of Nucl. Phys. at Intermediate Energy" (Trieste, May 1991), World Sci., Singapore, p. 224; E.N. Nikolov, M. Bergmann, C. Christov, K. Goeke, A.N. Antonov, S. Krewald: Phys. Lett. **281B**, 208 (1992)
9.70   U. Stroth, R. Hasse, P. Schuck: Phys. Lett. **171B**, 339 (1986)
9.71   U. Stroth, R. Hasse, P. Schuck: Nucl. Phys. **A462**, 45 (1987)
9.72   W.M. Alberico, P. Czerski, M. Ericson, A. Molinari: Nucl. Phys. **A462**, 269 (1987)
9.73   W.M. Alberico, G. Chanfray, M. Ericson, A. Molinari: Nucl. Phys. **A475**, 233 (1987)
9.74   L.S. Celenza, W.S. Pong, M.M. Rahman, C.M. Shakin: Phys. Rev. **C26**, 320 (1982)
9.75   L.S. Celenza, W.S. Pong, C.M. Shakin: Phys. Rev. **C27**, 2792 (1983)
9.76   R. Schiavilla, V.R. Pandharipande, A. Fabrocini: Phys. Rev. **C40**, 1484 (1989)
9.77   R. Schiavilla, V.R. Pandharipande: Phys. Rev. **C36**, 2221 (1987)
9.78   R. Schiavilla: Phys. Lett. **218B**, 1 (1989)
9.79   V.R. Pandharipande: Nucl. Phys. **A497**, 43 (1989)
9.80   S. Fantoni, V.R. Pandharipande: Nucl. Phys. **A473**, 234 (1987)
9.81   A. Fabrocini, S. Fantoni: University of Pisa Reprint IFUP-2H36/88, (1988)
9.82   S. Drożdż, M. Buballa, S. Krewald, J. Speth: Nucl. Phys. **A501**, 487 (1989); A. Mariano, E. Bauer, F. Krmpotić, A.F.R. de Toledo Piza: Phys. Lett. **268B**, 332 (1991)
9.83   C. Yannouleas, M. Dworzecka, J.J. Griffin: Nucl. Phys. **A379**, 256 (1982); **A397**, 239 (1983)
9.84   A.Yu. Korchin: Yad. Fiz. **35**, 1162 (1982); A.Yu. Korchin, E.L. Kuplennikov, A.V. Shebeko: Yad. Fiz. **44**, 932 (1986)
9.85   K. Wehrberger, F. Beck: Phys. Rev. **C35**, 298 (1987)
9.86   K. Wehrberger, F. Beck: Phys. Rev. **C37**, 1148 (1988)
9.87   K. Wehrberger, F. Beck: Nucl. Phys. **A491**, 587 (1989)
9.88   K. Wehrberger, F. Beck: Phys. Rev. **C39**, 2465 (1989); Phys. Lett. **266B**, 1 (1991); Phys. Lett. **270B**, 1 (1991)
9.89   G. Do Dang, P. Van Thieu: Phys. Rev. **C28**, 1845 (1983)
9.90   C.J. Horowitz, J. Piekarewicz: Phys. Rev. Lett. **62**, 391 (1989)
9.91   C.J. Horowitz, J. Piekarewicz: Nucl. Phys. **A511**, 461 (1990)
9.92   G. Do Dang, N. Van Giai: Phys. Rev. **C30**, 731 (1984)
9.93   S. Shlomo: Phys. Lett. **118B**, 233 (1982)
9.94   R. Cenni, C. Ciofi degli Atti, G. Salmé: Phys. Rev. **C39**, 1425 (1989)
9.95   J.M. Finn, R.W. Lourie, B.H. Cottman: Phys. Rev. **C29**, 2230 (1984)
9.96   F. Capuzzi, C. Giusti, F.D. Pacati: Nucl. Phys. **A524**, 681 (1991)
9.97   A.N. Antonov, I.Zh. Petkov: Bulg. J. Phys. **11**, 163 (1984)
9.98   U. Glawe, U. Strohbusch, J. Franz, P. Grosse-Wiesmann, G. Guzielski, G. Huber, G. Mecklenbrauck, W. Mecklenbrauck, E. Rössle: Phys. Lett. **89B**, 44 (1979)

9.99   C.V. Christov, E. Ruiz Arriola, K. Goeke: Phys. Lett. **225B**, 22 (1989); Nucl. Phys. **A510**, 689 (1990); Phys. Lett. **243B**, 333 (1990)
9.100   Y. Nambu, G. Jona-Lasinio: Phys. Rev. **122**, 354 (1961)
9.101   M. Bergmann, K. Goeke, S. Krewald: Phys. Lett. **243B**, 185 (1990)
9.102   M. Gari, W. Kruempelmann: Z. Phys. **A322**, 689 (1985)
9.103   L.B. Weinstein, H. Baghari, W. Bertozzi, J.M. Finn, J. Glickman, C.E. Hyde-Wright, N. Kalanter-Nayestanaki, R.W. Lourie, J.A. Nelson, W.W. Sapp, C.P. Sargent, P.E. Vemer, B.H. Cottman, L. Chedira, E.J. Winhold, J.R. Calarco, J. Wise, P. Boberg, C.C. Chang, D. Zhang, K. Aniol, M.B. Epstein, D.J. Margaziotis, C. Perdrisat, V. Punjabi: in "Proc. PANIC XII", MIT (1990), p. I–9
9.104   A. Zghiche, G.P. Capitani, E. De Sanctis, S. Frullani, F. Garibaldi, A. Magnon, C. Marchand, J. Morgenstern, J. Picard, D. Reffay-Pikeroen, M. Traini, S. Turck-Chieze, P. Vernin: in "Proc. PANIC XII", MIT (1990), p. I–8a
9.105   R. Schiavilla, D.S. Lewart, V.R. Pandharipande, S.C. Pieper, R.B. Wiringa, S. Fantoni: Nucl. Phys. **A473**, 267 (1987)
9.106   K. Takayanagi: Phys. Lett. **230B**, 11 (1989); W. Leidemann, G. Orlandini, M. Traini: Phys. Rev. **C44**, 1705 (1991); G. Orlandini and M. Traini: Rep. Progr. Phys. **54**, 257 (1991)
9.107   EM Collaboration (J.J. Aubert et al.): Phys. Lett. **105B**, 322 (1982); Phys. Lett. **123B**, 275 (1983)
9.108   A. Bodek, N. Giokaris, W.B. Atwood, D.H. Coward, D.J. Sherden, D.L. Dubin, J.E. Elias, J.I. Friedman, H.W. Kendall, J.S. Poucher, E.M. Riordan: Phys. Rev. Lett. **50**, 1431 (1983); A. Bodek, N. Giokaris, W.B. Atwood, D.H. Coward, D.L. Dubin, M. Breidenbach, J.E. Elias, J.I. Friedman, H.W. Kendall, J.S. Poucher, E.M. Riordan: Phys. Rev. Lett. **51**, 534 (1983)
9.109   SLAC Collaboration (R.G. Arnold et al.): Phys. Rev. Lett. **52**, 727 (1984)
9.110   A.E. Asratyan et al.: Preprint ITEP-110 (1983), Moscow
9.111   Hu Guoju, J.M. Irvine: J. Phys. **G15**, 147 (1989)
9.112   S.V. Akulinichev, S. Shlomo: Phys. Rev. **C33**, 1551 (1986)
9.113   E.L. Berger, F. Coester: Annu. Rev. Nucl. Part. Sci. **37**, 463 (1987)
9.114   A. Krzywicki: in Proc. of 3rd Workshop on Perspectives in Nuclear Physics at Intermediate Energies, Trieste (1987) (Eds. by S. Boffi, C. Ciofi degli Atti, M.M. Giannini, World Scientific, Singapore 1988) p. 159; E. Predazzi: in Proc 4th Workshop on Perspectives in Nuclear Physics at Intermediate Energies, Trieste (1989) (Eds. by S. Boffi, C. Ciofi degli Atti, M. Giannini, World Scientific, Singapore 1989), p. 91; D. Day: ibid. p. 165
9.115   R.P. Bickerstaff, A.W. Thomas: J. Phys. **G15**, 1523 (1989)
9.116   L. Kondratyuk, M. Shmatikov: Preprint ITEP-13 (1984), Moscow
9.117   Ma Bo-Qiang: Mod. Phys. Lett. **6**, 21 (1991)
9.118   Ma Bo-Qiang: Phys. Rev. **C43**, 2821 (1991); BIHEP-TH-90-5
9.119   R.L. Jaffe, F. Close, R. Roberts, G. Ross: Phys. Lett. **134B**, 449 (1984)
9.120   E.F. Hefter: Lett. Nuovo Cim. **44**, 99 (1985)
9.121   J.V. Noble: Phys. Rev. Lett. **46**, 412 (1981)
9.122   L.L. Frankfurt, M.I. Strikman: Preprint LINR No. 886 (1983), Leningrad
9.123   C.E. Carlson, T.J. Havens: Phys. Rev. Lett. **51**, 261 (1983)
9.124   O. Nachtmann, H.J. Pirner: Z. Phys. **C21**, 277 (1984)
9.125   C.H. Llewellyn Smith: Phys. Lett. **128B**, 107 (1983)
9.126   M. Ericson, A.W. Thomas: Phys. Lett. **128B**, 112 (1983)
9.127   J. Szwed: Phys. Lett. **128B**, 245 (1983)
9.128   A.I. Titov: Preprint JINR E2-83-460 (1983), Dubna
9.129   S.V. Akulinichev, S.A. Kulagin, G.M. Vagradov: Phys. Lett. **158B**, 485 (1985)
9.130   S.V. Akulinichev, S.A. Kulagin, G.M. Vagradov: JETP Lett. **42**, 105 (1985)
9.131   S.V. Akulinichev, S.A. Kulagin, G.M. Vagradov: J. Phys. **G11**, L245 (1985)
9.132   S.V. Akulinichev, S. Shlomo, S.A. Kulagin, G.M. Vagradov: Phys. Rev. Lett **55**, 2239 (1985)
9.133   R.P. Bickerstaff, A.W. Thomas: Phys. Rev. **D35**, 108 (1987)
9.134   L.L. Frankfurt, M.I. Strikman: Phys. Lett. **183B**, 254 (1987); Phys. Rep. **160**, 236 (1988)
9.135   B.L. Birbrair, E.M. Levin, A.G. Shuvaev: Nucl. Phys. **A491**, 618 (1989)

9.136   R.L. Jaffe: Nucl. Phys. **A478**, 3c (1988)
9.137   H. Jung, G.A. Miller: Phys. Lett. **200B**, 351 (1988)
9.138   G.L. Li, K.F. Liu, G.E. Brown: Phys. Lett. **213B**, 531 (1988)
9.139   C. Ciofi degli Atti, S. Liuti: Phys. Lett. **225B**, 215 (1989); Phys. Rev. **C44**, R1269 (1991)
9.140   C. Ciofi degli Atti, S. Liuti: Phys. Rev. **C41**, 1100 (1990); C. Ciofi degli Atti, S. Simula, L.L. Frankfurt, M.I. Strikman: Phys. Rev. **C44**, R7 (1991)
9.141   R.L. Jaffe: Comments Nucl. Part. Phys. **13**, 39 (1984)
9.142   B.-Q. Ma, J. Sun: J. Phys. **G16**, 823 (1990)
9.143   S. Shlomo, G.M. Vagradov: Phys. Lett. **232B**, 19 (1989)
9.144   Z. Cao, J. Shen, G. Li: Z. Phys. **A336**, 97 (1990)
9.145   J.G. Zabolitzky, W. Ey: Phys. Lett. **76B**, 527 (1978)
9.146   R. Jastrow: Phys. Rev. **98**, 1479 (1955)
9.147   V.V. Anisovich, S. Ohlsson, A.V. Sarantsev, V.E. Starodubsky: Nucl. Phys. **A474**, 662 (1987)
9.148   H. Araseki, T. Fujita: Nucl. Phys. **A439**, 681 (1985)
9.149   J. Rozynek, M.C. Birse: Phys. Rev. **C38**, 2201 (1988)
9.150   S. Fantoni, V.R. Pandharipande: Nucl. Phys. **A427**, 473 (1984)
9.151   S. Fantoni, B.L. Friman, V.R. Pandharipande: Nucl. Phys. **A399**, 51 (1983)
9.152   D.Yu. Bardin, V.Kh. Dodokhov, A.V. Efremov, N.G. Fadeev, V. Genchev, I.A. Golutvin, V.Yu. Karzhavin, M.Yu. Kazarinov, V.S. Khabarov, Yu.T. Kiryushin, V.S. Kisselev, I.G. Kosarev, V.G. Krivokhizin , V.V. Kukhtin, R. Lednicky, W. Lohmann, V.N. Lysyakov, S. Nemecek, Yu.A. Panebratsev, P. Reimer, I.A. Savin, N.B. Skachkov, A.I. Semenyushkin, G.I. Smirnov, D.A. Smolin, G. Sultanov, P. Todorov, A.V. Vishnevsky, N.V. Vlasov, A.G. Volodko, A.V. Zarubin, N.I. Zamyatin, S.P. Baranov, V.F. Grushin, A.A. Komar, A.A. Shikanyan, E.V. Telyukov, J. Cvach, J. Hladký, J. Strachota: Czech. J. Phys. **B36**, 1289 (1986)
9.153   A.N. Antonov, L.P. Kaptari, V.A. Nikolaev: JINR Rapid Communications No. 2 [41]-90 (1990), p. 14
9.154   A.N. Antonov, L.P. Kaptari, V.A. Nikolaev, A.Yu. Umnikov: Nuovo Cim. **A104**, 487 (1991)
9.155   F.E. Close, R.G. Roberts, G.G. Gross: Phys. Lett. **129B**, 346 (1983); R.L. Thews: Phys. Rev. **D29**, 398 (1984)
9.156   EM Collaboration (J. Ashman et al.): Phys. Lett. **202B**, 603 (1988)
9.157   BCDMS Collaboration (G. Bari et al.): Phys. Lett. **163B**, 282 (1985); BCDMS Collaboration (A.C. Benvenuti et al.): Phys. Lett. **189B**, 483 (1987)
9.158   B.L. Birbrair, E.M. Levin, A.G. Shuvaev: Phys. Lett. **222B**, 281 (1986)
9.159   J. Mougey, M. Bernheim, A. Bussiéré, A. Gillebert, P.X. Ho, M. Priou, D. Royer, I. Sick, G.J. Wagner: Nucl. Phys. **A262**, 461 (1976)
9.160   A.N. Antonov, V.A. Nikolaev, I.Zh. Petkov: Z. Phys. **A304**, 239 (1982)
9.161   A.N. Antonov, I.Zh. Petkov: Nuovo Cim. **A94**, 68 (1986)
9.162   A.C. Benvenuti, D. Bollini, G. Bruni, T. Camporesi, G. Heiman, L. Monari, F.L. Navarria, A. Argento, M. Bozzo, R. Brun, J. Cvach, H. Gennow, L. Piemontese, D. Schinzel, D.Yu. Bardin, N.G. Fadeev, I.A. Golutvin, Yu.T. Kiryushin, V.S. Kiselev, V.G. Krivokhizhin, V.V. Kuchtin, W. Lohman, P. Reimer, I.A. Savin, G.I. Smirnov, J. Strachota, G. Sultanov, P. Todorov, A.G. Volodko, D. Jamnik, R. Kopp, U. Meyer-Berkhout, A. Staude, K.-M. Teichert, R. Tirler, R. Voss, C. Zupančič, J. Feltesse, A. Milsztajn, A. Ouraou, J.F. Renardy, P. Rich-Hennion, Y. Sacquin, G. Smadja, P. Verrecchia, M. Virchaux: JINR, E1-87-549 (1987), Dubna
9.163   A.M. Baldin, V.K. Bondarev, N. Ghiordanescu, A.N. Khrenov, A.G. Litvinenko, A.N. Manyatovsky, S.N. Moroz, Yu.A. Panebratsev, M. Pentia, S.V. Rikhvitsky, V.S. Stavinsky: JINR, E2-82-472 (1982), Dubna
9.164   V.V. Burov, V.K. Lukyanov, A.I. Titov: Part. Nucl. **15**, 1249 (1984)
9.165   L.L. Frankfurt, M.I. Strikman: Phys. Rep. **76**, 215 (1981)
9.166   R.L. Jaffe: in *Relativistic Dynamics and Quark Nuclear Physics* (Ed. by M.B. Johnson and A. Picklesimer, Wiley, New York 1986)
9.167   E.L. Bratkovskaya, L.P. Kaptari, A.I. Titov, A.Yu. Umnikov: JINR, E2-89-306 (1989), Dubna

9.168   L.P. Kaptari, A.I. Titov, A.Yu. Umnikov: JINR, E2-89-564 (1989), Dubna
9.169   BCDMS Collaboration (A.C. Benvenuti et al.): CERN-EP/87-13
9.170   K. Nakano: Phys. Lett. **249B**, 169 (1990)
9.171   K. Nakano: Nucl. Phys. **A511**, 664 (1990)
9.172   S. Kumano, F.E. Close: Phys. Rev. **C41**, 1855 (1990)
9.173   A. Molinari, G. Vagradov: Z. Phys. **A332**, 119 (1989)
9.174   A.W. Thomas, A. Michels, A.W. Schreiber, P.A.M. Guichon: ADP-89-125/T73 (1989), Adelaide

Chapter 10

10.1   M.R. Sené, I. Anthony, D. Branford, A.G. Flowers, A.C. Shotter, C.H. Zimmerman, J.C. McGeorge, R.O. Owens, P.J. Thorley: Nucl. Phys. **A442**, 215 (1985)
10.2   J.L. Matthews: in "Proc. 3rd Workshop on Perspectives in Nuclear Physics at Intermediate Energies" (18–22 May 1987, Trieste) (Eds. S. Boffi, C. Ciofi degli Atti, M.M. Giannini, World Scientific, Singapore 1988), p. 611⁻
10.3   G. van der Steenhoven: in "Proc. 4th Workshop on Perspectives in Nuclear Physics at Intermediate Energies" (8–12 May 1989, Trieste), (Eds. S. Boffi, C. Ciofi degli Atti, M.M. Giannini, World Scientific, Singapore 1989), p. 469
10.4   J.L. Matthews, D.J.S. Findlay, S.N. Gardiner, R.O. Owens: Nucl. Phys. **A267**, 51 (1976)
10.5   G.M. Shklyarevskyi: JETP (Sov. Phys.) **9**, 1057 (1959)
10.6   J.L. Matthews, W. Bertozzi, S. Kowalski, C.P. Sargent, W. Turchinetz: Nucl. Phys. **A112**, 654 (1968)
10.7   R.C. Barrett, D.F. Jackson: *Nuclear Sizes and Structure* (Clarendon Press, Oxford 1977), Ch. 8.5
10.8   S.N. Gardiner: Ph.D. thesis, University of Glasgow (1971)
10.9   M. Fink, H. Hebach, H. Kümmel: Nucl. Phys. **A186**, 353 (1972)
10.10   D.J.S. Findlay, S.N. Gardiner, J.L. Matthews, R.O. Owens: J.Phys. **A7**, L157 (1974)
10.11   S. Boffi, C. Giusti, F.D. Pacati: Nucl. Phys. **A359**, 91 (1981)
10.12   M.J. Leitch, J.L. Matthews, W.W. Sapp, C.P. Sargent, S.A. Wood, D.J.S. Findlay, R.O. Owens, B.L. Roberts: Phys. Rev. **C31**, 1633 (1985)
10.13   G.M. Lotz. H.S. Sherif: Phys. Lett. **210B**, 45 (1988)
10.14   H. Ferdinande, D. Rickbosch, E. Kerkhove, P. Berkvens, R. van de Vyver, A. De Graeve, L. van Hoorebeke: Phys. Rev. **C39**, 253 (1989)
10.15   H. Göringer, B. Schoch, G. Lührs: Nucl. Phys. **A384**, 414 (1982)
10.16   H. Göringer, B. Schoch: Phys. Lett. **97B**, 41 (1980)
10.17   H. Schier, B. Schoch: Nucl. Phys. **A229**, 93 (1974); Lett. Nuovo Cim. **12**, 334 (1975)
10.18   H.G. Miller, W. Buss, J.A. Rawlins: Nucl. Phys. **A163**, 637 (1971)
10.19   J.L. Matthews, W. Bertozzi, M.J. Leitch, C.A. Peridier, B.L. Roberts, C.P. Sargent, W. Turchinetz, D.J.S. Findlay, R.O. Owens: Phys. Rev. Lett. **38**, 8 (1977)
10.20   M.J. Leitch: Ph.D. thesis, MIT (1979)
10.21   D.J.S. Findlay, R.O. Owens: Nucl. Phys. **A279**, 385 (1977)
10.22   H. Hebach, A. Wortberg, M. Gari: Nucl. Phys. **A267**, 425 (1976)
10.23   M. Marangoni, P.L. Ottaviani, A.M. Saruis: Report CNEN, RT/FI (76), 10, University Bologna (1976)
10.24   B. Schoch: Phys. Rev. Lett. **41**, 80 (1978)
10.25   D.J.S. Findlay, R.O. Owens: Phys. Rev. Lett. **37**, 674 (1976)
10.26   J.S. Levinger: Phys. Rev. **84**, 43 (1951); P.J. Carlos. H. Beil, R. Bergere, B.L. Berman, A. Laprétre, A. Veyssière: Nucl. Phys. **A378**, 317 (1982)
10.27   H. Hartmann, H. Hoffmann, B. Mecking, G.Nöldeke: in "Proc. Conf. on Photonuclear Reactions and Applications" (Asilomar 1973), p. 967
10.28   C.T. Noguchi, F. Prats: Phys. Rev. **C14**, 1133 (1976)

10.29   R. Bergère: in "Proc. Topical Conf. on Nucl. Phys. with Electromagn. Interactions", Mainz, 1979, Lecture Notes in Physics, **108** (Springer, Berlin 1979), p. 138

10.30   B. Ziegler: in"Proc. Topical Conf. on Nucl. Phys. with Electromagn. Interactions", Mainz, 1979, Lecture Notes in Physics, **108** (Springer, Berlin 1979), p. 148

10.31   M. Gari, H. Hebach: Phys. Rev. **C10**, 1629 (1974)

10.32   F. Prats: Phys. Lett. **88B**, 23 (1979)

10.33   M. Gari, H. Hebach: Phys. Rep. **72**, 1 (1981)

10.34   S.R. Cotanch: Phys. Lett. **76B**, 19 (1978)

10.35   J.L. Matthews: in "Proc. Topical Conf. on Nucl. Phys. with Electromagn. Interactions", Mainz, 1979, Lecture Notes in Physics, **108** (Springer, Berlin 1979), p. 369

10.36   T.W. Phillips, R.G. Johnson: Phys. Rev. **C20**, 1689 (1979)

10.37   M. Cavinato, M. Marangoni, P.L. Ottaviani, A.M. Saruis: Nucl. Phys. **A373**, 445 (1982); M. Cavinato, M. Marangoni, A.M. Saruis: Nuovo Cim. **76A**, 197 (1983); Nucl. Phys. **A422**, 237 (1984); Nucl. Phys. **A422**, 273 (1984)

10.38   J.T. Londergan, G.D. Nixon: Phys. Rev. **C19**, 998 (1979)

10.39   M.J. Leitch, F.C. Lin, J.L. Matthews, W.W. Sapp, C.P. Sargent, D.J.S. Findlay, R.O. Owens, B.L. Roberts: Phys. Rev. **C33**, 1511 (1986)

10.40   W. Weise, M.G. Huber: Nucl. Phys. **A162**, 330 (1971)

10.41   A. Małecki, P. Picchi: Lett. Nuovo Cim. **8**, 16 (1973)

10.42   B. Schoch, H. Göringer, P. Jennewein, F. Klein, G. Lührs, F. Zettl: Phys. Rev. **C22**, 2630 (1980)

10.43   G.M. Shklyarevskyi: JETP (Sov. Phys.) **14**, 324 (1962)

10.44   M. Gari, H. Hebach: Phys. Lett. **49B**, 29 (1974)

10.45   C. Ciofi degli Atti: *The Nuclear Many-Body Problem* v. 1, Ed. C. Ciofi degli Atti and F. Calogero (Editrice Compositori, Bologna 1973), p. 376

10.46   M. Bernheim, A. Bussiere, J. Mougey, D. Royer, D. Tarnowski, S. Turck-Chiese, S. Frullani, S. Boffi, C. Giusti, F.D. Pacati, G.P. Capitani, E.de Sanctis, G.J. Wagner: Nucl. Phys. **A375**, 381 (1982)

10.47   D.J.S. Findlay, R.O. Owens: Nucl. Phys. **A292**, 53 (1977)

10.48   J. Mougey, M. Bernheim, A. Bussiere, A. Gillebert, X.P. Ho, M. Priou, D. Royer, I. Sick, G.J. Wagner: Nucl. Phys. **A262**, 461 (1976)

10.49   A. Watt: Phys. Lett. **27B**, 190 (1968)

10.50   A.N. Gorbunov: Phys. Lett. **27B**, 436 (1968)

10.51   S.E. Kiergan, A.O. Hansen, L.J. Koester, Jr.: Phys. Rev. **C8**, 431 (1973)

10.52   V.A. Goldstein, E.L. Kuplennikov, R.I. Jibuti, R.Ya. Keserashvili: Nucl. Phys. **A355**, 333 (1981)

10.53   S. Frullani, J. Mougey: Advances in Nucl. Physics **14**, 1 (1984)

10.54   L.R.B. Elton, A. Swift: Nucl. Phys. **A94**, 52 (1967)

10.55   J.W. Negele: Phys. Rev. **C1**, 1260 (1970)

10.56   C. Ciofi degli Atti: Lett. Nuovo Cim. **1**, 590 (1971)

10.57   J.G. Zabolitzky, W. Ey: Phys. Lett. **76B**, 527 (1978)

10.58   S. Boffi, R. Cenni, C. Giusti, F.D. Pacati: Nucl. Phys. **A420**, 38 (1984)

10.59   J. Ryckebusch, M. Waroquier, K. Heyde, D. Ryckbosch: Phys. Lett. **194B**, 453 (1987); J. Ryckebusch, M. Waroquier, K. Heyde, J. Moreau, D. Ryckbosch: Nucl. Phys. **A476**, 237 (1988); J. Ryckebusch, K. Heyde, D. van Neck, M. Waroquier: Phys. Lett. **216B**, 252 (1989)

10.60   E.J. Beise: Ph.D. thesis, MIT (1988)

10.61   J.P. McDermott, E. Rost, J.R. Shepard, C.Y. Cheung: Phys. Rev. Lett. **61**, 814 (1988)

10.62   K. Maruyama: in "Proc. 4th Workshop on Perspectives in Nuclear Physics at Intermediate Energies" (8–12 May 1989, Trieste) (Eds. S. Boffi, C. Ciofi degli Atti, M.M. Giannini, World Scientific, Singapore 1989), p. 488

10.63   L. Boato, M.M. Giannini: J. Phys. **G15**, 1605 (1989)

10.64   K. Gottfried: Nucl. Phys. **5**, 557 (1958)

10.65   J.M. Vogt: Thesis, Mainz (1987)

10.66 J. Vogt, J. Annand, I. Anthony, R. Beck, D. Branford, G.I. Crawford, S. Dancer, J.C. McGeorge, J. Göbel, I.J.D. MacGregor, S.J. Hall, J.D. Kellie, G. Liesenfeld, R.O. Owens, B. Schoch, F. Zettl: in "Proc. 3rd Workshop on Perspectives in Nucl. Phys. at Intermediate Energies" (18–22 May 1987, Trieste) (Eds. S. Boffi, C. Ciofi degli Atti, M.M. Giannini, World Scientific, Singapore 1988), p. 633

10.67 S.N. Dancer, I.J.D. MacGregor, J.R.M. Annand, I. Anthony, G.I. Crawford, S.J. Hall, J.D. Kellie, J.C. McGeorge, R.O. Owens, P.A. Wallace, D. Branford, S.V. Springham, A.C. Shotter, B.Schoch, J.M. Vogt, R. Beck, G. Liesenfeld: Phys. Rev. Lett. **61**, 1170 (1988)

10.68 J. Friedrich: in "Proc. 4th Workshop on Perspectives in Nucl. Phys. at Intermediate Energies" (8–12 May 1989, Trieste) (Eds. S. Boffi, C. Ciofi degli Atti, M.M. Giannini, World Scientific, Singapore 1989), p. 479

10.69 S. Homma, M. Kanazawa, K. Maruyama, Y. Murata, H. Okuno, A. Sasaki, T. Taniguchi: Phys. Rev. Lett. **45**, 706 (1980)

10.70 J.W. Lightbody, Jr.: in "Proc. 3rd Workshop on Perspectives in Nucl. Phys. at Intermediate Energies" (18–22 May 1987, Trieste) Eds. S. Boffi, C. Ciofi degli Atti, M.M. Giannini, World Scientific, Singapore 1988), p. 455

Chapter 11

11.1   A.G. Sitenko: *Theory of Nuclear Reactions* (World Scientific, Singapore 1990)
11.2   R.J. Glauber: *Lectures in Theoretical Physics* v.1 (New York 1959), p. 315
11.3   A.G. Sitenko: Ukr. Phys. J. **4**, 152 (1959)
11.4   R.H. Bassel, C. Wilkin: Phys. Rev. **174**, 1179 (1968)
11.5   A.G. Sitenko: Particles and Nuclei **4**, 546 (1973)
11.6   J. Saudinos, C. Wilkin: Ann. Rev. Nucl. Sci. **24**, 341 (1974)
11.7   G.J. Igo: Rev. Mod. Phys. **50**, 523 (1978)
11.8   G.D. Alkhazov, S.L. Belostotsky, A.A. Vorobyov: Phys. Rep. **42**, 89 (1978)
11.9   R.D. Viollier, J.D. Walecka: Acta Phys. Pol. **B8**, 25 (1977)
11.10 L.L. Foldy, J.D. Walecka: Ann Phys. **54**, 447 (1969)
11.11 W. Czyż, L.C. Maximon: Ann. Phys. **52**, 59 (1969)
11.12 S.J. Wallace: Phys. Rev. **C12**, 179 (1975)
11.13 J.M. Eisenberg: Ann. Phys. **71**, 542 (1972)
11.14 C.W. Wong, S.K. Young: Phys. Rev. **C12**, 1301 (1975)
11.15 E. Bleszynski, M. Bleszynski, S. Hajisaeid, G.J. Igo, F. Irom, J.B. McClelland, G. Pauletta, A. Rahbar, A.T.M. Wang, C.A. Whitten, Jr., G. Adams, M. Barlett, G.W. Hoffmann, J.A. McGill, R. Boudrie, G. Kyle: Phys. Rev. **C25**, 2563 (1982)
11.16 D.R. Harrington, G.K. Varma: Phys. Lett. **74B**, 316 (1978)
11.17 J.P. Tabet, B. Bonin, A. Boudard, A. Chaumeaux, P. Couvert, J.L. Escudie, L. Farvacque, M. Garson, D. Legrand, J.C. Lugol, B. Nefkens, J. Saudinos, Y. Terrien: Nucl. Phys. **A386**, 552 (1982)
11.18 G.D. Alkhazov: Pis'ma ZETP **31**, 689 (1980)
11.19 J.P. Auger, R.J. Lombard: Ann. Phys. **115**, 442 (1978)
11.20 A.K. Kerman, H. McManus, R.M. Thaler: Ann. Phys. **8**, 551 (1959)
11.21 A. Chaumeaux, V. Layly, R. Schaeffer: Ann. Phys. **116**, 247 (1978)
11.22 H. Feshbach, J. Hüfner: Ann. Phys. **56**, 268 (1970)
11.23 H. Feshbach, A. Gal, J. Hüfner: Ann. Phys. **66**, 20 (1971)
11.24 E. Lambert, H. Feshbach: Ann. Phys. **76**, 80 (1973)
11.25 J. Ullo, H. Feshbach: Ann. Phys. **82**, 156 (1974)
11.26 E. Kujawski: Nucl. Phys. **A236**, 423 (1974)
11.27 E. Boridy, H. Feshbach: Ann. Phys. **109**, 468 (1977)
11.28 G.D. Alkhazov, S.L. Belostotsky, A.A. Vorobyov, O.A. Domchenkov, Yu.V. Dotsenko, N.P. Kuropatkin, V.N. Nikulin: Yad. Fiz. **42**, 8 (1985)
11.29 L. Ray: Phys. Rev. **C19**, 1855 (1979)

11.30   L. Ray, G.W. Hoffmann, M. Barlett, J. McGill, J. Amann, G. Adams, G. Pauletta, M. Gazzaly, G.S. Blanpied: Phys. Rev. **C23**, 828 (1981); G.K. Varma, L. Zamick: Phys. Rev. **C16**, 308 (1977); B.A. Brown, S.E. Massen, P.E. Hodgson: J. Phys. **G5**, 1655 (1979)

11.31   G.S. Blanpied, N.M. Hintz, G.S. Kyle, M.A. Franey, S.Y. Seestrom-Morris, R.K. Owen, J.W. Palm, D. Dennhard, M.L. Barlett, C.J. Harvey, G.W. Hoffmann, J.A. McGill, R.P. Liljestrand, L. Ray: Phys. Rev. **C25**, 422 (1982)

11.32   G.D. Alkhazov, S.L. Belostotsky, O.A. Domchenkov, Yu.V. Dotsenko, N.P. Kuropatkin, V.N. Nikulin, M. A. Shuvaev, A.A. Vorobyov: Nucl. Phys. **A381**, 430 (1982)

11.33   R.M. Lombard, G.D. Alkhazov, O.A. Domchenkov: Nucl. Phys. **A360**, 233 (1981)

11.34   G. Igo, G.S. Adams, T.S. Bauer, G. Pauletta, C.A. Whitten, Jr., A. Wreikat, G.W. Hoffmann, G.S. Blanpied, W.R. Coker, C. Harvey, R.P. Liljestrand, L. Ray, J.E. Spencer, H.A. Thiessen, C. Glashausser, N.M. Hintz, M.A. Oothoudt, H. Nann, K.K. Seth, B.E. Wood, D.K. McDaniels, M. Gazzaly: Phys. Lett. **81B**, 151 (1979)

11.35   G.S. Adams, Th.S. Bauer, G. Igo, G. Pauletta, C.A. Whitten, Jr., A. Wreikat, G.W. Hoffmann, G.R. Smith, M. Gazzaly: Phys. Rev. **C21**, 2485 (1980)

11.36   G. Fäldt, A. Ingemarsson: Nucl. Phys. **A392**, 249 (1983)

11.37   G.A. Moss, C.A. Davis, J.M. Greben, L.G. Greeniaus, G. Roy, J. Uegaki, R. Abegg, D.A. Hutcheon, C.A. Miller, W.T.H. van Oers: Nucl. Phys. **A392**, 361 (1983)

11.38   G.A. Moss, L.G. Greeniaus, J.M. Cameron, D.A. Hutcheon, R.P. Liljestrand, C.A. Miller, G. Roy, B.K.S. Koene, W.T.H. van Oers, A.W. Stetz, A. Willis, N. Willis: Phys. Rev. **C21**, 1932 (1980)

11.39   A.M. Kobos, E.D. Cooper, J.R. Rook, W. Haider: Nucl. Phys. **A435**, 677 (1985)

11.40   G. Igo: Nucl. Phys. **A374**, 253c (1982)

11.41   G.D. Alkhazov, S.L. Belostotsky, A.A. Vorobyov, V.T. Grachev, O.A. Domchenkov, Yu.V. Dotsenko, N.P. Kuropatkin, S.A. Semenov, M.A. Shuvaev: Preprint LINP No. 93, Leningrad (1974)

11.42   G.D. Alkhazov, T. Bauer, R. Beurtey, A. Boudard, G. Bruge, A. Chaumeaux, P. Couvert, C. Cvijanovich, H.H. Duhm, J.M. Fontaine, D. Garreta, A.V. Kulikov, D. Legrand, J.C. Lugol, J. Saudinos, J. Thiriou, A.A. Vorobyov: Nucl. Phys. **A274**, 443 (1976)

11.43   G.N. Afanasiev, A.G. Galperin, V.M. Shilov: JINR, E4-10495, Dubna (1977)

11.44   J.P. Auger, R.J. Lombard: Phys. Lett. **90B**, 200 (1980)

11.45   V.E. Starodubsky, O.A. Domchenkov: Preprint LINP No. 3, Leningrad (1972)

11.46   Y. Abgrall, J. Labarsouque, B. Morand: Nucl. Phys. **A271**, 477 (1976)

11.47   C. Ciofi degli Atti, R. Guardiola: Phys. Lett. **36B**, 287 (1971)

11.48   D.R. Harrington, G.K. Varma: Nucl. Phys. **A306**, 477 (1978)

11.49   E.J. Moniz, G.D. Nixon: Ann. Phys. **67**, 58 (1971)

11.50   V.N. Gribov: ZETP **57**, 1306 (1969)

11.51   V.E. Starodubsky: Yad. Fiz. **16**, 946 (1972)

11.52   V.E. Starodubsky: Nucl. Phys. **A219**, 525 (1974)

11.53   O. Kofoed-Hansen, C. Wilkin: Ann. Phys. **63**, 309 (1971)

11.54   R. Guardiola, E. Oset: Nucl. Phys. **A234**, 458 (1974)

11.55   I. Ahmad: Nucl. Phys. **A247**, 418 (1975)

11.56   G.D. Alkhazov: Nucl. Phys. **A280**, 330 (1977)

11.57   R.D. Viollier: Ann. Phys. **93**, 335 (1975)

11.58   Z.A. Khan: Z.Phys. **A303**, 161 (1981)

11.59   I. Ahmad, J.P. Auger: Nucl. Phys. **A352**, 425 (1981)

11.60   I. Ahmad, J.P. Auger: Preprint ICTP IC/81/230, Trieste (1981)

11.61   D.R. Harrington, V. Tutunjian: Nucl. Phys. **A375**, 453 (1982)

11.62   L. Ray: Phys. Rev. **C20**, 1857 (1979)

11.63   L. Ray: Nucl. Phys. **A335**, 443 (1980)

11.64   J.D. Lumpe, L. Ray: Phys. Rev. **C35**, 1040 (1987)

11.65   I. Ahmad: Phys. Lett. **36B**, 301 (1971)

11.66   A.N. Antonov, E.V. Inopin: Yad. Fiz. **16**, 74 (1972)

11.67   A.N. Antonov: Bulg. J. Phys. **2**, 287 (1975)

11.68   G.D. Alkhazov: Preprint LINP No. 75, Leningrad (1974)

11.69   Yu.A. Berezhnoy, V.V. Pilipenko, G.A. Khomenko: Izv. AN USSR, ser. fiz. **44**, 1950 (1980)

11.70   Yu.A. Berezhnoy, V.V. Pilipenko, G.A. Khomenko: Izv. AN USSR, ser. fiz. **45**, 1953 (1981)

11.71   I. Ahmad, Z.A. Khan: Nucl. Phys. **A274**, 519 (1976)

11.72   V.A. Karmanov: Yad. Fiz. **35**, 848 (1982)

11.73   S.J. Wallace: Nucl. Phys. **A374**, 203c (1982)

11.74   J.P. Auger, C. Lazard, R.J. Lombard: J. Phys. **G9**, 719 (1983)

11.75   L. Veg, B.Z. Kopeliovich, L.I. Lapidus: Yad. Fiz. **35**, 1514 (1982)

11.76   B.G. Zakharov, B.Z. Kopeliovich: Yad. Fiz. **42**, 1073 (1985)

11.77   S. Forte, E. Predazzi: Nuovo Cim. **A88**, 391 (1985); S. Forte: Nucl. Phys. **A467**, 665 (1987)

11.78   J.R. Shepard, J.A. McNeil, S.J. Wallace: Phys. Rev. Lett. **50**, 1443 (1983)

11.79   B.C. Clark, S. Hama, R.L. Mercer, L. Ray, B.D. Serot: Phys. Rev. Lett. **50**, 1644 (1983)

11.80   M. Hynes, A. Picklesimer, P.C. Tandy, R.M. Thaler: Phys. Rev. Lett. **52**, 978 (1984); Phys. Rev. **C31**, 1438 (1985)

11.81   J.A. Tjon, S.J. Wallace: Phys. Rev. Lett. **54**, 1357 (1985)

11.82   L.G. Arnold, B.C. Clark: Phys. Lett. **84B**, 46 (1979)

11.83   A. Picklesimer, P.C. Tandy: Phys. Rev. **C35**, 1174 (1987)

11.84   S. Frankel, W. Frati, O. van Dyck, R. Werbeck, V. Highland: Phys. Rev. Lett **36**, 642 (1976)

11.85   S. Frankel, W. Frati, G. Blanpied, G.W. Hoffmann, T. Kozlowski, C. Morris, H.A. Thiessen, O. van Dyck, R. Ridge, C. Whitten: Phys. Rev. **C18**, 1375 (1978)

11.86   S. Frankel, W. Frati, R.M. Woloshyn, D. Yang: Phys. Rev. **C18**, 1379 (1978)

11.87   V.I. Komarov, G.E. Kosarev, H. Müller, D. Netzband, T. Stiehler: Phys. Lett. **69B**, 37 (1977)

11.88   V.I. Komarov, G.E. Kosarev, H. Müller, D. Netzband, T. Stiehler, S. Tesch: JINR E1-11513, Dubna (1978)

11.89   K.R. Cordell, S.T. Thornton, L.C. Dennis, R.R. Doering, R.L. Parks, T.C. Schweizer: Nucl. Phys, **A352**, 485 (1981)

11.90   G. Roy, L.G. Greeniaus, G.A. Moss, D.A. Hutcheon, R.L. Liljestrand, R.M. Woloshyn, D.H. Boal, A.W. Stetz, K. Aniol, A. Willis, N. Willis, R. McCamis: Phys. Rev. **C23**, 1671 (1981)

11.91   V.I. Komarov, G.E. Kosarev, H. Müller, D. Netzband, T. Stiehler, S. Tesch, JINR E1-11354, Dubna (1978)

11.92   V.I. Komarov, G.E. Kosarev, H. Müller, D. Netzband, T. Stiehler, S.Tesch: JINR E1-12393, Dubna (1979)

11.93   M. Avan, A. Baldit, J. Castor, G. Chaigne, A. Devaux, J. Fargeix, P. Force, G. Landaud, G. Roche, J. Vicente, J.P. Didelez, F. Reide, S.A. Gurvitz: Phys. Rev. **C30**, 521 (1984)

11.94   A.N. Antonov, V.A. Nikolaev, I.Zh. Petkov: Preprint ICTP IC/78/152, Trieste (1978); Bulg. J. Phys. **6**, 151 (1979)

11.95   A.N. Antonov, V.A. Nikolaev, I.Zh. Petkov: Z. Phys. **A297**, 257 (1980)

11.96   A.N. Antonov, V.A. Nikolaev, I.Zh. Petkov: Izv. AN USSR, ser, fiz. **47**, 134 (1983); A.N. Antonov, I.Zh. Petkov: Particles and Nuclei **22**, 801 (1991); A.N. Antonov, L.I. Galanina, N.S. Zelenskaya: Yad. Fiz. **55**, 1868 (1992)

11.97   A.N. Antonov, P.E. Hodgson, I.Zh. Petkov: *Nucleon Momentum and Density Distributions in Nuclei* (Clarendon Press, Oxford 1988)

11.98   A.N. Antonov, V.A. Nikolaev, I.Zh. Petkov: Bulg. J. Phys. **10**, 42 (1983)

11.99   A.I. Akhiezer, I.Ya. Pomeranchuk: Usp. Fiz. Nauk **65**, 593 (1958)

11.100   A.N. Antonov, V.A. Nikolaev, I.Zh. Petkov: JINR E2-11283, Dubna (!978); Compt. rend. Acad. bulg. Sci. **31**, 409 (1978)

11.101   A.N. Antonov, I.S. Bonev, V.A. Nikolaev, I.Zh. Petkov: JINR R4-12634, Dubna (1979)

11.102   G.D. Alkhazov: Z. Phys. **A305**, 167 (1982)

11.103   A.N. Antonov, Chr.V. Christov, I. Zh. Petkov: Z. Phys. **A320**, 683 (1985)

11.104   A.N. Antonov, I.Zh. Petkov: in *"Diffraction Hadron-Nuclei Interaction"* (Naukova Dumka, Kiev 1987), p. 36

11.105   A.N. Antonov, Chr.V. Christov, E.N. Nikolov, I.Zh. Petkov, A.D. Polozov, A.M. Pushkash: Z. Phys. **A336**, 333 (1990)

11.106   I. Sick: Phys. Lett. **53B**, 15 (1974)

11.107 N.F. Golovanova, V. Iskra: Phys. Lett. **187B**, 7 (1987)

11.108 F. Shimizu, Y. Kubota, H. Koiso, F. Sai, S. Sakamoto, S.S. Yamamoto: Nucl. Phys. **A386**, 571 (1982)

11.109 F. Shimizu, H. Koiso, Y. Kubota, F. Sai, S. Sakamoto, S.S. Yamamoto: Nucl. Phys. **A389**, 445 (1982)

11.110 T.A. Murray, L. Riddiford, G.H. Grayer, T.W. Jones, Y. Tanimura: Nuovo Cim. **A49**, 261 (1967)

11.111 D.I. Blokhintzev: ZETP **33**, 989 (1957)

11.112 D.I. Blokhintzev, K.A. Toktarov: JINR R4-4018, Dubna (1968)

11.113 A.M. Baldin: in Proc. Int. Conf. on High Energy Physics and Nuclear Structure, Santa Fe and Los Alamos 1975 (Eds. D.E. Nagle, A.S. Goldhaber, C.K. Nargrave and R.L. Burman, AIP Conference Proceedings No. 26, New York 1975), p. 621

11.114 V.V. Burov, V.K. Lukyanov, A.I. Titov: Phys. Lett. **67B**, 46 (1977)

11.115 R.D. Amado, R.M. Woloshyn: Phys. Rev. Lett. **36**, 1435 (1976)

11.116 H.J. Weber, L.D. Miller: Phys. Rev. **C16**, 726 (1977)

11.117 A.N. Antonov, V.A. Nikolaev, I.Zh. Petkov: JINR R4-12207, Dubna (1979)

11.118 G. Jacob, Th.A. Maris: Rev. Mod. Phys. **38**, 121 (1966)

11.119 E.J. Moniz, I. Sick, R.R. Whitney, J.R. Ficenec, R.D. Kephart, W.P. Trower: Phys. Rev. Lett. **26**, 445 (1971)

11.120 G.D. Alkhazov, G.M. Amalsky. S.L. Belostotsky, A.A. Vorobyov, O.A.Domchenkov, Yu.V. Dotsenko, V.E.Starodubsky: Phys. Lett. **42B**, 121 (1972)

11.121 T. Fujita: Phys. Rev. Lett. **39**, 174 (1977)

11.122 J. Knoll: Phys. Rev. **C20**, 773 (1979)

11.123 V.I. Komarov, G.E. Kosarev, H. Müller, D. Netzband, V.D. Toneev, T. Stiehler, S. Tesch, K.K. Gudima, S.G. Mashnik: Nucl. Phys. **A326**, 297 (1979)

11.124 C.-M.Ko, M. To-Chung: Phys. Rev. Lett. **43**, 994 (1979)

11.125 Y. Haneishi, T. Fujita: Phys. Rev. **C33**, 260 (1986)

11.126 Y. Haneishi, T. Fujita: Phys. Rev. **C35**, 70 (1987)

11.127 V.I. Komarov, G.E. Kosarev, H. Müller, P. Netzband, T. Stiehler, S. Tesch: Phys. Lett. **80B**, 30 (1978)

11.128 Y. Miake, H. Hamagaki, S. Kadota, S. Nagamiya, S. Schnetzer, Y. Shida, H. Steiner, I. Tanihata: Phys. Rev. **C31**, 2168 (1985)

Chapter 12

12.1   A. Chaumeaux, G. Bruge, T. Bauer, R. Bertini, A. Boudard, H. Catz, P. Couvert, H.H. Duhm, J.M. Fontaine, D. Garreta, J.C. Lugol: Nucl. Phys. **A267**, 413 (1976)

12.2   G.D. Alkhazov, T. Bauer, R. Bertini, L. Bimbot, O. Bing, A. Boudard, G. Bruge, H. Catz, A. Chaumeaux, P. Couvert, J.M. Fontaine, F. Hibou, G.J. Igo, J.C. Lugol, M. Matoba: Nucl. Phys. **A280**, 365 (1977)

12.3   R.D. Viollier, E. Turtschi: Ann. Phys. **124**, 290 (1980)

12.4   A.K. Kerman, H. McManus, R.M. Thaler: Ann. Phys. **8**, 551 (1959)

12.5   V. Layly, R. Schaeffer: Phys. Rev. **C17**, 1145 (1978)

12.6   R.J. Glauber: *Lectures in Theoretical Physics* v. 1 (New York 1959), p. 315

12.7   A.G. Sitenko: Ukr. Fiz. J. **4**, 152 (1959)

12.8   A. Vitturi, F. Zardi: Fizika **9**, Suppl. No. 2, 79 (1977)

12.9   A. Vitturi, F. Zardi: Lett. Nuovo Cim, **20**, 640 (1977)

12.10 G.K. Varma: Nucl. Phys. **A294**, 465 (1978)

12.11 V. Franco, G.K. Varma: Phys. Rev. **C18**, 349 (1978)

12.12 I. Ahmad: J. Phys. **G4**, 1695 (1978)

12.13 A.S. Pak, A.V. Tarasov, V.V. Uzhinskyi, Ch. Tzeren: Pis'ma ZETP **28**, 314 (1978)

12.14 A.S. Pak, A.V. Tarasov, V.V. Uzhinskyi, Ch. Tzeren: Yad. Fiz. **30**, 102 (1979)

12.15 G. Fäldt, I. Hulthage: Nucl. Phys. **A316**, 253 (1979)

278   References

12.16   I. Ahmad: J. Phys. **G6**, 947 (1980)
12.17   M.A. El-Shabshiry: Ind. J. Phys. **60A**, 455 (1986)
12.18   W. Czyż, L.C. Maximon: Ann. Phys. **52**, 59 (1969)
12.19   G.D. Alkhazov: Preprint LINP No. 465, Leningrad (1979)
12.20   G.D. Alkhazov: Izv. AN USSR, ser. fiz. **43**, 2115 (1979)
12.21   A. Małecki, L. Satta: Preprint LNF-76/36(P), Frascati, (1976)
12.22   Y. Alexander, A.S. Rinat: Nucl. Phys. **A278**, 525 (1977)
12.23   D.C. Choudhury: Phys. Rev. **C22**, 1848 (1980)
12.24   A.N. Antonov, V.A. Nikolaev, I.Zh. Petkov: Bulg. J. Phys. **6**, 151 (1979)
12.25   A.N. Antonov, V.A. Nikolaev, I.Zh. Petkov: Z. Phys. **A297**, 257 (1980)
12.26   A.N. Antonov, P.E. Hodgson, I.Zh. Petkov: *Nucleon Momentum and Density Distributions in Nuclei* (Clarendon Press, Oxford 1988)
12.27   A.N. Antonov, V.A. Nikolaev, I.Zh. Petkov: Bulg. J. Phys. **10**, 42 (1983)
12.28   A.N. Antonov, V.A. Nikolaev, I.Zh. Petkov: Izv. AN USSR, ser. fiz. **47**, 134 (1983)
12.29   A.I. Akhiezer, I.Ya. Pomeranchuk: Usp. Fiz. Nauk **65**, 593 (1958)
12.30   V.V. Burov, Yu.N. Eldyshev, V.K. Lukyanov, Yu.S. Pol': JINR, E4-8029, Dubna (1974)
12.31   A.N. Antonov, I.Zh. Petkov: in *"Diffraction Hadron-Nuclei Interaction"* (Naukova Dumka, Kiev 1987), p. 36
12.32   A. Dar, Z. Kirzon: Phys. Lett. **37B**, 166 (1971)
12.33   V.P. Verbitzkyi, Yu.E. Penionzhkevich, V.N. Polyanskyi, K.O. Ternetzkyi: Yad. Fiz. **31**, 1134 (1980)
12.34   K.A. Brueckner, J.R. Buchler, M.M. Kelly: Phys. Rev. **173**, 944 (1968)
12.35   Ya.I. Delchev, I.Zh. Petkov: Compt. rend. Acad. bulg. Sci. **27**, 905 (1974)
12.36   D.T. Khoa, K.V. Shitikova: in "Particles and Nuclei", 10th Int. Conf. Heidelberg, 1984, Book of Abstracts, v. **2**, J3
12.37   V.S. Barashenkov, V.F. Nikitin: JINR R2-11440, Dubna (1978)
12.38   E.V. Inopin, A.V. Shebeko: Yad. Fiz. **11**, 140 (1970)
12.39   V.V. Kotlyar, A.V. Shebeko: Yad. Fiz. **34**, 370 (1981)
12.40   W.E. Frahn, D.H.E. Gross: Ann. Phys. **101**, 520 (1976)
12.41   W.E. Frahn: Nucl. Phys. **A302**, 267, 301 (1978)
12.42   Z. Kirzon, A. Dar: Nucl. Phys. **A237**, 319 (1975)
12.43   V.S. Barashenkov, E.S. Gavrilov, S.M. Eliseev: JINR R2-6423, Dubna (1972)
12.44   V.S. Barashenkov, Zh.Zh. Musulmanbekov: JINR R2-1453, Dubna (1978)
12.45   G.K. Varma: Phys. Rev. **C17**, 267 (1978)
12.46   S.A. Gurvitz: Phys. Rev. **C24**, 29 (1981)
12.47   A. Antonov, I. Bonev, V. Nikolaev, I. Petkov: JINR R4-12633, Dubna (1979)
12.48   A.I. Akhiezer, I.Ya. Pomeranchuk: ZETP **16**, 396 (1946)

Author Index

Subject Index

Springer-Verlag
and the Environment

We at Springer-Verlag firmly believe that an international science publisher has a special obligation to the environment, and our corporate policies consistently reflect this conviction.

We also expect our business partners – paper mills, printers, packaging manufacturers, etc. – to commit themselves to using environmentally friendly materials and production processes.

The paper in this book is made from low- or no-chlorine pulp and is acid free, in conformance with international standards for paper permanency.